AF850252

VOLUME THREE HUNDRED AND SEVENTY TWO

INTERNATIONAL REVIEW OF CELL AND MOLECULAR BIOLOGY

mRNA-Based Therapeutics

INTERNATIONAL REVIEW OF CELL AND MOLECULAR BIOLOGY

Series Editors

GEOFFREY H. BOURNE	*1949–1988*
JAMES F. DANIELLI	*1949–1984*
KWANG W. JEON	*1967–2016*
MARTIN FRIEDLANDER	*1984–1992*
JONATHAN JARVIK	*1993–1995*
LORENZO GALLUZZI	*2016–*

Editorial Advisory Board

AARON CIECHANOVER
SANDRA DEMARIA
SILVIA FINNEMANN
KWANG JEON
CARLOS LOPEZ-OTIN
WALLACE MARSHALL
SHIGEKAZU NAGATA
MOSHE OREN
ANNE SIMONSEN

VOLUME THREE HUNDRED AND SEVENTY TWO

International Review of CELL AND MOLECULAR BIOLOGY

mRNA-Based Therapeutics

Edited by

FERNANDO ARANDA
*Cima Universidad de Navarra,
Pamplona, Spain*

PEDRO BERRAONDO
*Cima Universidad de Navarra,
Pamplona, Spain*

LORENZO GALLUZZI
*Weill Cornell Medical College,
New York, NY, United States*

ELSEVIER

ACADEMIC PRESS
An imprint of Elsevier

Academic Press is an imprint of Elsevier
50 Hampshire Street, 5th Floor, Cambridge, MA 02139, United States
525 B Street, Suite 1650, San Diego, CA 92101, United States
The Boulevard, Langford Lane, Kidlington, Oxford OX5 1GB, United Kingdom
125 London Wall, London, EC2Y 5AS, United Kingdom

First edition 2022

Copyright © 2022 Elsevier Inc. All rights reserved.

No part of this publication may be reproduced or transmitted in any form or by any means, electronic or mechanical, including photocopying, recording, or any information storage and retrieval system, without permission in writing from the publisher. Details on how to seek permission, further information about the Publisher's permissions policies and our arrangements with organizations such as the Copyright Clearance Center and the Copyright Licensing Agency, can be found at our website: www.elsevier.com/permissions.

This book and the individual contributions contained in it are protected under copyright by the Publisher (other than as may be noted herein).

Notices
Knowledge and best practice in this field are constantly changing. As new research and experience broaden our understanding, changes in research methods, professional practices, or medical treatment may become necessary.

Practitioners and researchers must always rely on their own experience and knowledge in evaluating and using any information, methods, compounds, or experiments described herein. In using such information or methods they should be mindful of their own safety and the safety of others, including parties for whom they have a professional responsibility.

To the fullest extent of the law, neither the Publisher nor the authors, contributors, or editors, assume any liability for any injury and/or damage to persons or property as a matter of products liability, negligence or otherwise, or from any use or operation of any methods, products, instructions, or ideas contained in the material herein.

ISBN: 978-0-323-99401-9
ISSN: 1937-6448

For information on all Academic Press publications
visit our website at https://www.elsevier.com/books-and-journals

Publisher: Zoe Kruze
Acquisitions Editor: Leticia M. Lima
Developmental Editor: Jhon Michael Peñano
Production Project Manager: Vijayaraj Purushothaman
Cover Designer: Vicky Pearson

Typeset by STRAIVE, India

ELSEVIER Book Aid International
Working together to grow libraries in developing countries
www.elsevier.com • www.bookaid.org

Contents

Contributors

Estibaliz Arellano-Viera
Department of Internal Medicine II, University Hospital Würzburg, Würzburg, Germany

Matías A. Avila
Hepatology Program, Center for Applied Medical Research (CIMA), University of Navarra; Navarra Institute for Health Research (IDISNA), Pamplona; Centro de Investigación Biomédica en Red de Enfermedades Hepáticas y Digestivas (CIBERehd), Instituto de Salud Carlos III, Madrid, Spain

Marina Beraza-Millor
Pharmacokinetic, Nanotechnology and Gene Therapy Group (PharmaNanoGene), Faculty of Pharmacy, Centro de investigación Lascaray ikergunea, University of the Basque Country UPV/EHU; Bioaraba, Microbiology, Infectious Disease, Antimicrobial Agents, and Gene Therapy, Vitoria-Gasteiz, Spain

Pedro Berraondo
Navarra Institute for Health Research (IDISNA); Program of Immunology and Immunotherapy, CIMA-University of Navarra, Pamplona; Centro de Investigación Biomédica en Red de Cáncer (CIBERONC), Instituto de Salud Carlos III, Madrid, Spain

Karol M. Córdoba
Hepatology Program, Center for Applied Medical Research (CIMA), University of Navarra; Navarra Institute for Health Research (IDISNA), Pamplona, Spain

Ana del Pozo-Rodríguez
Pharmacokinetic, Nanotechnology and Gene Therapy Group (PharmaNanoGene), Faculty of Pharmacy, Centro de investigación Lascaray ikergunea, University of the Basque Country UPV/EHU; Bioaraba, Microbiology, Infectious Disease, Antimicrobial Agents, and Gene Therapy, Vitoria-Gasteiz, Spain

Ramon Diaz-Trelles
Arcturus Therapeutics, Inc, San Diego, CA, United States

Antonio Fontanellas
Hepatology Program, Center for Applied Medical Research (CIMA), University of Navarra; Navarra Institute for Health Research (IDISNA), Pamplona; Centro de Investigación Biomédica en Red de Enfermedades Hepáticas y Digestivas (CIBERehd), Instituto de Salud Carlos III, Madrid, Spain

Andoni Garitano-Trojaola
Department of Internal Medicine II, University Hospital Würzburg, Würzburg, Germany

Itziar Gómez-Aguado
Pharmacokinetic, Nanotechnology and Gene Therapy Group (PharmaNanoGene), Faculty of Pharmacy, Centro de investigación Lascaray ikergunea, University of the Basque Country UPV/EHU; Bioaraba, Microbiology, Infectious Disease, Antimicrobial Agents, and Gene Therapy, Vitoria-Gasteiz, Spain

Ruofan Connie Han
Nuffield Laboratory of Ophthalmology, Department of Clinical Neurosciences, University of Oxford, John Radcliffe Hospital, Oxford, United Kingdom

María J. Iraburu
Department of Biochemistry and Genetics. School of Sciences, University of Navarra, Pamplona, Spain

Daniel Jericó
Hepatology Program, Center for Applied Medical Research (CIMA), University of Navarra; Navarra Institute for Health Research (IDISNA), Pamplona, Spain

Lei Jiang
Moderna Inc, Cambridge, MA, United States

Sarah Elspeth Knudsen
Independent Scholar, Watertown, MA, United States

Mengyun Li
State Key Laboratory of Oncology in South China, Collaborative Innovation Center for Cancer Medicine, Sun Yat-sen University Cancer Center, Guangzhou, China

Kenneth Lundstrom
PanTherapeutics, Lutry, Switzerland

Robert E. MacLaren
Nuffield Laboratory of Ophthalmology, Department of Clinical Neurosciences, University of Oxford, John Radcliffe Hospital; Oxford Eye Hospital, John Radcliffe, Oxford, United Kingdom

Paolo G.V. Martini
Moderna Inc, Cambridge, MA, United States

Julia Mohr
BioNTech SE, Mainz, Germany

Carlos G. Perez-Garcia
Arcturus Therapeutics, Inc, San Diego, CA, United States

Musga Qureischi
BioNTech SE, Mainz, Germany

Julen Rodríguez-Castejón
Pharmacokinetic, Nanotechnology and Gene Therapy Group (PharmaNanoGene), Faculty of Pharmacy, Centro de investigación Lascaray ikergunea, University of the Basque Country UPV/EHU; Bioaraba, Microbiology, Infectious Disease, Antimicrobial Agents, and Gene Therapy, Vitoria-Gasteiz, Spain

Alicia Rodríguez-Gascón
Pharmacokinetic, Nanotechnology and Gene Therapy Group (PharmaNanoGene), Faculty of Pharmacy, Centro de investigación Lascaray ikergunea, University of the Basque Country UPV/EHU; Bioaraba, Microbiology, Infectious Disease, Antimicrobial Agents, and Gene Therapy, Vitoria-Gasteiz, Spain

Ana Sampedro
Hepatology Program, Center for Applied Medical Research (CIMA), University of Navarra, Pamplona, Spain

María Ángeles Solinís
Pharmacokinetic, Nanotechnology and Gene Therapy Group (PharmaNanoGene), Faculty of Pharmacy, Centro de investigación Lascaray ikergunea, University of the Basque Country UPV/EHU; Bioaraba, Microbiology, Infectious Disease, Antimicrobial Agents, and Gene Therapy, Vitoria-Gasteiz, Spain

Farrukh Vohidov
BioNTech US, Cambridge, MA, United States

Zining Wang
State Key Laboratory of Oncology in South China, Collaborative Innovation Center for Cancer Medicine, Sun Yat-sen University Cancer Center, Guangzhou, China

Xiaojun Xia
State Key Laboratory of Oncology in South China, Collaborative Innovation Center for Cancer Medicine, Sun Yat-sen University Cancer Center, Guangzhou, China

Chunyuan Xie
State Key Laboratory of Oncology in South China, Collaborative Innovation Center for Cancer Medicine, Sun Yat-sen University Cancer Center, Guangzhou, China

CHAPTER ONE

mRNA-based therapies: Preclinical and clinical applications

Musga Qureischi[a,†], Julia Mohr[a,†], Estibaliz Arellano-Viera[b], Sarah Elspeth Knudsen[c], Farrukh Vohidov[d], and Andoni Garitano-Trojaola[b,*]
[a]BioNTech SE, Mainz, Germany
[b]Department of Internal Medicine II, University Hospital Würzburg, Würzburg, Germany
[c]Independent Scholar, Watertown, MA, United States
[d]BioNTech US, Cambridge, MA, United States
*Corresponding author: e-mail address: garitano_a@ukw.de

Contents

Abstract

At the fundamental level, messenger RNA (mRNA)-based therapeutics involves the delivery of in vitro-transcribed (IVT) mRNA into the cytoplasm of a target cell, where it is translated into the desired protein. IVT mRNA presents various advantages compared to DNA and recombinant protein-based approaches that make it ideal for a broad range of therapeutic applications. IVT mRNA, which is translated in the cytoplasm after transfection into cells, can encode virtually any target protein. Notably, it does not enter the nucleus, which avoids its integration into the genome and the risk of insertional mutagenesis. The large-scale production of IVT mRNA is less complex than production

† Equal contribution.

International Review of Cell and Molecular Biology, Volume 372
ISSN 1937-6448
https://doi.org/10.1016/bs.ircmb.2022.04.007
Copyright © 2022 Elsevier Inc. All rights reserved.

of recombinant proteins, and Good Manufacturing Practice-compliant mRNA production is easily scalable, ideally poising mRNA for not only off-the-shelf, but more personalized treatment approaches. IVT mRNA's safety profile, pharmacokinetics, and pharmacodynamics, including its inherent immunostimulatory capacity, can be optimized for different therapeutic applications by harnessing a wide array of optimized sequence elements, chemical modifications, purification techniques, and delivery methods. The value of IVT mRNA was recently proved during the COVID-19 pandemic when mRNA-based vaccines outperformed the efficacy of established technologies, and millions of doses were rapidly deployed. In this review, we will discuss chemical modifications of IVT mRNA and highlight numerous preclinical and clinical applications including vaccines for cancer and infectious diseases, cancer immunotherapy, protein replacement, gene editing, and cell reprogramming.

1. Introduction

Messenger RNA (mRNA), and its central-dogma function as the carrier of genetic information from DNA to protein synthesis, was discovered in 1961 (Brenner et al., 1961). RNA has been investigated as a novel drug class to treat multiple diseases since 1990, when it was first demonstrated that RNA vectors could cause expression of the encoded protein in mouse muscle (Wolff et al., 1990).

IVT mRNA therapeutics have several notable differences from other nucleic acid-based therapies: First, IVT mRNA enters the cytoplasm and is translated directly into the desired protein without reaching the nucleus. Conversely, DNA therapeutics must enter the nucleus (when the nuclear envelope is broken down during cell division) to be transcribed into RNA. Second, unlike DNA and viral plasmids, IVT mRNA is not integrated into the genome, avoiding the risk of insertional mutagenesis (Meng et al., 2017). Third, IVT mRNA offers more stable and predictable protein expression than DNA-based therapeutics (Leonhardt et al., 2014; Ligon et al., 2014; Yamamoto et al., 2009). Fourth, IVT mRNA is only transiently expressed, which is advantageous for certain pharmaceutical applications that require brief protein expression (Granot and Peer, 2017). Finally, IVT mRNA production is cost-effective and provides easy scale-up, which makes this therapeutic approach particularly attractive.

In the last decade, several efforts have been done to utilize IVT mRNA as a new class of therapeutics. For example, both endogenous and synthetic mRNAs have a very short half-life due inherent instability and the degradation by environmental RNases (Wadhwa et al., 2020). Further, unmodified IVT mRNA has a strong immune activatory capacity and can lead to the production of type I interferons (IFNs) and other pro-inflammatory cytokines via multiple signaling receptors (Bessis et al., 2004; Fitzgerald and Kagan, 2020; Wu and Chen, 2014). These points have been primarily addressed by the introduction of chemical modification and the use of nanodrug delivery systems, which have not only improved IVT mRNA stability and cellular uptake efficiency but also increased the safety and therapeutic efficacy. These improvements have enabled the applications of IVT mRNA to various therapeutic areas such as vaccines, cancer immunotherapy, gene therapy, protein replacement, and cell reprogramming. In this review, we will primarily discuss effective chemical modifications for enhancing the stability and translation of IVT mRNA and then touch on the multiple therapeutic applications of IVT mRNAs.

2. Chemical modifications

The structure of mature eukaryotic mRNA consists of five elements: 5′-cap, polyadenylated tail (poly(A)), 5′- and 3′-untranslated region (5′-, 3′-UTR), and an open reading frame (ORF). IVT mRNA is prepared in a cell-free system, using a template of linearized DNA, which includes a phage promoter sequence (T7, T3, or SP6) and target gene sequence (Mignone et al., 2002). The poor stability and low translation efficiency of IVT mRNA result in suboptimal pharmacokinetic and pharmacodynamic activities; several chemical modifications have been established to overcome these limitations (Holtkamp et al., 2006; Kallen and Thess, 2014). Here, we will discuss what we consider to be the most effective approaches yet described.

2.1 Improvement of IVT mRNA stability and translation efficiency

The five common elements of mature eukaryotic mRNA and IVT mRNA (5′-cap, 3′-poly(A) tail, 5′-UTR, 3′-UTR, open reading frame (ORF)), as shown in Fig. 1A, can be chemically modified during mRNA synthesis and are discussed in the following sections.

2.1.1 5′-cap and 3′-poly(A) tail

The main functions of the 5′-cap are protecting nascent mRNA from exonucleases and assisting in ribosome binding during translation. The cap structure of natural mRNA consists of N7-methylated guanosine (m^7G) linked to the first nucleotide of the RNA through a $5' - 5'$ triphosphate

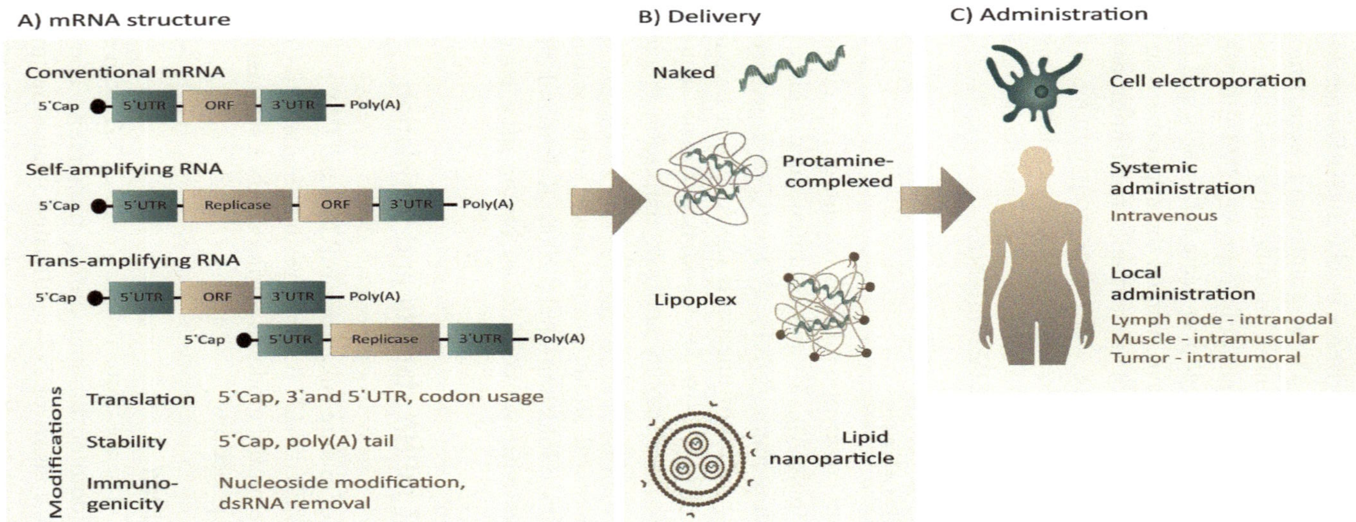

Fig. 1 Scheme of different mRNA structures, delivery methods and administration options. (A) mRNA structures: conventional mRNA harbors 5′cap, 5′ UTR, ORF, 3′UTR, and Poly (A) elements. Self-amplifying mRNA, a type of mRNA that encodes replicase. *Trans*-Amplifying mRNA, an advanced form of self-amplifying mRNA created by separating the target mRNA to be amplified and the replicase encoding mRNA. These mRNA structures can be modified chemically: The modification of 5′Cap, codon usage, 3′ and 5′UTR will change mRNA's translation activity. Additional modifications on 5′Cap and poly (A) will enhance the stability of mRNA. The nucleoside modification and dsRNA removal will reduce mRNA immunogenicity and this was essential to expand mRNA therapeutic applications. (B) Different mRNA delivery methods: naked mRNA, mRNA transfection by electroporation, nanoparticles, and protamine formulation. (C) Administration of mRNA into the organism: Local (intravenous) and systemic (Lymph node—intranodal, muscle—intramuscular, and tumor—intratumoral) administration of mRNA electroporated receptor cells into the organism. Local and systemic administration of naked and nanoparticles/protamine formulated mRNA into the organism. mRNA, messenger RNA; ORF, open reading frame; dsRNA, double-strand RNA; UTR, untranslated region.

bridge (Furuichi and Miura, 1975; Ramanathan et al., 2016). The binding of eukaryotic translation initiation factor 4E (eIF4E) to the 5′cap leads to efficient translation. However, the binding of decapping enzymes such as DCP1, DCP2, and DCPS induces mRNA decay (Li and Kiledjian, 2010; Ramanathan et al., 2016).

Different strategies have been developed to cap IVT mRNA. In post-transcription capping, recombinant vaccinia virus-derived capping enzymes are added to form the cap structure after mRNA is synthesized (Martin et al., 1975). The m^7GpppG cap generated in this approach is similar to those found on natural eukaryotic mRNAs. An alternative approach is to use a synthetic cap analog instead of m^7GpppG during the transcription reaction (Rydzik et al., 2017). This is a simple process and allows for diversity in the cap structure; however, it has two main drawbacks: decapping of mRNA results from competition between the cap analog and GDP nucleotide and reduced translation occurs due to random capping. Further, cap analogs can be attached to mRNA in two orientations, and substantial incorporation of a cap analog in the reverse orientation prevents recognition by translational machinery.

The anti-reverse cap analogs (ARCAs) $m_2{}^{7,O3'}$Gp(3)G, $m^7$3′dGp(3)G, $m^7$2′dGp4G, $m2^{7,2'-O}$Gp4G, and $m2^{7,3'-O}$Gp4G were developed to avoid issues arising from reverse capping (Jemielity et al., 2003; Stepinski et al., 2001). These ARCA-capped mRNAs show higher translational activity in some cell types (Mockey et al., 2006; Rabinovich et al., 2006). Additionally, a new generation of ARCAs with greater resistance to decapping enzymes, resulting in mRNAs with enhanced stability and translation efficiency, is being developed. These include phosphorothioate- (Grudzien-Nogalska et al., 2007), boranophosphate- (Kimchi-Sarfaty et al., 2007; Malone et al., 1989), phosphoroselenoate- (Wolff et al., 1990), methylene bisphosphonate- (Martinon et al., 1993), 1,2-dithiophosphate- (Strenkowska et al., 2016), and dihalogen methylene bisphosphonate-containing caps (Grudzien-Nogalska et al., 2007; Kowalska et al., 2008).

The interaction of a 3′-poly(A) tail with poly(A)-binding protein stabilizes an RNA molecule and prevents degradation by nucleases (Goldstrohm and Wickens, 2008). Two methods are commonly used to introduce a poly(A) tail into IVT mRNA: either a polyadenylation reaction catalyzed by recombinant poly(A) polymerase (Korner and Wahle, 1997; Martin and Keller, 1998) or direct transcription from the poly(T) sequence encoded in the DNA template (Holtkamp et al., 2006).

IVT mRNA tailed by recombinant poly(A) develops a mixture of mRNAs with different poly(A) tail lengths (Trepotec et al., 2019). In contrast,

IVT mRNAs tailed with the poly(T) method have a uniform poly(A) length (Murray and Schoenberg, 2008). In addition to affecting stability, the length of the poly(A) tail also affects the translational activity of mRNA (Weill et al., 2012). A longer poly(A) tail is associated with higher translation yields (Peng and Schoenberg, 2005). Besides the tail length, the addition of adenosine analogs such as 3′-deoxyadenosine or 8-aza adenosine protects mRNA from 3′exonuclease degradation, which can enhance protein (Rabinovich et al., 2006).

2.1.2 5′- and 3′-UTRs

The mRNA UTRs are non-coding regions located 5′ upstream and 3′ downstream of ORFs. These enable protein translation and provide stability by interacting with RNA-binding proteins (Mignone et al., 2002).

The 5′-UTR is essential for ribosome recruitment, choice of the start codon, and to provide a ribosomal entry site and facilitate effective mRNA translation even at low expression levels of eIF4E (Hinnebusch et al., 2016; Leppek et al., 2018; Mayr, 2017). Moreover, Meyer and colleagues have shown that mRNA containing an N6-methyladenosine- (m^6A) 5′-UTR can be translated in a cap-independent manner (Mayr, 2017).

The 3′-UTR controls the stability of mRNA and thus regulates the levels and duration of protein expression. For instance, the stability and translation of an IVT mRNA can be enhanced by the incorporation of several sequence elements from 3′-UTRs of α- and β-globin mRNAs (Kariko et al., 1999; Ross and Sullivan, 1985): The stability of mRNA increases in the presence of two human β-globin 3′-UTRs organized in a head-to-tail orientation (Holtkamp et al., 2006; Sahin et al., 2014). Moreover, some viral and cellular regions potentiate the stability and translational efficiency of mRNA. The 5′-UTR present in several orthopoxvirus mRNAs (Bergman et al., 2007) and the 3′-UTR of eukaryotic elongation factor 1α (*EEF1A1*) mRNA (Zinckgraf and Silbart, 2003) both prevented decapping and 3′–5′ exonucleolytic degradation. However, for some applications, it may be desirable to limit the duration of protein production. This can be achieved by adding adenylate/uridylate (AU)-rich elements to the 3′-UTR, which leads to rapid degradation of mRNA, reducing the duration of protein expression (Chen and Shyu, 1995).

2.1.3 The coding region

The base composition of mRNA directly impacts its translation activity (Novoa and Ribas de Pouplana, 2012; Plotkin and Kudla, 2011). For example, a reduction in the frequency of UU and UA dinucleotides increases stability and protein expression (Leppek et al., 2018). Further, the codon framework of mRNA regulates translation efficiency. For instance, replacing rare codons with synonymous but more common codons increases translation activity by allowing the same transfer RNA (tRNA) to be recycled by the ribosome (Cannarozzi et al., 2010). Historically, codon optimized IVT mRNA has shown promising results in vaccine studies targeting cancer, viral infections, and the expression of non-viral proteins (Benteyn et al., 2013; Kariko et al., 2012; Knights et al., 2009), but codon optimization may not always be necessary or desirable. For instance, for some IVT mRNA-encoded vaccines, slow translation from rare codons may be preferable to facilitate correct protein folding (Kimchi-Sarfaty et al., 2007).

2.2 Immunostimulatory properties of IVT mRNA

Compared to endogenous mRNA, IVT mRNA is highly immunogenic; both ORF and UTR sequences affect immunostimulation, and thus translational activity. Early studies applied IVT mRNAs composed of unmodified bases A, U, C, and G (Malone et al., 1989; Martinon et al., 1993; Wolff et al., 1990). U-rich sequences and several RNA structural features lead to the activation of innate immune cells through interaction with intracellular RNA sensors such as toll-like receptors (TLRs), retinoic acid-inducible gene I (RIG-I), and protein kinase R (PKR) (Alexopoulou et al., 2001; Diebold et al., 2004; Heil et al., 2004; Hornung et al., 2006; Kato et al., 2006; Nallagatla et al., 2007; Pichlmair et al., 2006; Rehwinkel et al., 2010). Consequently, the therapeutic applications of IVT mRNA were considered to be limited to vaccination, where its strong immune-stimulatory effect and intrinsic adjuvant activity lead to high antigen-specific cellular and humoral immune responses (Bourquin et al., 2007; Hoerr et al., 2000; Sander et al., 2011; Weissman et al., 2000).

The incorporation of modified nucleosides within IVT mRNAs such as pseudouridine (ψ), 2-thiouridines (s^2U), 5-methyluridine (m^5U), 5-methylcytidine, 5-methoxy uridine (mo^5U), N1-methylpseudouridine ($m^1\psi$), 5-methylcytidine (m^5C), 5-hydroxy methylcytidine (hm^5C), 5-methoxycytidine (mo^5C), N6-methyladenosine (m^6A) or 2-methylguanosine (m^2G), suppress the intrinsic adjuvant activity and increases stability and translational efficiency (Fig. 1A) (Kariko et al., 2005, 2008; Nelson et al., 2020). The three modified nucleosides which most significantly enhance protein translation of IVT mRNA are $m^1\psi$, mo^5U, and ψ

(Li and Kiledjian, 2010; Li et al., 2016). Notably, the combination of ψ and m^5C increases protein production relative to a single modification (Wroblewska et al., 2015; Zangi et al., 2013). Another study has shown that IVT mRNA modified by s^2U and m^5C reduces the activation of pathogen recognition receptors (PRRs) and extends the period of protein expression to 4 weeks (Kormann et al., 2011). The $m^1\psi$ modification reduces the immunostimulatory nature of IVT mRNA via several reported mechanisms, including by altering secondary RNA structures to be less like double-stranded RNA (dsRNA) structures commonly recognized by TLR3, disfavoring self-priming and subsequent formation of dsRNA, and reducing recognition via TLR7 due to the introduction of a steric incompatibility in the form of a methyl group (Nance and Meier, 2021). Because the dsRNA present in unpurified IVT mRNA enhances its immunogenicity, purification by liquid chromatography further reduces IVT mRNA's immunogenicity (Kariko et al., 2011; Nelson et al., 2020).

An alternative approach for effectively reducing the immunogenicity of IVT mRNA was reported by Thess et al. In their approach, GC-rich codons were used to code for each amino acid instead of modified nucleosides (Thess et al., 2015). They found this method enabled repeated dosing of unmodified IVT mRNA and expression of a therapeutic protein in large animals with no signs of immunogenicity.

The chemical modifications discussed in this section, and progress in sequence engineering and delivery methods (Fig. 1B), have led to the expansion of the potential therapeutic applications for IVT mRNA and are discussed in the next chapters (Fig. 2).

3. Preclinical and clinical applications of IVT mRNA

3.1 Vaccines against infectious diseases

Prophylactic vaccines against pathogens aim to generate immunological memory and thus protect individuals before exposure to the pathogen. As experienced during the COVID-19 pandemic, a framework for the fast development, regulatory approval, and global distribution of effective vaccines against emerging pathogens is essential.

Historically, most studies of IVT mRNA-based vaccines have been in the context of cancer immunotherapy (Lin et al., 2020; Schlake et al., 2012). However, the development of an mRNA-based vaccine for COVID-19 was accomplished in less than a year after the sequencing and publication of the severe acute respiratory syndrome coronavirus 2

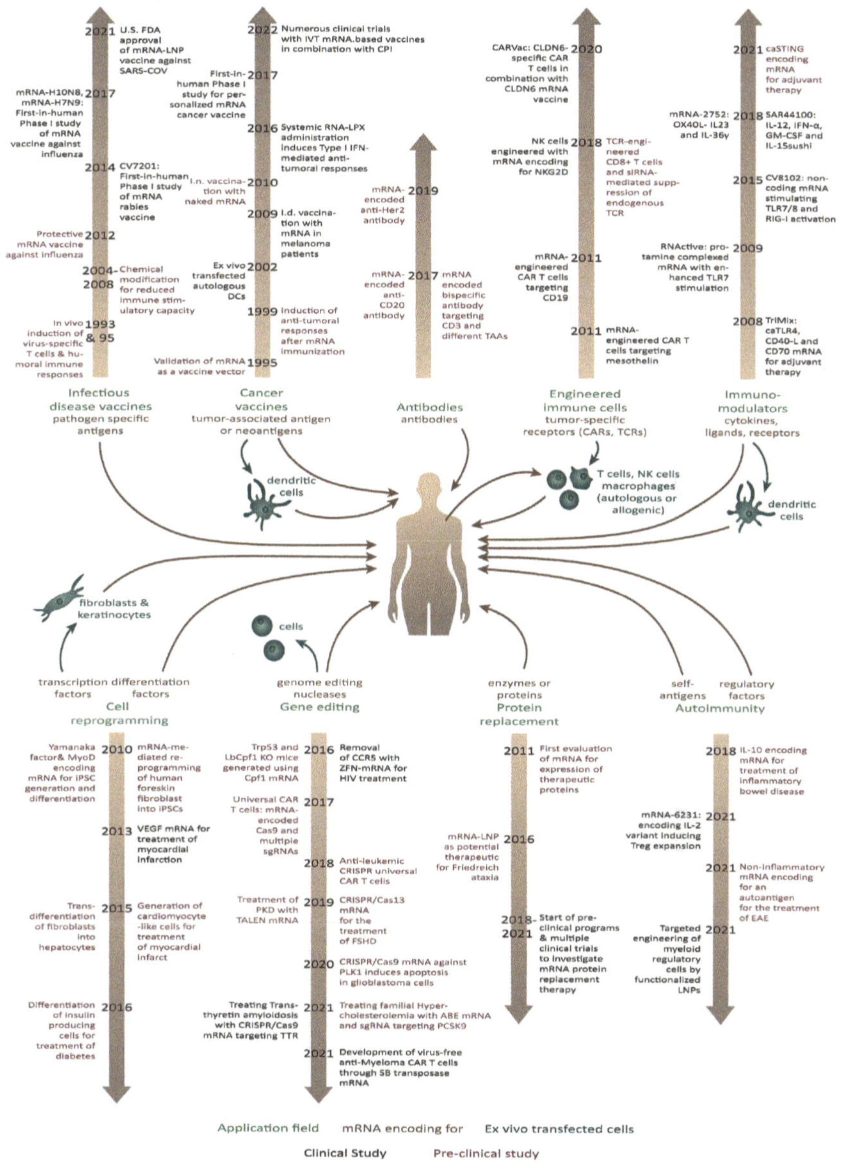

Fig. 2 Timeline of key preclinical and clinical mRNA therapeutics in infectious disease and cancer vaccines, cancer immunotherapy, protein replacement, gene editing, cell reprogramming and autoimmunity. Anti-infectious disease and anti-cancer vaccines: the direct injection of mRNA encoding pathogen-specific and tumor-associated antigens or neoantigens have shown successful results in different preclinical and clinical studies. The main hit of mRNA therapeutics until now has been the clinical approval of

(Continued)

(SARS-CoV-2) genome in 2019 (Zhu et al., 2020). The first approved mRNA-based vaccines, Spikevax® (mRNA-1273) and Comirnaty® (BNT162b) (Morais et al., 2021) consist of non-amplifying, m1ψ-modified IVT mRNA encoding for the full-length spike protein of SARS-CoV-2 formulated in lipid nanoparticles (LNPs). Both vaccines are administered

Fig. 2—Cont'd sars-cov-2 vaccines. As well, promising results have been observed with the administration of autologous mature dendritic cells (DC) loaded with mRNA encoding immunogenic neoantigens. Cancer immunotherapy: mRNA delivery of anti-Her2 and anti-CD3 and anti-CLDN6/CLD18.2/EpCAM bi-specific antibodies have shown promising preclinical results. Also, mRNA-based TCR, CAR T cells, and NKG2D engineered NK cells have shown excellent antitumor activity in different preclinical models. The intrinsic immune-activating capacity of mRNA makes its use advantageous in cancer immunotherapy to deliver co-stimulatory signals for vaccines or other mRNA therapies. Moreover, co-transfection of mRNA-encoded cytokines, ligands, and receptors together with DC-loaded neoantigens, through DC maturation and T cell activation, have shown strong anti-tumor activity in different preclinical models. For, autoimmune disorders, DC loaded with nucleoside modified mRNAs encoded self-antigens in non-inflammatory settings, led to T regs formation and suppression of multiple sclerosis. Another promising approach for autoimmune disease is mRNA encoded regulatory factors such as a low dose of IL-2. This cytokine has a higher affinity for T-regs than CD8 + T cells, being proposed as a therapy for lupus erythematosus. Protein replacement: mRNA encoded enzymes and protein, which are defective in rare genetic diseases, have shown promising results in preclinical models and clinical trials. Gene editing: Delivery of mRNA encoding DNA editing nucleases such asCas9, ZFNs, TALEN, SB transposase, and Cpf1 have shown successful results in different preclinical and early phase clinical trial studies. Recently, Cas13, which targets RNA instead of DNA, has opened new therapeutic options for infectious and genetic muscular muscle disorder disease. Cell reprogramming: mRNA encoding Yamanaka four factors and cardiac/pancreatic/hepatic reprogramming factors have shown successful results in reprogramming fibroblasts into pluripotent iPSCs cells and inducing the *trans*-differentiation of fibroblasts or pancreatic exocrine cells into cardiomyocyte-like cells/hepatocyte-like/insulin-secreting β cells, respectively. LNP, lipid nanoparticle; mRNA, messenger RNA; HER2, human epidermal growth factor receptor 2; NKG2D, natural killer group 2D; CLDN6, Claudin-6; TCR, T-cell receptor; CAR, chimeric antigen receptor; TLR, Toll like receptor; RIG-I retinoic acid-inducible gene I; DC, dendritic cells; GM-CSF, granulocyte-macrophage colony-stimulating factor; EAE experimental autoimmune encephalomyelitis; SB, Sleeping Beauty; Cas9, CRISPR associated protein 9; ZNFs, zinc finger nucleases; TALEN, transcription activator-like effector nuclease; Cpf1, CRISPR from Prevotella and Francisella; Trp53, transformation related protein 53; LbCpf1, lachnospiraceae bacterium N D2006; sgRNA, single guide RNA; CRISPR, clustered regularly interspaced short palindromic repeats; CCR5, C-C motif chemokine receptor 5; Cas13, CRISPR associated protein 13; FSHD, facioscapulohumeral muscular dystrophy; GVHD, graft-vs-host disease; iPSCs, induced pluripotent stem cells; PKD, pyruvate kinase deficiency; TTR, transthyretin; PLK1, polo like kinase 1; ABE, adenine base editors; PCSK9, proprotein convertase subtilisin/kexin type 9.

intramuscularly, are well tolerated (Anderson et al., 2020; Frenck Jr. et al., 2021; Walsh et al., 2020), and have shown more than 90% efficacy at preventing symptomatic COVID-19 (Baden et al., 2021; Polack et al., 2020; Thomas et al., 2021). Comirnaty® (BNT162b) not only induced neutralizing antibodies against the spike protein 21 days after a first-dose injection but also produced poly-specific T-cell responses (Sahin et al., 2021). These promising data regarding safety and efficacy demonstrate the potential of non-amplifying, nucleoside-modified IVT mRNA as a new safe and reliable platform for the rapid development and deployment of vaccines against emerging pathogens.

Another promising vaccine platform is the self-amplifying RNAs (saRNAs), a type of mRNA that autonomously replicates RNA molecules and is derived from positive- or negative-strand RNA viruses (Zimmer, 2010). The most well-studied saRNAs derive from alphavirus and have been evaluated as vectors for gene therapy (Polo et al., 2000) and as vaccines (Bloom et al., 2021). In contrast to conventional mRNA, saRNA contains two open reading frames (ORFs), with one encoding for an RNA polymerase (replicase) and the other one encoding for the gene of interest (Zhou et al., 1994) (Fig. 1A). The encoded replicase leads to amplification of the mRNA construct, and consequently, saRNA required a 65-fold lower dose than non-amplifying mRNA (for the same viral model antigen) to prolong the survival of mice infected with influenza (Vogel et al., 2018). However, due to the larger size of the saRNA transcript (10 vs 2–3 kb), in vitro transcription and delivery techniques will need to be further optimized to exploit the full potential of saRNA (Blakney et al., 2021). Based on promising preclinical results, trans-amplifying RNA (taRNA), could serve as another vector system for the induction of protective immune responses. The taRNA system is based on saRNA, but is separated into the RNA encoding for the gene of interest and another encoding for replicase (Fig. 1A). Beissert and colleagues showed that mice treated with only 50 ng of a naked taRNA system encoding a model antigen were protected against influenza—far lower than the 20 and 1.25 μg doses required for comparable optimized naked mRNA and saRNA systems, respectively (Beissert et al., 2020).

Sustained antigen availability, with kinetics similar to those observed in natural infection, has been shown to result in a T cell memory phenotype (Przybylski, 1989) and cause strong humoral responses by enabling more rounds of antibody affinity maturation in germinal centers (Cirelli and Crotty, 2017; Tam et al., 2016). In this setting, the tunable protein/antigen expression profile of IVT mRNA can be another advantage compared to conventional subunit vaccine platforms. Indeed, researchers have taken advantage of this property of IVT mRNA to provide sustained antigen kinetics, yielding high-affinity antibodies and durable protection (Pardi et al., 2018a). In accord with preclinical data, sustained antigen production in healthy recipients of Spikevax® vaccine was recently demonstrated in the clinic (Ogata et al., 2021). Using ultrasensitive single-molecule assays, Ogata et al. detected peak antigen levels an average of 5 days following the first injection, while no antigen could be detected on day 14, with concomitant detection of increased levels of antibodies induced against the SARS-CoV-2 antigen. Approved/Clinical trials using LNP-formulated mRNA vaccines for various infectious diseases are indicated in Table 1.

3.2 Cancer immunotherapy

3.2.1 Cancer vaccines

Conry et al. conducted the first proof of concept study using naked RNA to induce anti-tumoral immune responses in vivo in 1995 (Conry et al., 1995). Since then, mRNA-based vaccines have been intensively explored for cancer immunotherapy. Uptake of naked IVT mRNA primarily occurs via macropinocytosis by tissue-resident dendritic cells (DCs) (Diken et al., 2011; Selmi et al., 2016). The delivery of the sequence for a defined tumor antigen into DCs results in the translation and processing of the mRNA construct. The antigen is presented via major histocompatibility complex (MHC) molecules, which are recognized by the T cell receptors (TCRs) in the secondary lymphoid organs. In general, the translation of IVT mRNA leads to an accumulation of the antigen within the cytoplasm, which favors its processing and presentation via the MHC I pathway and activation of $CD8^+$ cytotoxic T lymphocytes (CTLs). The antigen can further be secreted and again uptaken by *endo*-or phagocytosis, resulting in the presentation of the antigen via MHC II leading to the activation of $CD4^+$ T cells. The presentation via MHC II can be further increased by the incorporation of lysosomal target peptides (Van Nuffel et al., 2012). DCs further deliver co-stimulation via B7-CD28 engagement, leading to the proliferation and differentiation of antigen-specific effector T cells (Jenkins et al., 1988, 1991). IVT mRNA further exhibits a self-adjuvant effect due to engagement of endosomal TLRs leading to activation and maturation of DCs. This was associated with the production of type I IFNs and an increased anti-tumoral efficacy even against self-antigens after administration of RNA-LPX in preclinical mouse models (Selmi et al., 2016). It should

Table 1 Overview of clinical trials using LNP-formulated mRNA vaccines for various infectious diseases.

Product candidate	Delivery system/method	mRNA encoding	Disease	Clinical trials phase or approved	Reference	Sponsor
BNT162b	LNP, i.m.	Full-length spike protein	SARS-COV-2	Approved	NCT04368728, PMID: 33301246	BioNTech SE
mRNA-1273	LNP, i.m.	Full-length spike protein	SARS-COV-2	Approved	NCT04470427, PMID: 33268462,PMID: 34551225	Moderna TX. Inc.
	LNP, i.m.	Full-length RBD	SARS-COV-2	Phase III, recruiting	NCT04847102	Walvax Biotechnology Co. Ltd., Abogen Biosciences Co. Ltd.
mRNA-1647	LNP, i.m.	Subunits of the HCMV pentamer complex and the glycoprotein B (gB) protein	HCMV	Phase III, recruiting	NCT05085366	Moderna TX. Inc.
mRNA-1893	LNP, i.m.	prME envelope protein	Zika vaccine	Phase II, recruiting	NCT04917861	Moderna TX. Inc.
mRNA-H10N8, mRNA-H7N9	LNP, i.m.	Full length HA glycoprotein from H10N8 and H7N9 strain	Influenza	Phase I, completed	NCT03076385, NCT03345043, PMID: 31079849	Moderna TX. Inc.
mRNA-1345	LNP, i.m.	Prefusion F protein	RSV	Phase I, recruiting	NCT04528719	Moderna TX. Inc.
mRNA-1653	LNP, i.m.	F protein from hMPV and PIV3	Human metapneumovirus and parainfluenza virus 3 (hMPV/PIV3)	Phase Ib, recuiting	NCT04144348, NCT03392389	Moderna TX. Inc.
mRNA-1189	n.a.	EBV gylcoproteins	EBV	Phase I, recruiting	NCT05164094	Moderna TX. Inc.

EBV, Epstein-barr virus; LNP, lipid nanoparticles; RBD, receptor binding domain; HA, hemagglutinin; HCM, human cytomegalovirus; RSV, respiratory syncytial virus; SARS-COV-2, severe acute respiratory syndrome coronavirus type 2.

From ClinicalTrial.gov and Nitika, Wei, J., & Huo, A.-M. (2021). The development of mRNA vaccines for infectious diseases: Recent updates. *Infection and Drug Resistance*, *14*, 5271–5285.

be noted here that high levels of type I IFNs can be counterproductive in infectious disease; for example, they may block humoral responses and induce immunoregulatory cytokines such as interleukin-10 (IL-10) (McNab et al., 2015). Similar parallels have been reported in the field of vaccine development, where it has been observed that the blockade of IFN signaling can result in increased vaccine potency (De Beuckelaer et al., 2016; Pollard et al., 2013). This is particularly relevant for IVT mRNA-based vaccines because, depending on the route of administration, type I IFNs have been shown to inhibit cells' translation machinery, limiting antigen expression (Pollard et al., 2013; Van Hoecke et al., 2020). Therefore, the immunostimulatory nature of IVT mRNAs, along with any delivery vehicles, must be tuned to achieve optimal translation and self-adjuvant strength for a given application and route of administration.

The antigens encoded for mRNA-based cancer vaccines can be tumor-associated self-antigens (TAAs) or tumor-specific antigens (TSAs; also called neoantigens). TAAs are abundant in malignant tissue but can also be expressed on healthy cells, TSAs, in contrast, are truly non-self-molecules, induced by patient-specific mutations (Schumacher and Schreiber, 2015). IVT mRNAs encoding for TAAs and/or TSAs have been used in two approaches for cancer immunotherapy: as a cell-based DC vaccine or as direct injection (Fig. 2). DCs are required to induce antigen-specific T cell responses, therefore the ex vivo transfection and re-infusion of autologous DCs deemed to be a feasible and safe approach for cancer immunotherapy, one which has been reviewed in detail previously (Beck et al., 2021; Benteyn et al., 2015; Pardi et al., 2018b). However, several factors changed the focus of the field toward the in vivo delivery of mRNA. First, the direct injection of IVT mRNA circumvents the labor-intensive and lengthy manufacturing process required for cell-based DC vaccines (reviewed by Perez and De Palma, 2019). Furthermore, the direct injection of mRNA and the release of soluble mediators such as cytokines and chemokines by the transfected DCs in the secondary lymphoid organs may further improve the efficacy of mRNA-based vaccines. The feasibility and safety of local injection of naked mRNA for cancer immunotherapy have been shown in several preclinical and clinical investigations (Table 2). Intradermal injection of globin UTR-stabilized mRNA combined with granulocyte-macrophage colony-stimulating factor (GM-CSF) (Carralot et al., 2004) and especially the intranodal injection of modified mRNA (Kreiter et al., 2010) demonstrated induction of antigen-specific T cell responses in preclinical tumor models. Consequently, a phase I/II trial assessed the intradermal injection of naked

Table 2 Overview of clinical trials with mRNA encoding for TAAs or neoantigens (mRNA cancer vaccines).

Product candidate	Delivery system/ method	mRNA encoding	Disease	Clinical trials phase	Reference	Sponsor
BNT111	LPX/i.v.	NY-ESO-1, MAGE-A3, tyrosinase and TPTE	Advanced melanoma	Phase I, Active, not recruiting	NCT02410733	BioNTech SE
BNT111	LPX/i.v.	NY-ESO-1, MAGE-A3, tyrosinase and TPTE, mono- and combination therapy with cemiplimab	Melanoma	Phase I, recruting	NCT04526899	BioNTech SE, Regeneron Pharmaceuticals
BNT112	LPX/i.v.	5 prostate cancer TAAs, mono- and combination therapy with cemiplimab	Prostate cancer	Phase I/II, recruiting	NCT04382898	BioNTech SE
BNT113	LPX/i.v.	HPV-16 oncoproteins E6/7 in combination with pembrolizumab	Head and Neck Squamous Cell Carcinoma (HNSCC)	Phase II, recruting	NCT04534205	BioNTech SE
BNT114	LPX/i.v.	TNB TAAs, up to 20 neoantigens and p53	TNBC	Phase I, Active, not recruiting	NCT02316457	BioNTech SE, Seventh Framework Programme
BNT115	LPX/i.v.	Three ovarian TAAs, in combination with neoadjuvant chemotherapy	Ovarian cancer	Phase I, recruiting	NCT04163094	University Medical Center Groningen, BioNTech SE

Continued

Table 2 Overview of clinical trials with mRNA encoding for TAAs or neoantigens (mRNA cancer vaccines).—cont'd

Product candidate	Delivery system/ method	mRNA encoding	Disease	Clinical trials phase	Reference	Sponsor
Autogene cevumeran, RO7198457	LPX/i.v.	Neoantigens, mono- and combination therapy with atezolizumab	Multiple solid tumors	Phase I, recruiting	NCT03289962	Genentech Inc., BioNTech SE
Autogene cevumeran, RO7198457	LPX/i.v.	Neoantigens, in combination with pembrolizumab	Advanced Melanoma	Phase II, recruting	NCT03815058	Genentech Inc., BioNTech SE
	mRNA/s..c.	Personalized neoantigen vaccine	Advanced Digestive System Neoplasms	Recruitment status unknown	NCT03468244	Changhai Hospital, Stemirna Therapeutics
RNActive®, CV9104	Protamine-formulated, i.d.	PSMA, PSCA, STEAP1, PAP and MUC1	Prostate cancer	Phase I/II, terminated	NCT01817738	CureVac AG
ECI-006	mRNA/i.n.	MAGE-A3, MAGE-C2 and PRAME and TriMix (CD40L, CD70 and caTLR4)	Melanoma	Phase I, terminated	NCT03394937	eTherna immunotherapies
BI 1361849	RNActive/i.d.	NY-ESO-1, MAGE-C1, MAGE-C2, TPBG, survivin, MUC1, in combination with Durvalumab and Tremelimumab	NSCLC	Phase I/II, completed	NCT03164772	Ludwig Institute for Cancer Research and collaborators

mRNA-5671/V941	LNP/i.m.	KRAS-mutations	KRAS-mutated neoplasms, NSCLC	Phase I, recruiting	NCT03948763	Merck Sharp & Dohme Corp.
mRNA-4157	LNP/i.m.	Neoantigens, in combination with pembrolizumab	Solid tumors	Phase I, recruiting	NCT03313778	Moderna TX. Inc., Merck Sharp & Dohme Corp.
mRNA-4157	LNP/i.v.	Neoantigens, in combination with pembrolizumab	Melanoma	Phase II, active not recruiting	NCT03897881	Moderna TX. Inc., Merck Sharp & Dohme Corp.
	Lipid-particles/i.v.	Autologues total tumor mRNA and LAMP-fused HCMV pp65	Glioblastoma	Phase I, recruting	NCT04573140	University of Florida and collaborators

HPV, human papilloma virus; i.d., intradermal; i.m., intramuscular; i.n., intranodal; i.v., intravenous; LAMP, lysosome-associated membrane glycoprotein; MAGE, melanoma-associated antigen; MUC1, mucin-1; neoAg, neoantigen; NY-ESO-1, New York esophageal squamous cell carcinoma-1; NSCLC, non-small-cell lung cancer; PD-1, programmed cell death protein 1; PD-L1, Programmed cell death ligand 1; PSA, prostate-specific antigen; PSCA, prostate stem cell antigen; PSMA, prostatespecific membrane antigen; 7RNActive, protamine-formulated mRNA; LPX, liposome-formulated mRNA; s.c., subcutanous; TAA, tumor-associated antigen; td, tetanus-diphtheria toxoid; TLR, toll-like receptor; TNBC, triple-negative breast cancer.

From ClinicalTrial.gov and adapted from Beck, J. D., Reidenbach, D., Salomon, N., Sahin, U., Tureci, O., Vormehr, M., et al. (2021). mRNA therapeutics in cancer immunotherapy. *Molecular Cancer*, *20*(1), 69; Miao, L., Zhang, Y., & Huang, L. (2021). mRNA vaccine for cancer immunotherapy. *Molecular Cancer*, *20*(1), 41.

IVT mRNA encoding for the TAAs mucin 1 (MUC1), carcinoembryonic antigen (CEA), human epidermal growth factor receptor 2 (Her-2/neu), telomerase, survivin, and melanoma-associated antigen 1 (MAGE-A1) in combination GM-CSF as an adjuvant. Vaccination was well-tolerated in advanced renal cell carcinoma (RCC) patients and induced antigen-specific T cell responses to the encoded TAAs (Rittig et al., 2011).

The concept of inducing anti-tumoral responses against neoantigens has been demonstrated in multiple preclinical studies (Castle et al., 2012; Kreiter et al., 2015). Based on these encouraging results, a pipeline for the development of a personalized neoantigen-specific mRNA vaccine was developed by BioNTech. The development of such vaccines is described as entailing the acquisition of tumor biopsies, next-generation sequencing, identification of mutations, computational prediction of neoepitopes, and the design and manufacturing of two patient-specific mRNA vectors encoding for up to 20 individual neoepitopes. In one study using this method, naked, unmodified IVT mRNAs were injected intranodally and led to the induction of poly-neoepitope-specific effector T cell responses in 13 melanoma patients (NCT02035956) (Sahin et al., 2017).

Independent of the administration route, mRNA is highly susceptible to degradation by endogenous RNases and cannot passively cross the cell membrane due to its negative charge (Gupta et al., 2013). Thus, the application of naked mRNA is mostly restricted to local injection. Protamine, a poly-cationic polymer, has been shown to increase RNA stability in vitro and to protect mRNA from degradation by RNases (Hoerr et al., 2000). In one study, in vivo administration of a two-component mRNA vaccine, consisting of free and protamine-complexed mRNA, induced Th1 and CTL responses in tumor-bearing mice (Fotin-Mleczek et al., 2011). In this context, the protamine formulation serves as an adjuvant, activating TLR7, and has been shown to induce cellular and humoral immune responses in preclinical models of influenza (Kallen et al., 2013). The RNActive® vaccine technology, developed by CureVac, has offered some of the first results showing intradermal injection with protamine-protected mRNA-encoding for TAAs (with GM-CSF, and KLH as adjuvant), is feasible and can induce anti-tumoral responses in metastatic melanoma patients (Weide et al., 2009) (NCT00204607). To improve therapeutic efficacy, the RNActive® platform is currently being evaluated in phase I/II trial as a combination therapy with durvalumab (anti-PD-L1) and tremelimumab (anti-CTLA-4) in non-small cell lung cancer (NSCLC) patients. Intradermally injected protamine-protected mRNA seems to be safe; however, the antigen presentation seems to be restricted to draining lymph nodes nearest the injection site (Kowalczyk et al., 2016). Other formulations suitable for intravenous administration will be necessary to increase vaccine efficacy the systematic targeting of DCs.

One promising delivery vehicle for mRNA vaccines is cationic liposomes introduced by Malone et al. in 1989 as an in vitro RNA transfection reagent (Malone et al., 1989). mRNA formulated in cationic lipoplexes (LPX) can be administered via different injection routes, and importantly, lead to selective uptake by $CD11c^+$ DCs in the spleen and secondary lymphoid organs after intravenous administration (Kranz et al., 2016) (Fig. 1C). Furthermore, the systemic administration of $m^1\psi$-modified RNA-LPX induces strong effector T cell responses against TAAs or TSAs in preclinical mouse models in a type I IFN-dependent manner (Kranz et al., 2016). The results of phase I clinical study utilizing RNA-LPX against the TAAs NYE-ESO-1, tyrosinase, MAGE-A3, and TPTE mediated durable objective responses and induced $CD4^+$ and $CD8^+$ T cell responses in checkpoint inhibitor-experienced melanoma patients (Sahin et al., 2020) (NCT02410733). Further phase I/II clinical trials were initiated to investigate the vaccination against non-mutated shared tumor antigens in TNBC (NCT02316457), ovarian (NCT04163094), and prostate cancer (NCT 04382898), and in combination with anti-PD-1 in melanoma patients (NCT04526899). The same RNA-LPX technology has also been utilized for a personalized vaccine approach in the individualized neoantigen-specific immunotherapy (iNeST) program. Unmodified mRNA encoding for 20 patient-specific neoantigens is currently being investigated in phase I and II clinical trials for the treatment of multiple solid tumors as mono- and combination therapy with pembrolizumab (NCT03289962, NCT038 15058, and NCT04486378).

The approval of patisiran in 2018, a siRNA-based therapeutic which is formulated in LNPs and accumulates in the liver after intravenous administration (Akinc et al., 2019), has facilitated the adaptation of LNPs for the delivery of mRNA molecules much larger than siRNA. LNP formulations enable high in vitro transfection efficiencies for various cell types, including DCs, and cause in vivo protein expression after systemic administration, while protecting their IVT mRNA cargo from degradation (Pardi et al., 2015). Notably, both approved mRNA-based COVID-19 vaccines consist of $m^1\psi$-modified mRNA encapsulated in LNPs. A similar approach entailing the intramuscular injection of LNP-formulated nucleoside-modified mRNA coding for up to 20 mutated neoepitopes is currently being investigated as a treatment for patients with solid cancer (NCT 03313778) and melanoma (NCT03897881). In another trial, mRNA encoding the most common mutations of KRAS oncogene is being evaluated as monotherapy or in combination with anti-PD-1 for the treatment of solid tumors (NCT03948763). LNPs typically consist of a mixture of cationic/ionizable and "helper" lipids, cholesterol, and polyethylene glycol (PEG) lipids. Decoration of the LNPs surface with various targeting molecules and tuning of LNPs' chemical composition is an active area of research and promises to enable target protein/antigen expression in organs other than the liver, and even the cell type-specific delivery of mRNAs following systemic administration (Hou et al., 2021). Of particular interest for vaccine development are LNPs with efficient accumulation in the spleen, which has been achieved by modulation of the surface charges in preclinical studies (Dilliard et al., 2021). Lastly, components of LNPs can act as additional adjuvants, as recently shown for heterocyclic lipids that activate stimulator of interferon genes (STING), enhancing the anti-tumor efficacy of a model vaccine (Miao et al., 2019).

3.2.2 mRNA-encoded immunomodulators

The intrinsic innate immune activating capacity of mRNA (Diebold et al., 2004; Heil et al., 2004; Hornung et al., 2006) can be utilized in cancer immunotherapy to provide adjuvant signals to vaccines; however, as discussed in Section 2.2, its immunogenic properties may also have an influence on translational efficiency in a context-depend manner. To optimize the translational efficiency of mRNA and provide additional co-stimulation for strengthened T cell responses for vaccination, CureVac developed the RNActive platform for enhanced adjuvant activity through TLR7 activation (discussed in Section 3.2.1). CureVac also developed the non-coding, unmodified mRNA CV8102, which is intended to induce TLR7/8 and RIG-I signaling after intratumoral injection, leading to enhanced activation and maturation of antigen-presenting cells (APCs). The combination of CV8102 with protein- or peptide-based vaccination resulted in increased antigen-specific T cell responses, heightened antibody levels in serum, and stronger tumor control compared to the vaccine alone (Heidenreich et al., 2015). Based on promising preclinical data, CureVac initiated a phase I dose-escalation study of CV8102 as mono- or combination therapy with anti-PD-1; initial results showed an increase of inflammatory cytokines in patients' serum. Furthermore, regression of injected and distant lesions was observed in some patients (NCT03291002, Abstract LB-021 AACR

Annual Meeting 2019, Abstract 800 SITC Annual Meeting 2021, Abstract 473 SITC Annual Meeting 2021).

Immunomodulatory molecules such as cytokines, costimulatory ligands, and their receptors, which augment DC maturation or increase T cell activation, can be co-transfected with mRNA-based cancer vaccines. Such a co-transfection approach was first described for a combination of constitutively active (ca) TLR4, CD40L, and CD70, called TriMix, which was used preclinically and clinically for ex-vivo transfection of DCs, as well as for direct in vivo injection (Table 3). caTLR4- and CD40L-induced DC maturation as well as CD70-dependent co-stimulation of naïve T cells led to increased antigen-specific T cell responses, which were protected against suppressive effects by regulatory T cells (Tregs) in preclinical models (Bonehill et al., 2008, 2009; Pen et al., 2013). Safety and immunogenicity were later demonstrated in melanoma patients, with objective tumor responses after combined intradermal and intravenous administration of the unmodified mRNA (De Keersmaecker et al., 2020; Wilgenhof et al., 2011, 2013) (NCT01066390, NCT01302496). The simultaneous, intranodal injection of non-formulated TriMix mRNA with antigen-encoding mRNA is currently being investigated in a phase I trial (Van Lint et al., 2012) (NCT03394937). Further, beneficial effects on DC maturation, T cell proliferation, and lytic activity of cytotoxic T cells have been observed for co-transfection of DCs with additional mRNAs encoding co-stimulatory molecules such as OX40L, 4-1BB, GITRL, CD40L, ICOSL, and the cytokines IL-12p70, GM-CSF, and IL-15 in preclinical studies (Amin et al., 2015; Bontkes et al., 2007; Daneshmandi et al., 2018; Grunebach et al., 2005; Levin et al., 2017; Minkis et al., 2008; Naka et al., 2008; Song et al., 2012; Tcherepanova et al., 2008; Tuyaerts et al., 2007; Van den Bergh et al., 2015, 2017) (NCT01582672). As the activation of STING in DCs is an essential step for the development of adaptive T cell responses (Vatner and Janssen, 2019), researchers developed mRNA encoding caSTING. It has been shown to be an effective adjuvant for the induction of antigen-specific $CD8^+$ T cell response (Tse et al., 2021) and it was demonstrated in a recent publication, which has not been peer-reviewed yet, to reactivate STING signaling in cancer types (Liu et al., 2022).

Cytokines are another class of immunomodulators and proved to be potent inducers of anti-tumoral responses in preclinical studies (Berraondo et al., 2019). However, recombinant cytokines have previously exhibited dose-limiting toxicities (Conlon et al., 2015; Leonard et al., 1997; Motzer et al., 1998) and promising results from preclinical studies have not been

Table 3 Overview of clinical trials with mRNA-based cancer immunotherapies excluding vaccination.

Product candidate	Delivery system/method	mRNA encoding	Disease	Clinical trials phase	Reference	Sponsor
RNActive, CV9103	50% not formulated 50% protamine-complexed/i.d.	mRNA encoding 4–6 tumor-specific antigens, half of the mRNA is protamine-complexed for TLR7 activation	Hormone Refractory Prostate Cancer	Phase I/IIa, Terminated	NCT00906243	Unviersity of Florida, CureVac AG
RNActive, CV9104			Prostate Carcinoma	Phase II, Terminated	NCT02140138	CureVac AG
RNActive, CV9202 (+radiotherapy)			Non Small Cell Lung Cancer	Phase I/IIa, Terminated	NCT01915524	CureVac AG
RNActive, CV9202 (+durvalumab/tremelimumab)			Mestatatic and non metastatic Non-small Cell Lung Cancer	Phase I/II, Completed	NCT03164772	Ludwig Institute for Cancer and collaborators
CV8102 (+ anti-PD-1)	LPX, i.t.	Non-coding mRNA inducing TLR7/8 and RIG-I activation	Multiple solid tumors	Phase I, Active	NCT03291002	CureVac AG and collaborators
TriMix-DC	Transfection ex vivo, cells i.v.	TriMix: CD40L, caTLR4, CD70L* tumor antigen either mRNA encoded or given as peptide	Melanoma	Phase I, Completed	NCT01066390	Bart Neyns, Universitair Ziekenhuis Brussel
TriMix-DC (+Ipilimumab)	Transfection ex vivo, cells i.v.		Malignant Melanoma	Phase II, Completed	NCT01302496	Bart Neyns, Vrije Universiteit Brussel
ECI-006	PBS/Ringer, i.n.	TriMix: CD40L, caTLR4, CD70L and tumor-associated antigen	Melanoma	Phase I, Terminated	NCT03394937	eTheRNA immunotherapies

TriMix	PBS/Ringer, i.t.	Trimix: CD40L, caTLR4, CD70L	Breast Cancer	Phase I, Recruiting	NCT03788083	Universitair Ziekenhuis Brussel, eTheRNA immunotherapies
Meso-CAR T cells	Transfection ex vivo, cells i.v.	CAR targeting mesothelin	Malignant Pleural Mesothelioma	Phase I, Completed	NCT01355965	University of Pennsylvania
Meso CAR T cells	Transfection ex vivo, cells i.v.	CAR targeting mesothelin	Metastatic Pancreatic Ductal Adenocarcinoma	Phase I, Completed	NCT01897415	University of Pennsylvania
Anti-MESO CAR T cells	Transfection ex vivo, cells i.v.	CAR targeting mesothelin	Refractory Malignant Solid Neoplasm	Phase I, Recruiting	NCT04981691	Ruijin Hospital, UTC Therapeutic Inc.
cMet RNA CAR T cells	Transfection ex vivo, cells i.t.	CAR targeting c-Met	Breast Cancer	Phase I, Completed	NCT01837602	University of Pennsylvania
anti-cMET CAR T cells	Transfection ex vivo, cells i.v.	CAR targeting c-Met	Malignant Melanoma and Breast Cancer	Early Phase I, Terminated	NCT03060356	University of Pennsylvania
RNA anti-CD19 CAR T cells	Transfection ex vivo, cells i.v.	CAR targeting CD19	Hodgkin Lymphoma	Early Phase I, Terminated	NCT02277522	University of Pennsylvania
RNA CART19 cells	Transfection ex vivo, cells i.v.	CAR targeting CD19	Hodgkin Lymphoma	Early Phase I, Terminated	NCT02624258	University of Pennsylvania, Children's Hospital of Philadelphia

Continued

Table 3 Overview of clinical trials with mRNA-based cancer immunotherapies excluding vaccination.—cont'd

Product candidate	Delivery system/method	mRNA encoding	Disease	Clinical trials phase	Reference	Sponsor
Anti-CD123 CAR TCR/4-1BB-expressing T lymphocytes	Transfection ex vivo, cells i.v.	CAR targeting CD123	Relapsed or Refractory Acute Myeloid Leukemia	Early Phase I, Terminated	NCT02623582	University of Pennsylvania
Descartes-08	Transfection ex vivo, cells i.v.	CAR targeting BCMA	Multiple Myeloma	Phase I/II, Recruiting	NCT03448978	Cartesian Therapeutics
CAR-NK cells targeting NKG2D ligands	Transfection ex vivo, cells i.v.	CARs targeting NKG2D ligands	Multiple solid tumors	Phase I, Status unknown	NCT03415100	The Third Affiliated Hospital of Guangzhou Medical University
BNT221	LPX/i.v.	CAR targeting Claudin 6 and Claudin6 in vivo vaccine	Multiple solid tumors	Phase I/IIa, Recruiting	NCT04503278	BioNTech Cell & Gene Therapies GmbH

DC, dendritic cells; i.v., intravenous; i.d., intradermal; i.t., intratumoral; TLR, toll-like receptor; LPX, liposome-formulated mRNA; CAR, chimeric antigen receptor; RIG-I, retinoic acid-inducible gene I; BCMA, B cell maturation antigen; NKG2D, natural killer group 2 member D.

From ClinicalTrial.gov, Beck, J.D., Reidenbach, D., Salomon N., Sahin, U., Tureci, O., Vormehr, M, et al., 2021. mRNA therapeutics in cancer immunotherapy. Mol. Cancer 20 (1), 69, and Wei et al., 2021.

validated in clinical trials. To date, only IL-2 and IFN-α were approved by the FDA and induced only limited clinical benefits (Atkins et al., 1999; Fyfe et al., 1995; Golomb et al., 1986; Groopman et al., 1984; Kirkwood et al., 1996; Solal-Celigny et al., 1993). mRNA-encoded cytokines offer the benefit of prolonged cytokine production with reduced peak serum levels, and mRNAs encoding for IL-12p70, IL-2, IL-7, IFNβ, and TGFβ were investigated in preclinical studies mainly for intratumoral administration (Hewitt et al., 2020; Lai et al., 2018; Li et al., 2020; Nguyen et al., 2020; Van der Jeught et al., 2014). The preclinical results sparked recent clinical studies for intratumoral administration of IL-12 and intravenous application of IL-2 and IL-7-encoding mRNAs (NCT03946800, NCT04455620, NCT04710043) (Table 3).

Importantly, mRNA-based formats allow the combination of multiple cytokines. Three combinatorial approaches are currently being tested in both preclinical and clinical studies. SAR441000 consists of $m^1\psi$-modified mRNAs encoding for IL-12, IFN-α, GM-CSF, and IL-15 sushi, and is being evaluated as mono- and combination therapy with cemiplimab in a phase I study for multiple solid tumors. The anti-tumoral efficacy of this combination was demonstrated in preclinical studies where the intratumoral injection of the aforementioned mRNA-encoded cytokine mix led to tumor regression of injected and distant lesions in several syngeneic mouse tumor models and induced an immunological memory protection which protected the mice from tumor rechallenge (Hotz et al., 2021) (NCT03739931). A combination of modified mRNA encoding OX40L, IL-23, and IL-36γ, mRNA-2752, exhibited promising preclinical anti-tumoral efficacy in several syngeneic mouse tumor models by inducing systemic immunity and memory formation following intratumoral administration (Hewitt et al., 2019). mRNA-2752 is currently being tested in a dose-escalation study in patients with solid tumor malignancies or lymphoma (NCT03739931). Furthermore, caTLR4, caCD40, IL-2, IL-12, and IL-15, which as single agents were previously proven to induce co-stimulation of T cells, are being developed as a combination therapy. In preclinical studies, this combination exhibited superior antigen-specific killing of autologous melanoma cells associated with IFNγ responses in vitro (Kariko et al., 2005; Levin et al., 2018; Weinstein-Marom et al., 2016, 2019).

mRNA-encoded signaling molecules such as MyD88, TNF, IRF5, or IKK can also be used to provide an additional intracellular stimulus to APCs. This approach holds substantial promise as a strategy to selectively reprogram suppressive myeloid cells, such as M2-like macrophages, in the tumor microenvironment (Seif et al., 2017; Zhang et al., 2019).

3.2.3 mRNA-engineered immune cells

Adoptive cell transfer (ACT) of immune cells has been extensively investigated for cancer immunotherapy. Patient-derived tumor-infiltrating lymphocytes can be expanded ex vivo, and furthermore, be modified by co-transfection with mRNA encoding for immunomodulators. For instance, the ex vivo transfection of T cells with mRNA encoding for IL-12 was shown to elicit stronger anti-tumoral responses after re-injection of the cells into the tumor (Etxeberria et al., 2019; Pohl-Guimaraes et al., 2019). Peripheral patient-derived or allogeneic T cells can further be engineered with tumor-specific receptors. So far, most studies utilize retro- or lentiviral transduction systems for T cell engineering, which has been associated with systemic toxicity due to cytokine release syndrome and on-target-off-tumor effects (Kochenderfer et al., 2012; Lamers et al., 2013).

TCR-engineered T cells targeting MHC-bound tumor antigens were first transcribed using mRNA in 2004 (Smits et al., 2004). A recent study demonstrated the feasibility of transfection of $CD8^+$ T cells with antigen-specific TCRs for cancer immunotherapy. This study sought to avoid potential mispairing between the endogenous and exogenous TCR, which could lead to new, potentially toxic target specificities, by suppressing the expression of endogenous TCR with siRNA and transfecting afterward engineered siRNA-resistant TCR (Campillo-Davo et al., 2018). In contrast to TCRs, the binding of chimeric antigen receptors (CARs) to the target antigens expressed on the tumor cell surface is independent of MHC and leads to potent T cell activation and tumor killing. CARs consist of a single-chain variable fragment (scFv) at N-terminus fused to spacer or hinge domains, a transmembrane domain, and an intracellular domain of CD3ζ (first-generation CARs). Additionally, second-generation CARs are built from CD3ζ and co-stimulatory domains CD28, OX40, or 4-1BB which improved significantly the efficacy, expansion, and persistence of CAR T cells (Freitag et al., 2020). Third-generation CARs carry two co-stimulatory domains in the same receptor, e.g., 4-1BB and CD28, while fourth-generation CAR-T-cells consist of one antigen-specific receptor with the addition of other transgenes that enhance anti-tumor function (Maucher et al., 2021).

T cells transfected with mRNA-encoded CARs were developed for several therapeutic targets mainly targeting CD19 on malignant B cells CD19 (Almasbak et al., 2011; Barrett et al., 2011; Svoboda et al., 2018). CD19 is expressed on tumor cells, but also on healthy B cells, which are targeted to the same extent by CD19-specific CAR T cells leading to cases of hypogammaglobulinemia during treatment. However, these on-target-off-tumor effects can be controlled via IgG reconstitution, CD19 has been widely used as a target for CAR T cell therapy (Anagnostou et al., 2020). The University of Pennsylvania started two pilot trials of an mRNA-encoded CD19 third-generation CAR, and no toxicity but limited efficacy was observed, with one of five enrolled patients showing a transient complete response (NCT02624258, NCT02277522). Other targets have been investigated for hematological malignancy treatment includes CD20, CD33, CD37, CD123, and B-cell maturation antigen (BCMA) (Foster et al., 2019; Kenderian et al., 2015; Koksal et al., 2019; Panjwani et al., 2016; Tasian et al., 2017). It was started a clinical study of CD123 but was discontinued due to a lack of funding, and an anti-BCMA CAR is currently under investigation for multiple myeloma patients in the phase I/II trial (NCT02623582, NCT03448978).

Mesothelin is expressed by several cancer types but also at low levels on healthy mesothelial surfaces and was one of the first evaluated CAR T cell targets for solid malignancies. Potent anti-tumoral responses of locally administered mesothelin-targeting mRNA-transfected CAR T cells were observed in preclinical studies (Beatty et al., 2014, 2018; Hung et al., 2018; Maus et al., 2013; Zhao et al., 2010). Two phase I studies administering the cells by i.v. infusion showed dose-limiting serositis likely induced by on-target-off-tumor effects on mesothelial cells, but anti-tumoral efficacy was also indicated by two cases of complete responses (NCT01355965, NCT01897415). Currently, patients with refractory malignant solid neoplasms are being infused with anti-mesothelin CAR T cells in a phase I study in China (NCT04981691).

Multiple targets for the treatment of solid tumors are under either preclinical or clinical investigation as binding targets for mRNA-transfected CAR T cells; these include c-Met, GD2, folate receptor, melanoma-associated chondroitin sulfate proteoglycan, vascular endothelial growth factor receptor 2, epithelial adhesion molecule, NKG2D, epidermal growth factor receptor and Her2/neu (NCT01837602, NCT03060356).

Despite their attractive safety profile, mRNA-based therapies for ACT face some limitations. While virally transfected receptors are long lasting, TCRs or CARs introduced via mRNA persist for only a few days. Hence, repeated infusions are performed, which can pose safety concerns, like the severe case of an IgE-mediated anaphylactic reaction observed in a clinical trial targeting mesothelin (Maus et al., 2013). Another limitation of mRNA-based ACT might be the challenge in reaching solid tumors within the short time of mRNA expression. In hematological malignancies, CAR T cells find their target in the periphery and are also activated by circulating B cells expressing the tumor antigen, leading to rapid proliferation and persistence of the adoptively transferred cells (Kalos et al., 2011; Porter et al., 2015). For solid tumors, i.v.-infused T cells must actively infiltrate the tumor, and to circumvent this challenge, many mRNA-based CAR T cells are administered intratumorally. However, Reinhard and colleagues recently published a new concept to prolong the persistence of i.v.-injected, mRNA-transfected CAR T cell in vivo for the treatment of solid tumors. They generated mRNA-transfected second-generation CAR T cells targeting the tight junction protein claudin 6 (CLDN6), which is overexpressed in several human solid cancer but not in healthy adult tissue. Following an adoptive transfer of the CLDN6-specific CAR T cells, a CAR-T cell amplifying RNA vaccine (CARVac) consisting of LPX-formulated CLDN6-mRNA was administered. The in vivo transfected DCs served as APCs to the CAR T cells, which improved their systemic persistence and expansion. With this approach, one sub-optimal, therapeutically non-efficient dose of CAR T cells followed by repeated CARVac treatments was sufficient for tumor regression (NCT04503278) (Reinhard et al., 2020).

mRNA-based engineering of y/δ T cells, NK cells, NKT cells, and macrophages pose no risk for mismatching of the endogenous and exogenous TCR and are not allogenic since they are MHC I independent. y/δ T cells, transfected with mRNA-TCRs, have exhibited expression and lysis capacity comparable to α/β T cells (Dorrie et al., 2014; Harrer et al., 2017). In a preclinical study, NK cells have been transfected with mRNA encoding for CD19, CD20, or BCMA CARs (Boissel et al., 2009, 2012; Chu et al., 2015; Gotley et al., 2009; Li et al., 2010). NKG2D CAR NK cells engineered using mRNA were locally infused to patients with colorectal cancer in a Phase I study, resulting in regression of primary tumors and liver metastases (NCT03415100). The approach has since been advanced by co-electroporation of CXCR-1 mRNA to enhance migration of the CAR NK cells to the tumor site when administered i.v. (Ng et al., 2020; Xiao et al., 2019).

3.2.4 mRNA-encoded antibodies

Aside from induction of ADCC, ADCP, and CDC, antibodies can be used in immunotherapy to block receptors involved in tumor survival or immunosuppression, as agonists for immunostimulatory receptors, as a vehicle for the targeted delivery of toxic/immunomodulatory payloads, and to direct immune cells toward tumor tissue via antibodies with multiple specificities (Zahavi and Weiner, 2020). Most of the currently approved antibodies are recombinant products, which face manufacturing challenges such as poor stability and a tendency toward aggregation (Sanchez-Martin et al., 2011). Moreover, full-sized antibodies exhibit poor tissue and tumor penetration, which led to the development of smaller antibody formats and nanobodies. The latter derives from heavy chain-only camelid antibodies, have a shorter half-live due to rapid clearance, making continuous administration via infusion pumps necessary (Schlake et al., 2019).

Two preclinical studies demonstrated the feasibility of mRNA-encoded antibodies for cancer immunotherapy in 2017. Thran and colleagues administered 50 μg of LNP-formulated, mRNA-encoding rituximab, binding to CD20 into tumor-bearing mice and observed superior tumor growth control compared to 200 μg recombinant rituximab (Thran et al., 2017). In another preclinical study, bispecific T cell-engaging antibodies targeting the T cell receptor-associated molecule CD3 and one of the tumor-associated antigens CLDN6, CLDN18.2, or EPCAM were encoded in modified mRNA (known as RiboMAB®s). RiboMAB®s showed in vitro efficient target lysis and high T cell activation induction, comparable to the corresponding recombinant antibodies. Complete tumor regression in a murine xenograft model could be observed upon i.v. injection of the CD3 × CLDN6 RiboMAB (Stadler et al., 2017). Furthermore, trastuzumab, an antibody targeting HER2 has been evaluated as mRNA-LNP targeting the liver and exhibited anti-tumoral efficacy following i.v. administration (Rybakova et al., 2019).

3.3 Autoimmunity

3.3.1 Non-inflammatory autoantigen vaccines

During homeostasis, immature DCs continuously uptake self-derived antigens. In the absence of pathogen- or danger-associated molecular patterns (PAMP/DAMP), DCs do not undergo maturation, presenting antigens while in a steady state with no costimulatory signals. Under such conditions, antigen presentation to lymphocytes leads to their anergy and formation of Tregs. In this setting, $CD11c^+$ DCs play a vital role in the maintenance of peripheral tolerance, as evidenced by the development of spontaneous autoimmunity when these DCs are ablated in mice (Ohnmacht et al., 2009). Therefore, delivery of self-antigens to steady-state dendritic cells under non-inflammatory conditions has been long-sought as an approach for treating some underlying causes of autoimmune diseases—i.e., depletion of autoreactive T cells and autoantibodies (Cifuentes-Rius et al., 2021; Streeter and Wraith, 2021). The mRNAs lipoplexes targeting splenic $CD11c^+$ APCs described earlier (Section 3.2.1) were recently repurposed to achieve suppression of autoreactive T cells against myelin-specific autoantigens in mouse models of multiple sclerosis. Krienke et al. used $m^1\psi$-modified, purified mRNA to effectively suppress the PAMP character of mRNAs, thus avoiding triggering TLR7, and so providing self-antigen expression and presentation in DCs in non-inflammatory settings. This approach led to the formation of Tregs that suppress disease via both cognate antigen and via bystander immunosuppression. These autoantigen-encoding mRNAs promise great potential for reversing autoimmune disorders without global suppression of the immune system (Krienke et al., 2021).

3.3.2 mRNA-encoded regulatory factors

In early clinical trials using IL-2 aimed at the expansion of $CD8^+$ T cells in cancer patients, a counterintuitive expansion of immunosuppressive Tregs was also observed (Sim et al., 2014). Tregs constitutively express the IL2Rα chain and therefore have a higher affinity for IL2 compared to naive $CD4^+$ and $CD8^+$ T cells; thus, low-dose IL-2 therapy has been proposed as a strategy for selective Treg expansion in vivo, with feasibility demonstrated in autoimmune diseases such as lupus erythematosus (Grasshoff et al., 2021). Recently, an mRNA LNP (mRNA-6231) encoding human serum albumin IL-2 mutein fusion protein entered a phase I clinical trial. mRNA-6231 is administered subcutaneously and encodes an IL2 variant which preferentially binds CD25, leading to selective expansion of Tregs.

Stephan and colleagues developed another IVT mRNA-based approach for the treatment of autoimmune diseases that relies on the in-situ engineering of myeloid regulatory cells (Parayath et al., 2021). Using intravenously administered surface-functionalized LNPs for selectively targeting CD64 expressed on circulating monocytes, they delivered IVT mRNA encoding an anti-inflammatory mediator glucocorticoid-induced leucine zipper. Repeated administration of the treatment was safe, and importantly, was shown to effectively reduce pathogenesis and prolong survival in a model of systemic lupus erythematosus. A related approach was reported by Veiga et al., who used an LNP-mRNA encoding IL-10 targeted to $Ly6c^+$ inflammatory leukocytes, resulting in the successful treatment of signs of inflammatory bowel disease (Veiga et al., 2018).

While promising, the use of encoded regulatory factors alleviates the symptoms of autoimmune diseases, and does treat the underlying cause; for example, such treatment does not lead to the depletion of autoreactive T cells. Therefore, in the clinic, this approach will most likely require repeated administration for long-term disease control, and its long-term effects on systemic immunosuppression remain to be seen.

3.4 mRNA-encoded protein replacement

Protein replacement therapies are intended to compensate for a defective or non-functional protein in rare genetic diseases. Most such therapies require repeated injections of recombinant products, which are both laborious to produce and expensive. IVT mRNA has the potential to serve as a new platform for this therapeutic area (Table 4). Kormann et al. performed the first proof-of-concept study using chemically modified mRNA with reduced binding to PRRs for a non-immunotherapeutic treatment in 2011 (Kormann et al., 2011). Balakrishnan and colleagues recently published a study using mRNA-LNPs to restore galactose-1 phosphate uridylyltransferase (GALT) expression and activity in a mouse model of classic galactosemia, a rare metabolic disorder that leads to accumulation of toxic galactose-1 phosphate (gap-1P). They observed a significant reduction of gap-1P in the plasma of GALT-deficient mice after single and repeated intravenous injection of a murine GalT encoding mRNA-LNPs (Balakrishnan et al., 2020). However, the safety and tolerability, especially of the repeated injections, was not a primary objective of this study, and some level of gal-1P remained after treatment. Propionic acidemia, a deficiency of propionyl-CoA carboxylase (PCC), is another rare metabolic disorder. PCC is involved in the metabolism of proteins and lipids, and PCC deficiency leads to the accumulation of toxic metabolites in the blood. In a recent preclinical study, PCC deficiency was compensated by dual mRNAs encoding the alpha and beta subunit of the enzyme to restore metabolic function in the liver. A single i.v. injection of dual mRNAs led to therapeutic-level expression of PCCs (hPCCa & hPCCB) in murine livers. Repeated dosing was well-tolerated in long-term (3–6-month) studies in these mice (Jiang et al., 2020) and an ongoing clinical trial conducted by Moderna (NCT04159103). An et al. offered further evidence for the feasibility of mRNA-based enzyme replacement therapies in a study of methylmalonic acidemia (MMA). 5-methocyU-modified codon-optimized mRNA encoding for the human methylmalonyl-CoA mutase (hMUT), a mitochondrial

Table 4 Overview of clinical trials with mRNA-based protein replacement and gene editing therapies.

Product candidate	Delivery system/ method	mRNA encoding	Disease	Pre-/-clinical trials phase or approved	Reference	Sponsor
AZD8601	mRNA/local tissue injection	VEGF-A	CABG patients	Phase II, completed	NCT03370887	AstraZeneca
MRT5005	Saline/nebulization	CFTR	Cystic fibrosis	Phase I/II, recruiting	NCT03375047	Translate Bio, Inc.
MRT5201	LNP/i.v.	OTC	OTC deficiency	Phase I/II, withdrawn	NCT03767270	Translate Bio, Inc.
mRNA-3704	LNP/i.v.	MUT	Methylmalonic Acidemia	Phase I/II, withdrawn due to business decision	NCT03810690, EU 2019–001061-32	Moderna TX. Inc.
mRNA-3927	LNP/i.v.	PCCA and PCCB	Propionic Acidemia	Phase I/II, recruiting	NCT04159103	Moderna TX. Inc.
ARCT-810	LNP/i.v.	OTC	OTC deficiency	Phase 1, recruiting	NCT04416126, NCT04442347	Arcturs Therapeutics, Inc.
NTLA-2001	LNP/i.v.	Cas9	Gene editing, target TTR, transthyretin amyloidosis	Phase I, recruiting	NCT04601051	Intellia Therapeutics and Regeneron Pharmaceuticals
SB-728mR	Transfection ex vivo, cells i.v.	ZFNs	HIV	Phase I, active, not recruiting	NCT02500849	Sangamo BioSciences Inc
SB-728mR	Transfection ex vivo, cells i.v.	ZFNs	HIV	Phase I, active, not recruiting	NCT03617198	Sangamo BioSciences Inc
SB100X transposase	Transfection ex vivo, cells i.v.	SB transposase	Multiple Myeloma	Phase I/IIA	EudraCT: 2019–001264-30	University Hospital of Würzburg

CABG, coronary artery bypass surgery; CFTR, cystic fibrosis trans-membrane conductance regulator; MUT, methylmalonyl-CoA mutase; OTC, ornithine transcarbamylase; PCCA and PCCB, propionyl CoA carboxylase; VEGF-A, vascular endothelial growth factor; SB, Sleeping Beauty; Cas9, CRISPR associated protein 9; ZNFs, zinc finger nucleases

From ClinicalTrial.gov and Vlatkovic, I., 2021. Non-immunotherapy application of LNP-mRNA: maximizing efficacy and safety. Biomedicine 9 (5).

enzyme responsible for the terminal step of fatty acid oxidation, was formulated into LNPs and administered by intravenous injection in mouse models of MMA (An et al., 2017). Repeated injections were well-tolerated, and the treatment reduced systemic levels of toxic metabolites. Based on the promising preclinical data, Moderna recently initiated a phase I study (NCT04899310).

Another promising example for a non-immunotherapeutic application of mRNA is the activation of regeneration pathways. This approach's feasibility has been demonstrated by the delivery of nucleoside-modified mRNA-LNPs encoding for hepatocyte growth factor (HGF) and epidermal growth factor (EFD) in a preclinical chronic liver injury model (Rizvi et al., 2021). Moderna and AstraZeneca recently completed two clinical trials with local injection of modified mRNA encoding for human vascular endothelial factor A (VEGF-A) to activate blood vessel regeneration in cardiovascular disease (NCT03370887), and for the treatment of ulcers in patients with type II diabetes mellitus (NCT02935712). Initial results showed that the treatment was well tolerated and that the local activation of regenerative pathways holds promise. Further preclinical studies showing the feasibility of mRNA-encoded protein replacement therapies include those targeting alpha-1-antitrypsin deficiency (systemic application of modified mRNA-LNPs encoding human alpha 1-antitrypsin in mice) (Karadagi et al., 2020), transient ischemia (Abe et al., 1991), hemophilia (intravenous injection of two LNP-formulated modified mRNA constructs encoding for human erythropoietin or factor IX) (DeRosa et al., 2016), and sensory nerve disorders (intranasal delivery of mRNA-loaded nanomicelles encoding for brain-derived neurotrophic factor) (Baba et al., 2015).

Despite the promising examples mentioned above, the development of mRNA-based protein replacement therapies still faces several challenges. mRNA-encoded proteins are only transiently expressed in vivo and the replacement of an intracellular protein, for instance, requires uptake of the LNP-encapsulated mRNA into the target cell of interest. Non-immune cells have a generally lower uptake rate compared to phagocytes. Furthermore, the absolute amount of translated protein, which correlates with the absolute amount of uptaken mRNA transcripts and the translation efficiency in target cells, will determine therapeutic efficacy (Sahin et al., 2014). Another important aspect (discussed in detail by Vlatkovic), is the safety and tolerability of both mRNA and the selected LNP (Vlatkovic, 2021). As mentioned earlier (Section 3.2.1), the mRNA, as well as the formulation itself, can mediate immune activation independent of the encoded antigen/protein. One potential method to avoid this was described by Miliotou and colleagues, who showed that mRNA could be delivered to cells via a protein transduction domain (PTD) (Miliotou et al., 2021). The requirement for repeated injections, which might lead to liver toxicity, is another challenge, along with anti-drug antibodies, cytokine storm, and immune cell and complement activation (Vlatkovic, 2021). Local injection or targeted delivery strategies might overcome these limitations in the future.

3.5 Gene editing

In the last decade, progress in the development of genome editing has revolutionized biomedical research on the human genome, which has empowered scientists to better understand the contribution of a single gene product to disease and therapeutic response in an organism. Custom-designed genome editing nucleases, such as zinc-finger nucleases (ZFNs), transcription activator-like effector nucleases (TALENs), and clustered regularly spaced short palindromic repeat–CRISPR-associated protein 9 (CRISPR-cas9) provide potent tools for achieving specific modification of the genomes (Gaj et al., 2013). However, a prolonged expression of these nucleases may present the risk of nonspecific editing. Thus, the nucleases translated from DNA vectors may lead to off-target mutagenesis (Fu et al., 2013). The transient expression feature of IVT mRNA can reduce the risk of off-target activity and may be an attractive approach to deliver such nucleases into human cells. In fact, IVT mRNAs encoding CRISPR-Cas9, TALENs, and ZFNs have been successfully deployed in both preclinical and early-phase clinical trials (Table 4). A major strength unique to CRISPR-Cas9 is that the Cas9 nuclease can specifically target any site in the genome simply by modifying a single guide RNA (sgRNA) sequence; it is also possible to simultaneously edit multiple targets using several sgRNAs (Wang et al., 2013). These features make the delivery of CRISPR-Cas9 via mRNA a natural fit.

Recent studies have described a new strategy to develop universal CAR T cells by editing their genome, co-delivering mRNA encoded Cas9 and multiple sgRNAs to target specifically immunogenic, and graft-versus-host disease-associated/immunogenic genes (Georgiadis et al., 2018; Ren et al., 2017). This strategy makes it possible to develop allogeneic CAR T cells and avoids several limitations of the traditional procedure—which requires the patient's own cells, and is costly, time-consuming, and sometimes not even possible (Chen et al., 2021; Liu et al., 2017). Moreover, Ren and colleagues have eliminated PD-1 from allogeneic CD19 and prostate stem cell antigen (PSCA) CAR T cells, resulting in the enhanced antitumor activity of PD-1 KO CAR T cells in leukemic and prostate cancer preclinical models (Ren et al., 2017). Also, the disruption of gene expression in vivo, through applying CRISPR-Cas9 has opened a new therapeutic strategy. Rosenblum et al. have shown that the delivery of LNPs-encapsulated Cas9 mRNA and sgRNA for PLK1 led to 70% gene editing in an aggressive orthotopic glioblastoma (GBM) in vivo model, which inhibited tumor growth and increased survival in orthotopic GBM (Rosenblum et al., 2020). Moreover, they develop EGFR-targeted sgPLK1-cLNPs, which strongly reduce tumor growth and increase overall survival in the metastatic ovarian adenocarcinoma model (Rosenblum et al., 2020). Another study has demonstrated the benefit of delivering LNPs-encapsulated m1Ψ-containing mRNA encoding adenine base editors (ABEs) and sgRNA to target PCSK9, a negative regulator of the low-density lipoprotein (LDL) receptor (Rothgangl et al., 2021). Following administration, plasma PCSK9 and LDL levels were significantly reduced in mice and macaques. Importantly, the ABE mRNA was cleared rapidly, and no off-target mutations in genomic DNA were found. Thus, this therapeutic approach offers promise for treating familial hypercholesterolemia and monogenic liver diseases (Rothgangl et al., 2021).

Intellia Therapeutics has developed LNPs encapsulating chemically modified sgRNA and mRNA Cas9 to target transthyretin (TTR) in the liver for the treatment of transthyretin amyloidosis (ATTR amyloidosis), alpha-1 antitrypsin deficiency transthyretin amyloidosis (AATD), and primary hyperoxaluria type 1 (PH1) (Finn et al., 2018). Currently, this product, under the NTLA-2001 name is being studied in a Phase I trial in adults with hereditary ATTR amyloidosis with polyneuropathy. In June 2021, Intellia and its collaborator Regeneron announced positive interim clinical results from the first two cohorts of this study (Gillmore et al., 2021).

Quintana-Bustamante et al. reported the use of an mRNA TALEN system to correct mutations in the pyruvate kinase L/R gene (PKLR) in HSPCs to treat pyruvate kinase deficiency (PKD). While they were able to edit PKLR in HSPCs in vitro and detected edited cells 4 months following the engraftment of edited HSPCs in immunodeficient mice, they were not present at levels that suggested therapeutic potential without the substantial further refinement of the technique (Quintana-Bustamante et al., 2019).

IVT mRNA-based ZFNs have been applied to treat HIV disease. The homozygous deletion of CCR5, the major co-receptor required for

HIV-1 infection, in hematopoietic stem cells (HSPCs) leads to resistance to HIV-1 infection (Lopalco, 2010). Allogeneic transplantation of cells harboring a natural CCR5 homozygous deletion has shown evidence as a promising therapeutic strategy to cure HIV (Li, 2020). In fact, Sangamo Therapeutics has developed SB-728mR, an mRNA that encodes ZFNs to target CCR5 for HIV treatment (DiGiusto et al., 2016) (SB-728mR-HSPC; NCT02500849); SB728-mR was delivered into HSPCs via electroporation. The same therapeutic mRNA is being used safely in a separate phase I clinical trial (SB-728mR-T; NCT03617198) to disrupt the CCR5 receptor on $CD4^+$ T cells.

Other IVT mRNAs encoding DNA-editing machinery—such as the application of CRISPR from *Prevotella* and *Francisella* (Cpf1) and transposases from Sleeping Beauty (SB)—have been successfully applied in vivo and in an early phase clinical trial, respectively (Dupuy et al., 2002; Kim et al., 2016; Wilber et al., 2006). In specific examples, Kim et al. used mRNA encoding Cpf1 to generate transformation-related protein 53 (Trp53) and *Lachnospiraceae bacterium ND2006* (LbCpf1) knockout mice (Kim et al., 2016), and Prommersberger et al. used IVT mRNA-encoded SB transposase to develop virus-free CAR T cells for the treatment of myeloma, currently in phase I/II trial (Prommersberger et al., 2021).

Recently, Abudayyeh and colleagues showed that Cas13 can target RNA instead of DNA, meaning CRISPR/Cas13 may offer even more therapeutic options than previously thought (Abudayyeh et al., 2017). In fact, Blanchard et al. co-delivered sgRNAs for highly conserved regions of influenza virus and against the replicase/nucleocapsid genes of SARS-CoV-2 with polymer-formulated Cas13a mRNA and found it degraded influenza RNA in murine lung tissue and reduced SARS-CoV-2 replication and symptoms in hamsters, respectively (Blanchard et al., 2021). And Rashnonejad and co-workers treated facioscapulohumeral muscular dystrophy by co-administration of multiple sgRNAs for the DUX4 gene and mRNA encoding Cas13b, which they found drastically reduced DUX4 protein expression both in vitro and in vivo (Rashnonejad et al., 2019).

In summary, the application of modified IVT mRNA in gene editing shows enormous potential to translate to the clinic for the treatment of genetic diseases.

3.6 Cell reprogramming and therapeutic cell generation

In 2006, Yamanaka and colleagues first demonstrated that adult somatic cells or specialized cells can be reprogrammed into induced pluripotent stem cells (iPSCs) by transfection of retroviral vectors encoding four transcription factors (Oct4, Sox2, Klf4, and c-Myc; also known as Yamanaka factors) (Takahashi and Yamanaka, 2006). The retroviral delivery strategy used to generate iPSCs confers the risk of genomic integration, which has limited its use in the clinic (Anson, 2004). However, several features of IVT modified mRNA, such as transient expression and direct translation into transcription factors within the cytoplasm, eliminate the risk of insertional mutagenesis/tumorogenesis, and lead to higher transfection efficiency than is possible with DNA plasmids. These advantages make modified IVT mRNA a promising tool in regenerative medicine.

In 2010, Yakubov et al. were the first to reprogram human foreskin fibroblasts into iPSCs using modified mRNAs encoding Yamanaka factors (Yakubov et al., 2010). Another study has generated iPSCs using these four factors and subsequently introduced MyoD using modified mRNA to differentiate iPSCs into myogenic cells. Notably, the researchers tested multiple mRNA modifications, all with 3′-0-Me-m^7G(5′)ppp(5′)G ARCAs: otherwise unmodified-, m^5C triphosphate-, Ψ triphosphate-, and m^5C triphosphate/Ψ triphosphate-modified mRNA (Warren et al., 2010). The modified mRNA-based approach showed cell reprogramming efficiency greater than 1%, surpassing the reprogramming efficiency observed with viruses and DNA vectors (Warren et al., 2010). Also, heart progenitor cells have been reprogrammed using mRNA. Zangi et al. used a murine model for myocardial infarction to examine the effects of intramyocardial injection of modified mRNA encoding human VEGF-A and found it resulted in the differentiation of heart progenitor cells into endothelial cells, and improved heart function and long-term survival (Zangi et al., 2013).

In addition, somatic cells have been transdifferentiated and transformed directly into other types of the mature somatic cells through the transfection of mRNA encoding specific factors (Corritore et al., 2016; Koblas et al., 2016). For instance, Lee and colleagues treated myocardial infarction with functional cardiomyocytes using cardiac fibroblasts transformed into cardiomyocyte-like cells by modified IVT mRNAs encoding three cardiac reprogramming factors (Gata4, Mef2c, and Tbx5) (Lee et al., 2015). Two different groups have transdifferentiated pancreatic exocrine cells and human pancreatic duct-derived cells into insulin-secreting β cells through the administration of modified mRNA encoding three pancreatic transcription factors (Pdx1, Ngn3, and MafA) (Corritore et al., 2016; Koblas et al., 2016). And Simeonov and Uppal demonstrated transdifferentiation of human fibroblasts into the hepatocyte-like cells using mRNA encoding

HNF1A plus any two of the factors FOXA1, FOXA3, or HNF4A in the presence of an optimized hepatic growth medium (Simeonov and Uppal, 2014).

IVT modified mRNAs have been applied to engineer mesenchymal stem cells (MSCs). These cells may hold promise as cellular therapies to treat multiple diseases. However, minimally invasive and efficient methods must first be developed for delivering MSCs to the organ of interest. In work to that end, Nowakowski et al. developed a modified mRNA encoding integrin α4 subunit (ITGA4) in MSCs, which resulted in high surface expression of adhesion molecule ITGA4 and enhanced MSCs' migration capabilities (Nowakowski et al., 2017). Besides, Levy and colleagues harnessed mRNA transfection to develop MSCs that simultaneously expressed functional rolling machinery (P-selectin glycoprotein ligand-1 [PSGL-1] and Sialyl-Lewis^x [SLeX]) to rapidly target inflamed tissues and that express the potent immunosuppressive cytokine interleukin-10 (IL-10), which is not inherently produced by MSCs and showed an anti-inflammatory effect in vivo (Levy et al., 2013). Despite the rapid clearance of MSCs in vivo, engineered MSCs can be harnessed via a "hit-and-run" action for the targeted delivery of potent immunomodulatory factors to treat distant sites of inflammation (Levy et al., 2013).

The application of IVT mRNAs for the production of engineered-stem cells is a nascent but exciting field. Taken together, the studies discussed above show the massive potential of IVT modified mRNA in cellular reprogramming, which will bring the dream of truly regenerative medicine closer to reality.

4. Conclusions and future perspectives

The urgent need to respond to the COVID-19 pandemic, and subsequent rapid development and deployment of mRNA-based vaccines, has shown the promise of mRNA technologies to the world and marked a new era in medicine. The potential of mRNA to prevent and treat a wide range of diseases stretches far beyond COVID-19. On the therapeutic side, IVT mRNA offers tremendous potential due to its transient expression after delivery to the cytoplasm, and importantly, its degradation through physiological metabolic pathways, which eliminates the risk of genomic integration and oncogenic mutagenesis inherent in many other genetic therapies. In the last decade, substantial progress has been made in the chemical modification of mRNA, which has been key to increasing stability, reducing immunogenicity, and enabling its use as a new class of therapeutics for multiple applications. Moreover, one of the central advantages of mRNA therapeutics hinges on the rapidity of manufacture which is independent of the encoding antigen/protein. Clinical batches can be developed within only a few weeks after the sequence encoding a target immunogen becomes available, and the process is cell-free and easily scalable. The further study of IVT mRNAs for cancer immunotherapy, vaccination for cancer & infectious diseases, protein replacement, gene editing, and cell reprogramming, will lead to significant growth of mRNA-based therapies in clinical trials and subsequent approval by regulatory bodies.

Conflict of interest disclosure

M.Q., J.M., F.V. are employees of BioNTech SE. The publication reflects the opinion of the authors and not of BioNTech. This publication is not a BioNTech publication and the content is not endorsed by BioNTech.

References

Abe, K., Tanzi, R.E., Kogure, K., 1991. Induction of HSP70 mRNA after transient ischemia in gerbil brain. Neurosci. Lett. 125 (2), 166–168.

Abudayyeh, O.O., Gootenberg, J.S., Essletzbichler, P., Han, S., Joung, J., Belanto, J.J., et al., 2017. RNA targeting with CRISPR-Cas13. Nature 550 (7675), 280–284.

Akinc, A., Maier, M.A., Manoharan, M., Fitzgerald, K., Jayaraman, M., Barros, S., et al., 2019. The Onpattro story and the clinical translation of nanomedicines containing nucleic acid-based drugs. Nat. Nanotechnol. 14 (12), 1084–1087.

Alexopoulou, L., Holt, A.C., Medzhitov, R., Flavell, R.A., 2001. Recognition of double-stranded RNA and activation of NF-kappaB by Toll-like receptor 3. Nature 413 (6857), 732–738.

Almasbak, H., Rian, E., Hoel, H.J., Pule, M., Walchli, S., Kvalheim, G., et al., 2011. Transiently redirected T cells for adoptive transfer. Cytotherapy 13 (5), 629–640.

Amin, A., Dudek, A.Z., Logan, T.F., Lance, R.S., Holzbeierlein, J.M., Knox, J.J., et al., 2015. Survival with AGS-003, an autologous dendritic cell-based immunotherapy, in combination with sunitinib in unfavorable risk patients with advanced renal cell carcinoma (RCC): phase 2 study results. J. Immunother. Cancer 3, 14.

An, D., Schneller, J.L., Frassetto, A., Liang, S., Zhu, X., Park, J.S., et al., 2017. Systemic messenger RNA therapy as a treatment for methylmalonic acidemia. Cell Rep. 21 (12), 3548–3558.

Anagnostou, T., Riaz, I.B., Hashmi, S.K., Murad, M.H., Kenderian, S.S., 2020. Anti-CD19 chimeric antigen receptor T-cell therapy in acute lymphocytic leukaemia: a systematic review and meta-analysis. Lancet Haematol. 7 (11), e816–e826.

Anderson, E.J., Rouphael, N.G., Widge, A.T., Jackson, L.A., Roberts, P.C., Makhene, M., et al., 2020. Safety and immunogenicity of SARS-CoV-2 mRNA-1273 vaccine in older adults. N. Engl. J. Med. 383 (25), 2427–2438.

Anson, D.S., 2004. The use of retroviral vectors for gene therapy-what are the risks? A review of retroviral pathogenesis and its relevance to retroviral vector-mediated gene delivery. Genet. Vaccines Ther. 2 (1), 9.

Atkins, M.B., Lotze, M.T., Dutcher, J.P., Fisher, R.I., Weiss, G., Margolin, K., et al., 1999. High-dose recombinant interleukin 2 therapy for patients with metastatic melanoma:

analysis of 270 patients treated between 1985 and 1993. J. Clin. Oncol. 17 (7), 2105–2116.
Baba, M., Itaka, K., Kondo, K., Yamasoba, T., Kataoka, K., 2015. Treatment of neurological disorders by introducing mRNA in vivo using polyplex nanomicelles. J. Control. Release 201, 41–48.
Baden, L.R., El Sahly, H.M., Essink, B., Kotloff, K., Frey, S., Novak, R., et al., 2021. Efficacy and safety of the mRNA-1273 SARS-CoV-2 vaccine. N. Engl. J. Med. 384 (5), 403–416.
Balakrishnan, B., An, D., Nguyen, V., DeAntonis, C., Martini, P.G.V., Lai, K., 2020. Novel mRNA-based therapy reduces toxic galactose metabolites and overcomes galactose sensitivity in a mouse model of classic galactosemia. Mol. Ther. 28 (1), 304–312.
Barrett, D.M., Zhao, Y., Liu, X., Jiang, S., Carpenito, C., Kalos, M., et al., 2011. Treatment of advanced leukemia in mice with mRNA engineered T cells. Hum. Gene Ther. 22 (12), 1575–1586.
Beatty, G.L., Haas, A.R., Maus, M.V., Torigian, D.A., Soulen, M.C., Plesa, G., et al., 2014. Mesothelin-specific chimeric antigen receptor mRNA-engineered T cells induce anti-tumor activity in solid malignancies. Cancer Immunol. Res. 2 (2), 112–120.
Beatty, G.L., O'Hara, M.H., Lacey, S.F., Torigian, D.A., Nazimuddin, F., Chen, F., et al., 2018. Activity of mesothelin-specific chimeric antigen receptor T cells against pancreatic carcinoma metastases in a phase 1 trial. Gastroenterology 155 (1), 29–32.
Beck, J.D., Reidenbach, D., Salomon, N., Sahin, U., Tureci, O., Vormehr, M., et al., 2021. mRNA therapeutics in cancer immunotherapy. Mol. Cancer 20 (1), 69.
Beissert, T., Perkovic, M., Vogel, A., Erbar, S., Walzer, K.C., Hempel, T., et al., 2020. A trans-amplifying RNA vaccine strategy for induction of potent protective immunity. Mol. Ther. 28 (1), 119–128.
Benteyn, D., Anguille, S., Van Lint, S., Heirman, C., Van Nuffel, A.M., Corthals, J., et al., 2013. Design of an optimized Wilms' tumor 1 (WT1) mRNA construct for enhanced WT1 expression and improved immunogenicity in vitro and in vivo. Mol. Ther. Nucleic Acids 2, e134.
Benteyn, D., Heirman, C., Bonehill, A., Thielemans, K., Breckpot, K., 2015. mRNA-based dendritic cell vaccines. Expert Rev. Vaccines 14 (2), 161–176.
Bergman, N., Moraes, K.C., Anderson, J.R., Zaric, B., Kambach, C., Schneider, R.J., et al., 2007. Lsm proteins bind and stabilize RNAs containing 5' poly(A) tracts. Nat. Struct. Mol. Biol. 14 (9), 824–831.
Berraondo, P., Sanmamed, M.F., Ochoa, M.C., Etxeberria, I., Aznar, M.A., Perez-Gracia, J.L., et al., 2019. Cytokines in clinical cancer immunotherapy. Br. J. Cancer 120 (1), 6–15.
Bessis, N., GarciaCozar, F.J., Boissier, M.C., 2004. Immune responses to gene therapy vectors: influence on vector function and effector mechanisms. Gene Ther. 11 (Suppl. 1), S10–S17.
Blakney, A.K., Ip, S., Geall, A.J., 2021. An update on self-amplifying mRNA vaccine development. Vaccines (Basel) 9 (2).
Blanchard, E.L., Vanover, D., Bawage, S.S., Tiwari, P.M., Rotolo, L., Beyersdorf, J., et al., 2021. Treatment of influenza and SARS-CoV-2 infections via mRNA-encoded Cas13a in rodents. Nat. Biotechnol. 39 (6), 717–726.
Bloom, K., van den Berg, F., Arbuthnot, P., 2021. Self-amplifying RNA vaccines for infectious diseases. Gene Ther. 28 (3-4), 117–129.
Boissel, L., Betancur, M., Wels, W.S., Tuncer, H., Klingemann, H., 2009. Transfection with mRNA for CD19 specific chimeric antigen receptor restores NK cell mediated killing of CLL cells. Leuk. Res. 33 (9), 1255–1259.
Boissel, L., Betancur, M., Lu, W., Wels, W.S., Marino, T., Van Etten, R.A., et al., 2012. Comparison of mRNA and lentiviral based transfection of natural killer cells with

chimeric antigen receptors recognizing lymphoid antigens. Leuk. Lymphoma 53 (5), 958–965.
Bonehill, A., Tuyaerts, S., Van Nuffel, A.M., Heirman, C., Bos, T.J., Fostier, K., et al., 2008. Enhancing the T-cell stimulatory capacity of human dendritic cells by co-electroporation with CD40L, CD70 and constitutively active TLR4 encoding mRNA. Mol. Ther. 16 (6), 1170–1180.
Bonehill, A., Van Nuffel, A.M., Corthals, J., Tuyaerts, S., Heirman, C., Francois, V., et al., 2009. Single-step antigen loading and activation of dendritic cells by mRNA electroporation for the purpose of therapeutic vaccination in melanoma patients. Clin. Cancer Res. 15 (10), 3366–3375.
Bontkes, H.J., Kramer, D., Ruizendaal, J.J., Kueter, E.W., van Tendeloo, V.F., Meijer, C.J., et al., 2007. Dendritic cells transfected with interleukin-12 and tumor-associated antigen messenger RNA induce high avidity cytotoxic T cells. Gene Ther. 14 (4), 366–375.
Bourquin, C., Schmidt, L., Hornung, V., Wurzenberger, C., Anz, D., Sandholzer, N., et al., 2007. Immunostimulatory RNA oligonucleotides trigger an antigen-specific cytotoxic T-cell and IgG2a response. Blood 109 (7), 2953–2960.
Brenner, S., Jacob, F., Meselson, M., 1961. An unstable intermediate carrying information from genes to ribosomes for protein synthesis. Nature 190, 576–581.
Campillo-Davo, D., Fujiki, F., Van den Bergh, J.M.J., De Reu, H., Smits, E., Goossens, H., et al., 2018. Efficient and non-genotoxic RNA-based engineering of human T cells using tumor-specific T cell receptors with minimal TCR mispairing. Front. Immunol. 9, 2503.
Cannarozzi, G., Schraudolph, N.N., Faty, M., von Rohr, P., Friberg, M.T., Roth, A.C., et al., 2010. A role for codon order in translation dynamics. Cell 141 (2), 355–367.
Carralot, J.P., Probst, J., Hoerr, I., Scheel, B., Teufel, R., Jung, G., et al., 2004. Polarization of immunity induced by direct injection of naked sequence-stabilized mRNA vaccines. Cell. Mol. Life Sci. 61 (18), 2418–2424.
Castle, J.C., Kreiter, S., Diekmann, J., Lower, M., van de Roemer, N., de Graaf, J., et al., 2012. Exploiting the mutanome for tumor vaccination. Cancer Res. 72 (5), 1081–1091.
Chen, C.Y., Shyu, A.B., 1995. AU-rich elements: characterization and importance in mRNA degradation. Trends Biochem. Sci. 20 (11), 465–470.
Chen, Q., Zhang, Y., Yin, H., 2021. Recent advances in chemical modifications of guide RNA, mRNA and donor template for CRISPR-mediated genome editing. Adv. Drug Deliv. Rev. 168, 246–258.
Chu, Y., Hochberg, J., Yahr, A., Ayello, J., van de Ven, C., Barth, M., et al., 2015. Targeting CD20+ aggressive B-cell non-Hodgkin lymphoma by anti-CD20 CAR mRNA-modified expanded natural killer cells in vitro and in NSG mice. Cancer Immunol. Res. 3 (4), 333–344.
Cifuentes-Rius, A., Desai, A., Yuen, D., Johnston, A.P.R., Voelcker, N.H., 2021. Inducing immune tolerance with dendritic cell-targeting nanomedicines. Nat. Nanotechnol. 16 (1), 37–46.
Cirelli, K.M., Crotty, S., 2017. Germinal center enhancement by extended antigen availability. Curr. Opin. Immunol. 47, 64–69.
Conlon, K.C., Lugli, E., Welles, H.C., Rosenberg, S.A., Fojo, A.T., Morris, J.C., et al., 2015. Redistribution, hyperproliferation, activation of natural killer cells and CD8 T cells, and cytokine production during first-in-human clinical trial of recombinant human interleukin-15 in patients with cancer. J. Clin. Oncol. 33 (1), 74–82.
Conry, R.M., LoBuglio, A.F., Wright, M., Sumerel, L., Pike, M.J., Johanning, F., et al., 1995. Characterization of a messenger RNA polynucleotide vaccine vector. Cancer Res. 55 (7), 1397–1400.
Corritore, E., Lee, Y.S., Pasquale, V., Liberati, D., Hsu, M.J., Lombard, C.A., et al., 2016. V-Maf musculoaponeurotic fibrosarcoma oncogene homolog A synthetic modified

mRNA drives reprogramming of human pancreatic duct-derived cells into insulin-secreting cells. Stem Cells Transl. Med. 5 (11), 1525–1537.

Daneshmandi, S., Pourfathollah, A.A., Forouzandeh-Moghaddam, M., 2018. Enhanced CD40 and ICOSL expression on dendritic cells surface improve anti-tumor immune responses; effectiveness of mRNA/chitosan nanoparticles. Immunopharmacol. Immunotoxicol. 40 (5), 375–386.

De Beuckelaer, A., Pollard, C., Van Lint, S., Roose, K., Van Hoecke, L., Naessens, T., et al., 2016. Type I interferons interfere with the capacity of mRNA lipoplex vaccines to elicit cytolytic T cell responses. Mol. Ther. 24 (11), 2012–2020.

De Keersmaecker, B., Claerhout, S., Carrasco, J., Bar, I., Corthals, J., Wilgenhof, S., et al., 2020. TriMix and tumor antigen mRNA electroporated dendritic cell vaccination plus ipilimumab: link between T-cell activation and clinical responses in advanced melanoma. J. Immunother. Cancer 8 (1).

DeRosa, F., Guild, B., Karve, S., Smith, L., Love, K., Dorkin, J.R., et al., 2016. Therapeutic efficacy in a hemophilia B model using a biosynthetic mRNA liver depot system. Gene Ther. 23 (10), 699–707.

Diebold, S.S., Kaisho, T., Hemmi, H., Akira, S., Reis e Sousa C., 2004. Innate antiviral responses by means of TLR7-mediated recognition of single-stranded RNA. Science 303 (5663), 1529–1531.

DiGiusto, D.L., Cannon, P.M., Holmes, M.C., Li, L., Rao, A., Wang, J., et al., 2016. Preclinical development and qualification of ZFN-mediated CCR5 disruption in human hematopoietic stem/progenitor cells. Mol. Ther. Methods Clin. Dev. 3, 16067.

Diken, M., Kreiter, S., Selmi, A., Britten, C.M., Huber, C., Tureci, O., et al., 2011. Selective uptake of naked vaccine RNA by dendritic cells is driven by macropinocytosis and abrogated upon DC maturation. Gene Ther. 18 (7), 702–708.

Dilliard, S.A., Cheng, Q., Siegwart, D.J., 2021. On the mechanism of tissue-specific mRNA delivery by selective organ targeting nanoparticles. Proc. Natl. Acad. Sci. U. S. A. 118 (52).

Dorrie, J., Krug, C., Hofmann, C., Muller, I., Wellner, V., Knippertz, I., et al., 2014. Human adenovirus-specific gamma/delta and CD8+ T cells generated by T-cell receptor transfection to treat adenovirus infection after allogeneic stem cell transplantation. PLoS One 9 (10), e109944.

Dupuy, A.J., Clark, K., Carlson, C.M., Fritz, S., Davidson, A.E., Markley, K.M., et al., 2002. Mammalian germ-line transgenesis by transposition. Proc. Natl. Acad. Sci. U. S. A. 99 (7), 4495–4499.

Etxeberria, I., Bolanos, E., Quetglas, J.I., Gros, A., Villanueva, A., Palomero, J., et al., 2019. Intratumor adoptive transfer of IL-12 mRNA transiently engineered antitumor CD8(+) T cells. Cancer Cell 36 (6), 613–29 e7.

Finn, J.D., Smith, A.R., Patel, M.C., Shaw, L., Youniss, M.R., van Heteren, J., et al., 2018. A single administration of CRISPR/Cas9 lipid nanoparticles achieves robust and persistent in vivo genome editing. Cell Rep. 22 (9), 2227–2235.

Fitzgerald, K.A., Kagan, J.C., 2020. Toll-like receptors and the control of immunity. Cell 180 (6), 1044–1066.

Foster, J.B., Barrett, D.M., Kariko, K., 2019. The emerging role of in vitro-transcribed mRNA in adoptive T cell immunotherapy. Mol. Ther. 27 (4), 747–756.

Fotin-Mleczek, M., Duchardt, K.M., Lorenz, C., Pfeiffer, R., Ojkic-Zrna, S., Probst, J., et al., 2011. Messenger RNA-based vaccines with dual activity induce balanced TLR-7 dependent adaptive immune responses and provide antitumor activity. J. Immunother. 34 (1), 1–15.

Freitag, F., Maucher, M., Riester, Z., Hudecek, M., 2020. New targets and technologies for CAR-T cells. Curr. Opin. Oncol. 32 (5), 510–517.

Frenck Jr., R.W., Klein, N.P., Kitchin, N., Gurtman, A., Absalon, J., Lockhart, S., et al., 2021. Safety, immunogenicity, and efficacy of the BNT162b2 Covid-19 vaccine in adolescents. N. Engl. J. Med. 385 (3), 239–250.

Fu, Y., Foden, J.A., Khayter, C., Maeder, M.L., Reyon, D., Joung, J.K., et al., 2013. High-frequency off-target mutagenesis induced by CRISPR-Cas nucleases in human cells. Nat. Biotechnol. 31 (9), 822–826.

Furuichi, Y., Miura, K., 1975. A blocked structure at the 5' terminus of mRNA from cytoplasmic polyhedrosis virus. Nature 253 (5490), 374–375.

Fyfe, G., Fisher, R.I., Rosenberg, S.A., Sznol, M., Parkinson, D.R., Louie, A.C., 1995. Results of treatment of 255 patients with metastatic renal cell carcinoma who received high-dose recombinant interleukin-2 therapy. J. Clin. Oncol. 13 (3), 688–696.

Gaj, T., Gersbach, C.A., Barbas 3rd., C.F., 2013. ZFN, TALEN, and CRISPR/Cas-based methods for genome engineering. Trends Biotechnol. 31 (7), 397–405.

Georgiadis, C., Preece, R., Nickolay, L., Etuk, A., Petrova, A., Ladon, D., et al., 2018. Long terminal repeat CRISPR-CAR-coupled "universal" T cells mediate potent anti-leukemic effects. Mol. Ther. 26 (5), 1215–1227.

Gillmore, J.D., Gane, E., Taubel, J., Kao, J., Fontana, M., Maitland, M.L., et al., 2021. CRISPR-Cas9 in vivo gene editing for transthyretin amyloidosis. N. Engl. J. Med. 385 (6), 493–502.

Goldstrohm, A.C., Wickens, M., 2008. Multifunctional deadenylase complexes diversify mRNA control. Nat. Rev. Mol. Cell Biol. 9 (4), 337–344.

Golomb, H.M., Jacobs, A., Fefer, A., Ozer, H., Thompson, J., Portlock, C., et al., 1986. Alpha-2 interferon therapy of hairy-cell leukemia: a multicenter study of 64 patients. J. Clin. Oncol. 4 (6), 900–905.

Gotley, L.M., Blanch, A., Kimble, R., Frawley, K., Acworth, J.P., 2009. Pyloric stenosis: a retrospective study of an Australian population. Emerg. Med. Australas. 21 (5), 407–413.

Granot, Y., Peer, D., 2017. Delivering the right message: challenges and opportunities in lipid nanoparticles-mediated modified mRNA therapeutics-An innate immune system standpoint. Semin. Immunol. 34, 68–77.

Grasshoff, H., Comduhr, S., Monne, L.R., Muller, A., Lamprecht, P., Riemekasten, G., et al., 2021. Low-dose IL-2 therapy in autoimmune and rheumatic diseases. Front. Immunol. 12, 648408.

Groopman, J.E., Gottlieb, M.S., Goodman, J., Mitsuyasu, R.T., Conant, M.A., Prince, H., et al., 1984. Recombinant alpha-2 interferon therapy for Kaposi's sarcoma associated with the acquired immunodeficiency syndrome. Ann. Intern. Med. 100 (5), 671–676.

Grudzien-Nogalska, E., Jemielity, J., Kowalska, J., Darzynkiewicz, E., Rhoads, R.E., 2007. Phosphorothioate cap analogs stabilize mRNA and increase translational efficiency in mammalian cells. RNA 13 (10), 1745–1755.

Grunebach, F., Kayser, K., Weck, M.M., Muller, M.R., Appel, S., Brossart, P., 2005. Cotransfection of dendritic cells with RNA coding for HER-2/neu and 4-1BBL increases the induction of tumor antigen specific cytotoxic T lymphocytes. Cancer Gene Ther. 12 (9), 749–756.

Gupta, S.K., Haigh, B.J., Griffin, F.J., Wheeler, T.T., 2013. The mammalian secreted RNases: mechanisms of action in host defence. Innate Immun. 19 (1), 86–97.

Harrer, D.C., Simon, B., Fujii, S.I., Shimizu, K., Uslu, U., Schuler, G., et al., 2017. RNA-transfection of gamma/delta T cells with a chimeric antigen receptor or an alpha/beta T-cell receptor: a safer alternative to genetically engineered alpha/beta T cells for the immunotherapy of melanoma. BMC Cancer 17 (1), 551.

Heidenreich, R., Jasny, E., Kowalczyk, A., Lutz, J., Probst, J., Baumhof, P., et al., 2015. A novel RNA-based adjuvant combines strong immunostimulatory capacities with a favorable safety profile. Int. J. Cancer 137 (2), 372–384.

Heil, F., Hemmi, H., Hochrein, H., Ampenberger, F., Kirschning, C., Akira, S., et al., 2004. Species-specific recognition of single-stranded RNA via toll-like receptor 7 and 8. Science 303 (5663), 1526–1529.

Hewitt, S.L., Bai, A., Bailey, D., Ichikawa, K., Zielinski, J., Karp, R., et al., 2019. Durable anticancer immunity from intratumoral administration of IL-23, IL-36gamma, and OX40L mRNAs. Sci. Transl. Med. 11 (477).

Hewitt, S.L., Bailey, D., Zielinski, J., Apte, A., Musenge, F., Karp, R., et al., 2020. Intratumoral IL12 mRNA therapy promotes TH1 transformation of the tumor microenvironment. Clin. Cancer Res. 26 (23), 6284–6298.

Hinnebusch, A.G., Ivanov, I.P., Sonenberg, N., 2016. Translational control by 5'-untranslated regions of eukaryotic mRNAs. Science 352 (6292), 1413–1416.

Hoerr, I., Obst, R., Rammensee, H.G., Jung, G., 2000. In vivo application of RNA leads to induction of specific cytotoxic T lymphocytes and antibodies. Eur. J. Immunol. 30 (1), 1–7.

Holtkamp, S., Kreiter, S., Selmi, A., Simon, P., Koslowski, M., Huber, C., et al., 2006. Modification of antigen-encoding RNA increases stability, translational efficacy, and T-cell stimulatory capacity of dendritic cells. Blood 108 (13), 4009–4017.

Hornung, V., Ellegast, J., Kim, S., Brzozka, K., Jung, A., Kato, H., et al., 2006. 5'-Triphosphate RNA is the ligand for RIG-I. Science 314 (5801), 994–997.

Hotz, C., Wagenaar, T.R., Gieseke, F., Bangari, D.S., Callahan, M., Cao, H., et al., 2021. Local delivery of mRNA-encoded cytokines promotes antitumor immunity and tumor eradication across multiple preclinical tumor models. Sci. Transl. Med. 13 (610), eabc7804.

Hou, X., Zaks, T., Langer, R., Dong, Y., 2021. Lipid nanoparticles for mRNA delivery. Nat. Rev. Mater., 1–17.

Hung, C.F., Xu, X., Li, L., Ma, Y., Jin, Q., Viley, A., et al., 2018. Development of anti-human mesothelin-targeted chimeric antigen receptor messenger RNA-transfected peripheral blood lymphocytes for ovarian cancer therapy. Hum. Gene Ther. 29 (5), 614–625.

Jemielity, J., Fowler, T., Zuberek, J., Stepinski, J., Lewdorowicz, M., Niedzwiecka, A., et al., 2003. Novel "anti-reverse" cap analogs with superior translational properties. RNA 9 (9), 1108–1122.

Jenkins, M.K., Ashwell, J.D., Schwartz, R.H., 1988. Allogeneic non-T spleen cells restore the responsiveness of normal T cell clones stimulated with antigen and chemically modified antigen-presenting cells. J. Immunol. 140 (10), 3324–3330.

Jenkins, M.K., Taylor, P.S., Norton, S.D., Urdahl, K.B., 1991. CD28 delivers a costimulatory signal involved in antigen-specific IL-2 production by human T cells. J. Immunol. 147 (8), 2461–2466.

Jiang, L., Park, J.S., Yin, L., Laureano, R., Jacquinet, E., Yang, J., et al., 2020. Dual mRNA therapy restores metabolic function in long-term studies in mice with propionic acidemia. Nat. Commun. 11 (1), 5339.

Kallen, K.J., Thess, A., 2014. A development that may evolve into a revolution in medicine: mRNA as the basis for novel, nucleotide-based vaccines and drugs. Ther. Adv. Vaccines 2 (1), 10–31.

Kallen, K.J., Heidenreich, R., Schnee, M., Petsch, B., Schlake, T., Thess, A., et al., 2013. A novel, disruptive vaccination technology: self-adjuvanted RNActive((R)) vaccines. Hum. Vaccin. Immunother. 9 (10), 2263–2276.

Kalos, M., Levine, B.L., Porter, D.L., Katz, S., Grupp, S.A., Bagg, A., et al., 2011. T cells with chimeric antigen receptors have potent antitumor effects and can establish memory in patients with advanced leukemia. Sci. Transl. Med. 3 (95), 95ra73.

Karadagi, A., Cavedon, A.G., Zemack, H., Nowak, G., Eybye, M.E., Zhu, X., et al., 2020. Systemic modified messenger RNA for replacement therapy in alpha 1-antitrypsin deficiency. Sci. Rep. 10 (1), 7052.

Kariko, K., Kuo, A., Barnathan, E., 1999. Overexpression of urokinase receptor in mammalian cells following administration of the in vitro transcribed encoding mRNA. Gene Ther. 6 (6), 1092–1100.
Kariko, K., Buckstein, M., Ni, H., Weissman, D., 2005. Suppression of RNA recognition by Toll-like receptors: the impact of nucleoside modification and the evolutionary origin of RNA. Immunity 23 (2), 165–175.
Kariko, K., Muramatsu, H., Welsh, F.A., Ludwig, J., Kato, H., Akira, S., et al., 2008. Incorporation of pseudouridine into mRNA yields superior nonimmunogenic vector with increased translational capacity and biological stability. Mol. Ther. 16 (11), 1833–1840.
Kariko, K., Muramatsu, H., Ludwig, J., Weissman, D., 2011. Generating the optimal mRNA for therapy: HPLC purification eliminates immune activation and improves translation of nucleoside-modified, protein-encoding mRNA. Nucleic Acids Res. 39 (21), e142.
Kariko, K., Muramatsu, H., Keller, J.M., Weissman, D., 2012. Increased erythropoiesis in mice injected with submicrogram quantities of pseudouridine-containing mRNA encoding erythropoietin. Mol. Ther. 20 (5), 948–953.
Kato, H., Takeuchi, O., Sato, S., Yoneyama, M., Yamamoto, M., Matsui, K., et al., 2006. Differential roles of MDA5 and RIG-I helicases in the recognition of RNA viruses. Nature 441 (7089), 101–105.
Kenderian, S.S., Ruella, M., Shestova, O., Klichinsky, M., Aikawa, V., Morrissette, J.J., et al., 2015. CD33-specific chimeric antigen receptor T cells exhibit potent preclinical activity against human acute myeloid leukemia. Leukemia 29 (8), 1637–1647.
Kim, Y., Cheong, S.A., Lee, J.G., Lee, S.W., Lee, M.S., Baek, I.J., et al., 2016. Generation of knockout mice by Cpf1-mediated gene targeting. Nat. Biotechnol. 34 (8), 808–810.
Kimchi-Sarfaty, C., Oh, J.M., Kim, I.W., Sauna, Z.E., Calcagno, A.M., Ambudkar, S.V., et al., 2007. A "silent" polymorphism in the MDR1 gene changes substrate specificity. Science 315 (5811), 525–528.
Kirkwood, J.M., Strawderman, M.H., Ernstoff, M.S., Smith, T.J., Borden, E.C., Blum, R.H., 1996. Interferon alfa-2b adjuvant therapy of high-risk resected cutaneous melanoma: the Eastern Cooperative Oncology Group Trial EST 1684. J. Clin. Oncol. 14 (1), 7–17.
Knights, A.J., Nuber, N., Thomson, C.W., de la Rosa, O., Jager, E., Tiercy, J.M., et al., 2009. Modified tumour antigen-encoding mRNA facilitates the analysis of naturally occurring and vaccine-induced CD4 and CD8 T cells in cancer patients. Cancer Immunol. Immunother. 58 (3), 325–338.
Koblas, T., Leontovyc, I., Loukotova, S., Kosinova, L., Saudek, F., 2016. Reprogramming of pancreatic exocrine cells AR42J into insulin-producing cells using mRNAs for Pdx1, Ngn3, and MafA transcription factors. Mol. Ther. Nucleic Acids 5, e320.
Kochenderfer, J.N., Dudley, M.E., Feldman, S.A., Wilson, W.H., Spaner, D.E., Maric, I., et al., 2012. B-cell depletion and remissions of malignancy along with cytokine-associated toxicity in a clinical trial of anti-CD19 chimeric-antigen-receptor-transduced T cells. Blood 119 (12), 2709–2720.
Koksal, H., Dillard, P., Josefsson, S.E., Maggadottir, S.M., Pollmann, S., Fane, A., et al., 2019. Preclinical development of CD37CAR T-cell therapy for treatment of B-cell lymphoma. Blood Adv. 3 (8), 1230–1243.
Kormann, M.S., Hasenpusch, G., Aneja, M.K., Nica, G., Flemmer, A.W., Herber-Jonat, S., et al., 2011. Expression of therapeutic proteins after delivery of chemically modified mRNA in mice. Nat. Biotechnol. 29 (2), 154–157.
Korner, C.G., Wahle, E., 1997. Poly(A) tail shortening by a mammalian poly(A)-specific 3'-exoribonuclease. J. Biol. Chem. 272 (16), 10448–10456.

Kowalczyk, A., Doener, F., Zanzinger, K., Noth, J., Baumhof, P., Fotin-Mleczek, M., et al., 2016. Self-adjuvanted mRNA vaccines induce local innate immune responses that lead to a potent and boostable adaptive immunity. Vaccine 34 (33), 3882–3893.

Kowalska, J., Zuberek, J., Darzynkiewicz, Z.M., Lukaszewicz, M., Darzynkiewicz, E., Jemielity, J., 2008. The first examples of mRNA cap analogs bearing boranophosphate modification. Nucleic Acids Symp. Ser. (Oxf) 52, 289–290.

Kranz, L.M., Diken, M., Haas, H., Kreiter, S., Loquai, C., Reuter, K.C., et al., 2016. Systemic RNA delivery to dendritic cells exploits antiviral defence for cancer immunotherapy. Nature 534 (7607), 396–401.

Kreiter, S., Selmi, A., Diken, M., Koslowski, M., Britten, C.M., Huber, C., et al., 2010. Intranodal vaccination with naked antigen-encoding RNA elicits potent prophylactic and therapeutic antitumoral immunity. Cancer Res. 70 (22), 9031–9040.

Kreiter, S., Vormehr, M., van de Roemer, N., Diken, M., Lower, M., Diekmann, J., et al., 2015. Erratum: mutant MHC class II epitopes drive therapeutic immune responses to cancer. Nature 523 (7560), 370.

Krienke, C., Kolb, L., Diken, E., Streuber, M., Kirchhoff, S., Bukur, T., et al., 2021. A non-inflammatory mRNA vaccine for treatment of experimental autoimmune encephalomyelitis. Science 371 (6525), 145–153.

Lai, I., Swaminathan, S., Baylot, V., Mosley, A., Dhanasekaran, R., Gabay, M., et al., 2018. Lipid nanoparticles that deliver IL-12 messenger RNA suppress tumorigenesis in MYC oncogene-driven hepatocellular carcinoma. J. Immunother. Cancer 6 (1), 125.

Lamers, C.H., Sleijfer, S., van Steenbergen, S., van Elzakker, P., van Krimpen, B., Groot, C., et al., 2013. Treatment of metastatic renal cell carcinoma with CAIX CAR-engineered T cells: clinical evaluation and management of on-target toxicity. Mol. Ther. 21 (4), 904–912.

Lee, K., Yu, P., Lingampalli, N., Kim, H.J., Tang, R., Murthy, N., 2015. Peptide-enhanced mRNA transfection in cultured mouse cardiac fibroblasts and direct reprogramming towards cardiomyocyte-like cells. Int. J. Nanomedicine 10, 1841–1854.

Leonard, J.P., Sherman, M.L., Fisher, G.L., Buchanan, L.J., Larsen, G., Atkins, M.B., et al., 1997. Effects of single-dose interleukin-12 exposure on interleukin-12-associated toxicity and interferon-gamma production. Blood 90 (7), 2541–2548.

Leonhardt, C., Schwake, G., Stogbauer, T.R., Rappl, S., Kuhr, J.T., Ligon, T.S., et al., 2014. Single-cell mRNA transfection studies: delivery, kinetics and statistics by numbers. Nanomedicine 10 (4), 679–688.

Leppek, K., Das, R., Barna, M., 2018. Functional 5' UTR mRNA structures in eukaryotic translation regulation and how to find them. Nat. Rev. Mol. Cell Biol. 19 (3), 158–174.

Levin, N., Pato, A., Cafri, G., Eisenberg, G., Peretz, T., Margalit, A., et al., 2017. Spontaneous activation of antigen-presenting cells by genes encoding truncated homo-oligomerizing derivatives of CD40. J. Immunother. 40 (2), 39–50.

Levin, N., Weinstein-Marom, H., Pato, A., Itzhaki, O., Besser, M.J., Eisenberg, G., et al., 2018. Potent activation of human T cells by mRNA encoding constitutively active CD40. J. Immunol. 201 (10), 2959–2968.

Levy, O., Zhao, W., Mortensen, L.J., Leblanc, S., Tsang, K., Fu, M., et al., 2013. mRNA-engineered mesenchymal stem cells for targeted delivery of interleukin-10 to sites of inflammation. Blood 122 (14), e23–e32.

Li, J., 2020. Advances toward a cure for HIV: getting beyond n = 2. Top Antivir. Med. 27 (4), 91–95.

Li, Y., Kiledjian, M., 2010. Regulation of mRNA decapping. Wiley Interdiscip. Rev. RNA 1 (2), 253–265.

Li, L., Liu, L.N., Feller, S., Allen, C., Shivakumar, R., Fratantoni, J., et al., 2010. Expression of chimeric antigen receptors in natural killer cells with a regulatory-compliant non-viral method. Cancer Gene Ther. 17 (3), 147–154.

Li, B., Luo, X., Dong, Y., 2016. Effects of chemically modified messenger RNA on protein expression. Bioconjug. Chem. 27 (3), 849–853.

Li, Y., Su, Z., Zhao, W., Zhang, X., Momin, N., Zhang, C., et al., 2020. Multifunctional oncolytic nanoparticles deliver self-replicating IL-12 RNA to eliminate established tumors and prime systemic immunity. Nat. Cancer 1 (9), 882–893.

Ligon, T.S., Leonhardt, C., Radler, J.O., 2014. Multi-level kinetic model of mRNA delivery via transfection of lipoplexes. PLoS One 9 (9), e107148.

Lin, Y.X., Wang, Y., Blake, S., Yu, M., Mei, L., Wang, H., et al., 2020. RNA nanotechnology-mediated cancer immunotherapy. Theranostics 10 (1), 281–299.

Liu, X., Zhang, Y., Cheng, C., Cheng, A.W., Zhang, X., Li, N., et al., 2017. CRISPR-Cas9-mediated multiplex gene editing in CAR-T cells. Cell Res. 27 (1), 154–157.

Liu, W., Alameh, M.-G., Yang, J.F., Xu, J.R., Lin, P.J., Tam, Y.K., et al., 2022. Lipid nanoparticles delivering constitutively active STING mRNA as a novel anti-cancer therapeutic approach. bioRxiv. 2022.01.08.475499.

Lopalco, L., 2010. CCR5: from natural resistance to a new anti-HIV strategy. Viruses 2 (2), 574–600.

Malone, R.W., Felgner, P.L., Verma, I.M., 1989. Cationic liposome-mediated RNA transfection. Proc. Natl. Acad. Sci. U. S. A. 86 (16), 6077–6081.

Martin, G., Keller, W., 1998. Tailing and 3'-end labeling of RNA with yeast poly(A) polymerase and various nucleotides. RNA 4 (2), 226–230.

Martin, S.A., Paoletti, E., Moss, B., 1975. Purification of mRNA guanylyltransferase and mRNA (guanine-7-) methyltransferase from vaccinia virions. J. Biol. Chem. 250 (24), 9322–9329.

Martinon, F., Krishnan, S., Lenzen, G., Magne, R., Gomard, E., Guillet, J.G., et al., 1993. Induction of virus-specific cytotoxic T lymphocytes in vivo by liposome-entrapped mRNA. Eur. J. Immunol. 23 (7), 1719–1722.

Maucher, M., Srour, M., Danhof, S., Einsele, H., Hudecek, M., Yakoub-Agha, I., 2021. Current limitations and perspectives of chimeric antigen receptor-T-cells in acute myeloid leukemia. Cancers (Basel) 13 (24).

Maus, M.V., Haas, A.R., Beatty, G.L., Albelda, S.M., Levine, B.L., Liu, X., et al., 2013. T cells expressing chimeric antigen receptors can cause anaphylaxis in humans. Cancer Immunol. Res. 1 (1), 26–31.

Mayr, C., 2017. Regulation by 3'-untranslated regions. Annu. Rev. Genet. 51, 171–194.

McNab, F., Mayer-Barber, K., Sher, A., Wack, A., O'Garra, A., 2015. Type I interferons in infectious disease. Nat. Rev. Immunol. 15 (2), 87–103.

Meng, Z., O'Keeffe-Ahern, J., Lyu, J., Pierucci, L., Zhou, D., Wang, W., 2017. A new developing class of gene delivery: messenger RNA-based therapeutics. Biomater. Sci. 5 (12), 2381–2392.

Miao, L., Li, L., Huang, Y., Delcassian, D., Chahal, J., Han, J., et al., 2019. Delivery of mRNA vaccines with heterocyclic lipids increases anti-tumor efficacy by STING-mediated immune cell activation. Nat. Biotechnol. 37 (10), 1174–1185.

Mignone, F., Gissi, C., Liuni, S., Pesole, G., 2002. Untranslated regions of mRNAs. Genome Biol. 3 (3), REVIEWS0004.

Miliotou, A.N., Pappas, I.S., Spyroulias, G., Vlachaki, E., Tsiftsoglou, A.S., Vizirianakis, I.S., et al., 2021. Development of a novel PTD-mediated IVT-mRNA delivery platform for potential protein replacement therapy of metabolic/genetic disorders. Mol. Ther. Nucleic Acids 26, 694–710.

Minkis, K., Kavanagh, D.G., Alter, G., Bogunovic, D., O'Neill, D., Adams, S., et al., 2008. Type 2 bias of T cells expanded from the blood of melanoma patients switched to type 1 by IL-12p70 mRNA-transfected dendritic cells. Cancer Res. 68 (22), 9441–9450.

Mockey, M., Goncalves, C., Dupuy, F.P., Lemoine, F.M., Pichon, C., Midoux, P., 2006. mRNA transfection of dendritic cells: synergistic effect of ARCA mRNA capping with Poly(A) chains in cis and in trans for a high protein expression level. Biochem. Biophys. Res. Commun. 340 (4), 1062–1068.

Morais, P., Adachi, H., Yu, Y.T., 2021. The critical contribution of pseudouridine to mRNA COVID-19 vaccines. Front. Cell Dev. Biol. 9, 789427.

Motzer, R.J., Rakhit, A., Schwartz, L.H., Olencki, T., Malone, T.M., Sandstrom, K., et al., 1998. Phase I trial of subcutaneous recombinant human interleukin-12 in patients with advanced renal cell carcinoma. Clin. Cancer Res. 4 (5), 1183–1191.

Murray, E.L., Schoenberg, D.R., 2008. Assays for determining poly(A) tail length and the polarity of mRNA decay in mammalian cells. Methods Enzymol. 448, 483–504.

Naka, T., Iwahashi, M., Nakamura, M., Ojima, T., Nakamori, M., Ueda, K., et al., 2008. Tumor vaccine therapy against recrudescent tumor using dendritic cells simultaneously transfected with tumor RNA and granulocyte macrophage colony-stimulating factor RNA. Cancer Sci. 99 (2), 407–413.

Nallagatla, S.R., Hwang, J., Toroney, R., Zheng, X., Cameron, C.E., Bevilacqua, P.C., 2007. 5'-triphosphate-dependent activation of PKR by RNAs with short stem-loops. Science 318 (5855), 1455–1458.

Nance, K.D., Meier, J.L., 2021. Modifications in an emergency: the role of N1-methylpseudouridine in COVID-19 vaccines. ACS Cent. Sci. 7 (5), 748–756.

Nelson, J., Sorensen, E.W., Mintri, S., Rabideau, A.E., Zheng, W., Besin, G., et al., 2020. Impact of mRNA chemistry and manufacturing process on innate immune activation. Sci. Adv. 6 (26), eaaz6893.

Ng, Y.Y., Tay, J.C.K., Wang, S., 2020. CXCR1 expression to improve anti-cancer efficacy of intravenously injected CAR-NK cells in mice with peritoneal xenografts. Mol. Ther. Oncolytics 16, 75–85.

Nguyen, K.G., Vrabel, M.R., Mantooth, S.M., Hopkins, J.J., Wagner, E.S., Gabaldon, T.A., et al., 2020. Localized interleukin-12 for cancer immunotherapy. Front. Immunol. 11, 575597.

Novoa, E.M., Ribas de Pouplana, L., 2012. Speeding with control: codon usage, tRNAs, and ribosomes. Trends Genet. 28 (11), 574–581.

Nowakowski, A., Andrzejewska, A., Boltze, J., Nitzsche, F., Cui, L.L., Jolkkonen, J., et al., 2017. Translation, but not transfection limits clinically relevant, exogenous mRNA based induction of alpha-4 integrin expression on human mesenchymal stem cells. Sci. Rep. 7 (1), 1103.

Ogata, A.F., Cheng, C.A., Desjardins, M., Senussi, Y., Sherman, A.C., Powell, M., et al., 2021. Circulating SARS-CoV-2 vaccine antigen detected in the plasma of mRNA-1273 vaccine recipients. Clin. Infect. Dis. 74 (4), 715–718.

Ohnmacht, C., Pullner, A., King, S.B., Drexler, I., Meier, S., Brocker, T., et al., 2009. Constitutive ablation of dendritic cells breaks self-tolerance of CD4 T cells and results in spontaneous fatal autoimmunity. J. Exp. Med. 206 (3), 549–559.

Panjwani, M.K., Smith, J.B., Schutsky, K., Gnanandarajah, J., O'Connor, C.M., Powell, D.J., Jr., et al., 2016. Feasibility and safety of RNA-transfected CD20-specific chimeric antigen receptor T cells in dogs with spontaneous B cell lymphoma. Mol. Ther. 24 (9), 1602–1614.

Parayath, N.N., Hao, S., Stephan, S.B., Koehne, A.L., Watson, C.E., Stephan, M.T., 2021. Genetic in situ engineering of myeloid regulatory cells controls inflammation in autoimmunity. J. Control. Release 339, 553–561.

Pardi, N., Tuyishime, S., Muramatsu, H., Kariko, K., Mui, B.L., Tam, Y.K., et al., 2015. Expression kinetics of nucleoside-modified mRNA delivered in lipid nanoparticles to mice by various routes. J. Control. Release 217, 345–351.

Pardi, N., Hogan, M.J., Naradikian, M.S., Parkhouse, K., Cain, D.W., Jones, L., et al., 2018a. Nucleoside-modified mRNA vaccines induce potent T follicular helper and germinal center B cell responses. J. Exp. Med. 215 (6), 1571–1588.

Pardi, N., Hogan, M.J., Porter, F.W., Weissman, D., 2018b. mRNA vaccines—a new era in vaccinology. Nat. Rev. Drug Discov. 17 (4), 261–279.

Pen, J.J., De Keersmaecker, B., Maenhout, S.K., Van Nuffel, A.M., Heirman, C., Corthals, J., et al., 2013. Modulation of regulatory T cell function by monocyte-derived dendritic cells matured through electroporation with mRNA encoding CD40 ligand, constitutively active TLR4, and CD70. J. Immunol. 191 (4), 1976–1983.

Peng, J., Schoenberg, D.R., 2005. mRNA with a <20-nt poly(A) tail imparted by the poly(A)-limiting element is translated as efficiently in vivo as long poly(A) mRNA. RNA 11 (7), 1131–1140.

Perez, C.R., De Palma, M., 2019. Engineering dendritic cell vaccines to improve cancer immunotherapy. Nat. Commun. 10 (1), 5408.

Pichlmair, A., Schulz, O., Tan, C.P., Naslund, T.I., Liljestrom, P., Weber, F., et al., 2006. RIG-I-mediated antiviral responses to single-stranded RNA bearing 5'-phosphates. Science 314 (5801), 997–1001.

Plotkin, J.B., Kudla, G., 2011. Synonymous but not the same: the causes and consequences of codon bias. Nat. Rev. Genet. 12 (1), 32–42.

Pohl-Guimaraes, F., Yang, C., Dyson, K.A., Wildes, T.J., Drake, J., Huang, J., et al., 2019. RNA-modified T cells mediate effective delivery of immunomodulatory cytokines to brain tumors. Mol. Ther. 27 (4), 837–849.

Polack, F.P., Thomas, S.J., Kitchin, N., Absalon, J., Gurtman, A., Lockhart, S., et al., 2020. Safety and efficacy of the BNT162b2 mRNA Covid-19 vaccine. N. Engl. J. Med. 383 (27), 2603–2615.

Pollard, C., Rejman, J., De Haes, W., Verrier, B., Van Gulck, E., Naessens, T., et al., 2013. Type I IFN counteracts the induction of antigen-specific immune responses by lipid-based delivery of mRNA vaccines. Mol. Ther. 21 (1), 251–259.

Polo, J.M., Gardner, J.P., Ji, Y., Belli, B.A., Driver, D.A., Sherrill, S., et al., 2000. Alphavirus DNA and particle replicons for vaccines and gene therapy. Dev. Biol. (Basel) 104, 181–185.

Porter, D.L., Hwang, W.T., Frey, N.V., Lacey, S.F., Shaw, P.A., Loren, A.W., et al., 2015. Chimeric antigen receptor T cells persist and induce sustained remissions in relapsed refractory chronic lymphocytic leukemia. Sci. Transl. Med. 7 (303), 303ra139.

Prommersberger, S., Reiser, M., Beckmann, J., Danhof, S., Amberger, M., Quade-Lyssy, P., et al., 2021. CARAMBA: a first-in-human clinical trial with SLAMF7 CAR-T cells prepared by virus-free Sleeping Beauty gene transfer to treat multiple myeloma. Gene Ther. 28 (9), 560–571.

Przybylski, Z., 1989. The epidemiology of the morbidity of respiratory system malignant neoplasm in Poland on the background of the increasing atmospheric air pollution by dust and vehicle exhaust emission. Rocz. Panstw. Zakl. Hig. 40 (4-6), 257–265.

Quintana-Bustamante, O., Fananas-Baquero, S., Orman, I., Torres, R., Duchateau, P., Poirot, L., et al., 2019. Gene editing of PKLR gene in human hematopoietic progenitors through 5' and 3' UTR modified TALEN mRNA. PLoS One 14 (10), e0223775.

Rabinovich, P.M., Komarovskaya, M.E., Ye, Z.J., Imai, C., Campana, D., Bahceci, E., et al., 2006. Synthetic messenger RNA as a tool for gene therapy. Hum. Gene Ther. 17 (10), 1027–1035.

Ramanathan, A., Robb, G.B., Chan, S.H., 2016. mRNA capping: biological functions and applications. Nucleic Acids Res. 44 (16), 7511–7526.

Rashnonejad, A., Amini Chermahini, G., Wallace, L., Harper, S., 2019. O.8DUX4 mRNA silencing with CRISPR-Cas13 gene therapy as a prospective treatment for Facioscapulohumeral muscular dystrophy. Neuromuscul. Disord. 29, S40.

Rehwinkel, J., Tan, C.P., Goubau, D., Schulz, O., Pichlmair, A., Bier, K., et al., 2010. RIG-I detects viral genomic RNA during negative-strand RNA virus infection. Cell 140 (3), 397–408.

Reinhard, K., Rengstl, B., Oehm, P., Michel, K., Billmeier, A., Hayduk, N., et al., 2020. An RNA vaccine drives expansion and efficacy of claudin-CAR-T cells against solid tumors. Science 367 (6476), 446–453.

Ren, J., Liu, X., Fang, C., Jiang, S., June, C.H., Zhao, Y., 2017. Multiplex genome editing to generate universal CAR T cells resistant to PD1 inhibition. Clin. Cancer Res. 23 (9), 2255–2266.

Rittig, S.M., Haentschel, M., Weimer, K.J., Heine, A., Muller, M.R., Brugger, W., et al., 2011. Intradermal vaccinations with RNA coding for TAA generate CD8+ and CD4+ immune responses and induce clinical benefit in vaccinated patients. Mol. Ther. 19 (5), 990–999.

Rizvi, F., Everton, E., Smith, A.R., Liu, H., Osota, E., Beattie, M., et al., 2021. Murine liver repair via transient activation of regenerative pathways in hepatocytes using lipid nanoparticle-complexed nucleoside-modified mRNA. Nat. Commun. 12 (1), 613.

Rosenblum, D., Gutkin, A., Kedmi, R., Ramishetti, S., Veiga, N., Jacobi, A.M., et al., 2020. CRISPR-Cas9 genome editing using targeted lipid nanoparticles for cancer therapy. Sci. Adv. 6 (47).

Ross, J., Sullivan, T.D., 1985. Half-lives of beta and gamma globin messenger RNAs and of protein synthetic capacity in cultured human reticulocytes. Blood 66 (5), 1149–1154.

Rothgangl, T., Dennis, M.K., Lin, P.J.C., Oka, R., Witzigmann, D., Villiger, L., et al., 2021. In vivo adenine base editing of PCSK9 in macaques reduces LDL cholesterol levels. Nat. Biotechnol. 39 (8), 949–957.

Rybakova, Y., Kowalski, P.S., Huang, Y., Gonzalez, J.T., Heartlein, M.W., DeRosa, F., et al., 2019. mRNA delivery for therapeutic anti-HER2 antibody expression in vivo. Mol. Ther. 27 (8), 1415–1423.

Rydzik, A.M., Warminski, M., Sikorski, P.J., Baranowski, M.R., Walczak, S., Kowalska, J., et al., 2017. mRNA cap analogues substituted in the tetraphosphate chain with CX2: identification of O-to-CCl2 as the first bridging modification that confers resistance to decapping without impairing translation. Nucleic Acids Res. 45 (15), 8661–8675.

Sahin, U., Kariko, K., Tureci, O., 2014. mRNA-based therapeutics—developing a new class of drugs. Nat. Rev. Drug Discov. 13 (10), 759–780.

Sahin, U., Derhovanessian, E., Miller, M., Kloke, B.P., Simon, P., Lower, M., et al., 2017. Personalized RNA mutanome vaccines mobilize poly-specific therapeutic immunity against cancer. Nature 547 (7662), 222–226.

Sahin, U., Oehm, P., Derhovanessian, E., Jabulowsky, R.A., Vormehr, M., Gold, M., et al., 2020. An RNA vaccine drives immunity in checkpoint-inhibitor-treated melanoma. Nature 585 (7823), 107–112.

Sahin, U., Muik, A., Vogler, I., Derhovanessian, E., Kranz, L.M., Vormehr, M., et al., 2021. BNT162b2 vaccine induces neutralizing antibodies and poly-specific T cells in humans. Nature 595 (7868), 572–577.

Sanchez-Martin, D., Sanz, L., Alvarez-Vallina, L., 2011. Engineering human cells for in vivo secretion of antibody and non-antibody therapeutic proteins. Curr. Opin. Biotechnol. 22 (6), 924–930.

Sander, L.E., Davis, M.J., Boekschoten, M.V., Amsen, D., Dascher, C.C., Ryffel, B., et al., 2011. Detection of prokaryotic mRNA signifies microbial viability and promotes immunity. Nature 474 (7351), 385–389.

Schlake, T., Thess, A., Fotin-Mleczek, M., Kallen, K.J., 2012. Developing mRNA-vaccine technologies. RNA Biol. 9 (11), 1319–1330.

Schlake, T., Thess, A., Thran, M., Jordan, I., 2019. mRNA as novel technology for passive immunotherapy. Cell. Mol. Life Sci. 76 (2), 301–328.

Schumacher, T.N., Schreiber, R.D., 2015. Neoantigens in cancer immunotherapy. Science 348 (6230), 69–74.
Seif, M., Hoppstadter, J., Breinig, F., Kiemer, A.K., 2017. Yeast-mediated mRNA delivery polarizes immuno-suppressive macrophages towards an immuno-stimulatory phenotype. Eur. J. Pharm. Biopharm. 117, 1–13.
Selmi, A., Vascotto, F., Kautz-Neu, K., Tureci, O., Sahin, U., von Stebut, E., et al., 2016. Uptake of synthetic naked RNA by skin-resident dendritic cells via macropinocytosis allows antigen expression and induction of T-cell responses in mice. Cancer Immunol. Immunother. 65 (9), 1075–1083.
Sim, G.C., Martin-Orozco, N., Jin, L., Yang, Y., Wu, S., Washington, E., et al., 2014. IL-2 therapy promotes suppressive ICOS+ Treg expansion in melanoma patients. J. Clin. Invest. 124 (1), 99–110.
Simeonov, K.P., Uppal, H., 2014. Direct reprogramming of human fibroblasts to hepatocyte-like cells by synthetic modified mRNAs. PLoS One 9 (6), e100134.
Smits, E., Ponsaerts, P., Lenjou, M., Nijs, G., Van Bockstaele, D.R., Berneman, Z.N., et al., 2004. RNA-based gene transfer for adult stem cells and T cells. Leukemia 18 (11), 1898–1902.
Solal-Celigny, P., Lepage, E., Brousse, N., Reyes, F., Haioun, C., Leporrier, M., et al., 1993. Recombinant interferon alfa-2b combined with a regimen containing doxorubicin in patients with advanced follicular lymphoma. Groupe d'Etude des Lymphomes de l'Adulte. N. Engl. J. Med. 329 (22), 1608–1614.
Song, Z., Guo, C., Li, Y., Tan, B., Fan, L., Xiao, J., 2012. Enhanced antitumor effects of a dendritic cell vaccine transfected with gastric cancer cell total RNA carrying the 4-1BBL gene in vitro. Exp. Ther. Med. 3 (2), 319–323.
Stadler, C.R., Bahr-Mahmud, H., Celik, L., Hebich, B., Roth, A.S., Roth, R.P., et al., 2017. Elimination of large tumors in mice by mRNA-encoded bispecific antibodies. Nat. Med. 23 (7), 815–817.
Stepinski, J., Waddell, C., Stolarski, R., Darzynkiewicz, E., Rhoads, R.E., 2001. Synthesis and properties of mRNAs containing the novel "anti-reverse" cap analogs 7-methyl (3'-O-methyl)GpppG and 7-methyl (3'-deoxy)GpppG. RNA 7 (10), 1486–1495.
Streeter, H.B., Wraith, D.C., 2021. Manipulating antigen presentation for antigen-specific immunotherapy of autoimmune diseases. Curr. Opin. Immunol. 70, 75–81.
Strenkowska, M., Grzela, R., Majewski, M., Wnek, K., Kowalska, J., Lukaszewicz, M., et al., 2016. Cap analogs modified with 1,2-dithiodiphosphate moiety protect mRNA from decapping and enhance its translational potential. Nucleic Acids Res. 44 (20), 9578–9590.
Svoboda, J., Rheingold, S.R., Gill, S.I., Grupp, S.A., Lacey, S.F., Kulikovskaya, I., et al., 2018. Nonviral RNA chimeric antigen receptor-modified T cells in patients with Hodgkin lymphoma. Blood 132 (10), 1022–1026.
Takahashi, K., Yamanaka, S., 2006. Induction of pluripotent stem cells from mouse embryonic and adult fibroblast cultures by defined factors. Cell 126 (4), 663–676.
Tam, H.H., Melo, M.B., Kang, M., Pelet, J.M., Ruda, V.M., Foley, M.H., et al., 2016. Sustained antigen availability during germinal center initiation enhances antibody responses to vaccination. Proc. Natl. Acad. Sci. U. S. A. 113 (43), E6639–E6648.
Tasian, S.K., Kenderian, S.S., Shen, F., Ruella, M., Shestova, O., Kozlowski, M., et al., 2017. Optimized depletion of chimeric antigen receptor T cells in murine xenograft models of human acute myeloid leukemia. Blood 129 (17), 2395–2407.
Tcherepanova, I.Y., Adams, M.D., Feng, X., Hinohara, A., Horvatinovich, J., Calderhead, D., et al., 2008. Ectopic expression of a truncated CD40L protein from synthetic post-transcriptionally capped RNA in dendritic cells induces high levels of IL-12 secretion. BMC Mol. Biol. 9, 90.

Thess, A., Grund, S., Mui, B.L., Hope, M.J., Baumhof, P., Fotin-Mleczek, M., et al., 2015. Sequence-engineered mRNA without chemical nucleoside modifications enables an effective protein therapy in large animals. Mol. Ther. 23 (9), 1456–1464.

Thomas, S.J., Moreira, E.D., Jr., Kitchin, N., Absalon, J., Gurtman, A., Lockhart, S., et al., 2021. Safety and efficacy of the BNT162b2 mRNA Covid-19 vaccine through 6 months. N. Engl. J. Med. 385 (19), 1761–1773.

Thran, M., Mukherjee, J., Ponisch, M., Fiedler, K., Thess, A., Mui, B.L., et al., 2017. mRNA mediates passive vaccination against infectious agents, toxins, and tumors. EMBO Mol. Med. 9 (10), 1434–1447.

Trepotec, Z., Geiger, J., Plank, C., Aneja, M.K., Rudolph, C., 2019. Segmented poly(A) tails significantly reduce recombination of plasmid DNA without affecting mRNA translation efficiency or half-life. RNA 25 (4), 507–518.

Tse, S.W., McKinney, K., Walker, W., Nguyen, M., Iacovelli, J., Small, C., et al., 2021. mRNA-encoded, constitutively active STING(V155M) is a potent genetic adjuvant of antigen-specific CD8(+) T cell response. Mol. Ther. 29 (7), 2227–2238.

Tuyaerts, S., Van Meirvenne, S., Bonehill, A., Heirman, C., Corthals, J., Waldmann, H., et al., 2007. Expression of human GITRL on myeloid dendritic cells enhances their immunostimulatory function but does not abrogate the suppressive effect of CD4+CD25+ regulatory T cells. J. Leukoc. Biol. 82 (1), 93–105.

Van den Bergh, J., Willemen, Y., Lion, E., Van Acker, H., De Reu, H., Anguille, S., et al., 2015. Transpresentation of interleukin-15 by IL-15/IL-15Ralpha mRNA-engineered human dendritic cells boosts antitumoral natural killer cell activity. Oncotarget 6 (42), 44123–44133.

Van den Bergh, J.M.J., Smits, E., Versteven, M., De Reu, H., Berneman, Z.N., Van Tendeloo, V.F.I., et al., 2017. Characterization of interleukin-15-transpresenting dendritic cells for clinical use. J. Immunol. Res. 2017, 1975902.

Van der Jeught, K., Joe, P.T., Bialkowski, L., Heirman, C., Daszkiewicz, L., Liechtenstein, T., et al., 2014. Intratumoral administration of mRNA encoding a fusokine consisting of IFN-beta and the ectodomain of the TGF-beta receptor II potentiates antitumor immunity. Oncotarget 5 (20), 10100–10113.

Van Hoecke, L., Roose, K., Ballegeer, M., Zhong, Z., Sanders, N.N., De Koker, S., et al., 2020. The opposing effect of type I IFN on the T cell response by non-modified mRNA-lipoplex vaccines is determined by the route of administration. Mol. Ther. Nucleic Acids 22, 373–381.

Van Lint, S., Goyvaerts, C., Maenhout, S., Goethals, L., Disy, A., Benteyn, D., et al., 2012. Preclinical evaluation of TriMix and antigen mRNA-based antitumor therapy. Cancer Res. 72 (7), 1661–1671.

Van Nuffel, A.M., Benteyn, D., Wilgenhof, S., Pierret, L., Corthals, J., Heirman, C., et al., 2012. Dendritic cells loaded with mRNA encoding full-length tumor antigens prime CD4+ and CD8+ T cells in melanoma patients. Mol. Ther. 20 (5), 1063–1074.

Vatner, R.E., Janssen, E.M., 2019. STING, DCs and the link between innate and adaptive tumor immunity. Mol. Immunol. 110, 13–23.

Veiga, N., Goldsmith, M., Granot, Y., Rosenblum, D., Dammes, N., Kedmi, R., et al., 2018. Cell specific delivery of modified mRNA expressing therapeutic proteins to leukocytes. Nat. Commun. 9 (1), 4493.

Vlatkovic, I., 2021. Non-immunotherapy application of LNP-mRNA: maximizing efficacy and safety. Biomedicine 9 (5).

Vogel, A.B., Lambert, L., Kinnear, E., Busse, D., Erbar, S., Reuter, K.C., et al., 2018. Self-amplifying RNA vaccines give equivalent protection against influenza to mRNA vaccines but at much lower doses. Mol. Ther. 26 (2), 446–455.

Wadhwa, A., Aljabbari, A., Lokras, A., Foged, C., Thakur, A., 2020. Opportunities and challenges in the delivery of mRNA-based vaccines. Pharmaceutics 12 (2).

Walsh, E.E., Frenck Jr., R.W., Falsey, A.R., Kitchin, N., Absalon, J., Gurtman, A., et al., 2020. Safety and immunogenicity of two RNA-based covid-19 vaccine candidates. N. Engl. J. Med. 383 (25), 2439–2450.

Wang, H., Yang, H., Shivalila, C.S., Dawlaty, M.M., Cheng, A.W., Zhang, F., et al., 2013. One-step generation of mice carrying mutations in multiple genes by CRISPR/Cas-mediated genome engineering. Cell 153 (4), 910–918.

Warren, L., Manos, P.D., Ahfeldt, T., Loh, Y.H., Li, H., Lau, F., et al., 2010. Highly efficient reprogramming to pluripotency and directed differentiation of human cells with synthetic modified mRNA. Cell Stem Cell 7 (5), 618–630.

Weide, B., Pascolo, S., Scheel, B., Derhovanessian, E., Pflugfelder, A., Eigentler, T.K., et al., 2009. Direct injection of protamine-protected mRNA: results of a phase 1/2 vaccination trial in metastatic melanoma patients. J. Immunother. 32 (5), 498–507.

Weill, L., Belloc, E., Bava, F.A., Mendez, R., 2012. Translational control by changes in poly(A) tail length: recycling mRNAs. Nat. Struct. Mol. Biol. 19 (6), 577–585.

Weinstein-Marom, H., Pato, A., Levin, N., Susid, K., Itzhaki, O., Besser, M.J., et al., 2016. Membrane-attached cytokines expressed by mRNA electroporation act as potent T-cell adjuvants. J. Immunother. 39 (2), 60–70.

Weinstein-Marom, H., Levin, N., Pato, A., Shmuel, N., Sharabi-Nov, A., Peretz, T., et al., 2019. Combined expression of genetic adjuvants via mRNA electroporation exerts multiple immunostimulatory effects on antitumor T cells. J. Immunother. 42 (2), 43–50.

Weissman, D., Ni, H., Scales, D., Dude, A., Capodici, J., McGibney, K., et al., 2000. HIV gag mRNA transfection of dendritic cells (DC) delivers encoded antigen to MHC class I and II molecules, causes DC maturation, and induces a potent human in vitro primary immune response. J. Immunol. 165 (8), 4710–4717.

Wilber, A., Frandsen, J.L., Geurts, J.L., Largaespada, D.A., Hackett, P.B., McIvor, R.S., 2006. RNA as a source of transposase for Sleeping Beauty-mediated gene insertion and expression in somatic cells and tissues. Mol. Ther. 13 (3), 625–630.

Wilgenhof, S., Van Nuffel, A.M., Corthals, J., Heirman, C., Tuyaerts, S., Benteyn, D., et al., 2011. Therapeutic vaccination with an autologous mRNA electroporated dendritic cell vaccine in patients with advanced melanoma. J. Immunother. 34 (5), 448–456.

Wilgenhof, S., Van Nuffel, A.M.T., Benteyn, D., Corthals, J., Aerts, C., Heirman, C., et al., 2013. A phase IB study on intravenous synthetic mRNA electroporated dendritic cell immunotherapy in pretreated advanced melanoma patients. Ann. Oncol. 24 (10), 2686–2693.

Wolff, J.A., Malone, R.W., Williams, P., Chong, W., Acsadi, G., Jani, A., et al., 1990. Direct gene transfer into mouse muscle in vivo. Science 247 (4949 Pt. 1), 1465–1468.

Wroblewska, L., Kitada, T., Endo, K., Siciliano, V., Stillo, B., Saito, H., et al., 2015. Mammalian synthetic circuits with RNA binding proteins for RNA-only delivery. Nat. Biotechnol. 33 (8), 839–841.

Wu, J., Chen, Z.J., 2014. Innate immune sensing and signaling of cytosolic nucleic acids. Annu. Rev. Immunol. 32, 461–488.

Xiao, L., Cen, D., Gan, H., Sun, Y., Huang, N., Xiong, H., et al., 2019. Adoptive transfer of NKG2D CAR mRNA-engineered natural killer cells in colorectal cancer patients. Mol. Ther. 27 (6), 1114–1125.

Yakubov, E., Rechavi, G., Rozenblatt, S., Givol, D., 2010. Reprogramming of human fibroblasts to pluripotent stem cells using mRNA of four transcription factors. Biochem. Biophys. Res. Commun. 394 (1), 189–193.

Yamamoto, A., Kormann, M., Rosenecker, J., Rudolph, C., 2009. Current prospects for mRNA gene delivery. Eur. J. Pharm. Biopharm. 71 (3), 484–489.

Zahavi, D., Weiner, L., 2020. Monoclonal antibodies in cancer therapy. Antibodies (Basel) 9 (3).

Zangi, L., Lui, K.O., von Gise, A., Ma, Q., Ebina, W., Ptaszek, L.M., et al., 2013. Modified mRNA directs the fate of heart progenitor cells and induces vascular regeneration after myocardial infarction. Nat. Biotechnol. 31 (10), 898–907.

Zhang, F., Parayath, N.N., Ene, C.I., Stephan, S.B., Koehne, A.L., Coon, M.E., et al., 2019. Genetic programming of macrophages to perform anti-tumor functions using targeted mRNA nanocarriers. Nat. Commun. 10 (1), 3974.

Zhao, Y., Moon, E., Carpenito, C., Paulos, C.M., Liu, X., Brennan, A.L., et al., 2010. Multiple injections of electroporated autologous T cells expressing a chimeric antigen receptor mediate regression of human disseminated tumor. Cancer Res. 70 (22), 9053–9061.

Zhou, X., Berglund, P., Rhodes, G., Parker, S.E., Jondal, M., Liljestrom, P., 1994. Self-replicating Semliki Forest virus RNA as recombinant vaccine. Vaccine 12 (16), 1510–1514.

Zhu, N., Zhang, D., Wang, W., Li, X., Yang, B., Song, J., et al., 2020. A novel coronavirus from patients with pneumonia in China, 2019. N. Engl. J. Med. 382 (8), 727–733.

Zimmer, G., 2010. RNA replicons—a new approach for influenza virus immunoprophylaxis. Viruses 2 (2), 413–434.

Zinckgraf, J.W., Silbart, L.K., 2003. Modulating gene expression using DNA vaccines with different 3'-UTRs influences antibody titer, seroconversion and cytokine profiles. Vaccine 21 (15), 1640–1649.

CHAPTER TWO

Messenger RNA as a personalized therapy: The moment of truth for rare metabolic diseases

Karol M. Córdoba[a,b,†], Daniel Jericó[a,b,†], Ana Sampedro[a], Lei Jiang[c], María J. Iraburu[d], Paolo G.V. Martini[c], Pedro Berraondo[b,e,f], Matías A. Avila[a,b,g], and Antonio Fontanellas[a,b,g,]*

[a]Hepatology Program, Center for Applied Medical Research (CIMA), University of Navarra, Pamplona, Spain
[b]Navarra Institute for Health Research (IDISNA), Pamplona, Spain
[c]Moderna Inc, Cambridge, MA, United States
[d]Department of Biochemistry and Genetics. School of Sciences, University of Navarra, Pamplona, Spain
[e]Program of Immunology and Immunotherapy, CIMA-University of Navarra, Pamplona, Spain
[f]Centro de Investigación Biomédica en Red de Cáncer (CIBERONC), Instituto de Salud Carlos III, Madrid, Spain
[g]Centro de Investigación Biomédica en Red de Enfermedades Hepáticas y Digestivas (CIBERehd), Instituto de Salud Carlos III, Madrid, Spain
*Corresponding author: e-mail address: afontanellas@unav.es

Contents

† These two authors contributed equally: K.M.C. and D.J.

International Review of Cell and Molecular Biology, Volume 372
ISSN 1937-6448
https://doi.org/10.1016/bs.ircmb.2022.03.005
Copyright © 2022 Elsevier Inc.
All rights reserved.

Abstract

Inborn errors of metabolism (IEM) encompass a group of monogenic diseases affecting both pediatric and adult populations and currently lack effective treatments. Some IEM such as familial hypercholesterolemia or X-linked protoporphyria are caused by gain of function mutations, while others are characterized by an impaired protein function, causing a metabolic pathway blockage. Pathophysiology classification includes intoxication, storage and energy-related metabolic disorders. Factors specific to each disease trigger acute metabolic decompensations. IEM require prompt and effective care, since therapeutic delay has been associated with the development of fatal events including severe metabolic acidosis, hyperammonemia, cerebral edema, and death. Rapid expression of therapeutic proteins can be achieved hours after the administration of messenger RNAs (mRNA), representing an etiological solution for acute decompensations. mRNA-based therapy relies on modified RNAs with enhanced stability and translatability into therapeutic proteins. The proteins produced in the ribosomes can be targeted to specific intracellular compartments, the cell membrane, or be secreted. Non-immunogenic lipid nanoparticle formulations have been optimized to prevent RNA degradation and to allow safe repetitive administrations depending on the disease physiopathology and clinical status of the patients, thus, mRNA could be also an effective chronic treatment for IEM. Given that the liver plays a key role in most of metabolic pathways or can be used as bioreactor for excretable proteins, this review focuses on the preclinical and clinical evidence that supports the implementation of mRNA technology as a promising personalized strategy for liver metabolic disorders such as acute intermittent porphyria, ornithine transcarbamylase deficiency or glycogen storage disease.

1. Introduction

Inborn errors of metabolism (IEM) are a subtype of rare inherited diseases caused by an enzyme defect in a metabolic pathway. Depending on the pathophysiology, IEM can be classified into three different groups (Solares et al., 2021): (i) Intoxication disorders, characterized by the accumulation of potentially toxic metabolites, triggered by specific factors for each disease and that lead to acute decompensations. In this group, patients usually alternate asymptomatic periods with symptomatic acute decompensations. Such is the case with patients with methylmalonic and propionic acidemias, classic galactosemia or acute intermittent porphyria. (ii) Energy metabolism disorders, such as most of the glycogenosis or defects in the beta oxidation of fatty acids, result from the disruption of either production or utilization of energy by tissues. The heart, skeletal muscle, or brain, the organs with the highest energy consumption, are involved in the diverse spectrum of these disorders. (iii) Storage disorders, such as Gaucher and Fabry diseases, are caused by an enzyme deficiency involved in the synthesis or catabolism of large molecules. Thus, the body is unable to store or break down complex molecules. In some cases, glucose homeostasis is compromised whereas in others, complex molecules are accumulated in cell organelles, resulting in lysosomal or peroxisomal malfunction, and cell dysfunction. Acute decompensations rarely happen in this group, although they occur in strokes in patients with Fabry disease (Solares et al., 2021).

Due to the genetic origin of rare metabolic diseases, gene therapy (GT) emerged as a promising therapeutic alternative for these disorders in the early 1990s. The liver plays an essential role in metabolism and protein synthesis. Thereby, it is the main target organ for the treatment of many IEM. Liver-directed GT strategies using non-viral and viral vectors have been assayed as an etiological treatment for these disorders. Viral-based vector strategies are leading to a slow but constant approval of advanced medicines by regulatory agencies. However, they are not advantageous for treating acute metabolic decompensations as their therapeutic action is delayed since it requires nuclear delivery, DNA transcription and RNA transport to the cytoplasm before translation into the therapeutic protein. Furthermore, most of these diseases have a childhood onset, and thus, the use of non-integrative viral vectors would require re-administrations due to the dilution of the therapeutic effect caused by rapid normal organ growth. Moreover, viral vectors induce an immune activation that prevents re-administration. Finally, due to the complexity of the production process and regulatory procedures, it is unlikely that therapeutic products based on viral vectors for diseases with low or very low prevalence will be commercialized. Thus, there is an unmet need for therapeutic alternatives for these rare and ultrarare diseases.

RNA therapeutics can be classified into four different groups (Fig. 1). The first group includes RNA molecules that range from 15 to 30 base pairs in length and target nucleic acids, either DNA or RNA, regulating gene expression in the cell. This group can be subdivided into two categories: single-stranded antisense RNAs and double-stranded RNAs. The second group comprises RNA aptamers (56–120 nucleotides) that interact with proteins and modulate their function. The third group consists of suppressor transfer RNAs (tRNAs), molecules of 70–80 nucleotides long that avoid translation of premature stop codons; they appear due to mutations in the nucleotide sequence and are responsible for more than 10% of all genetic

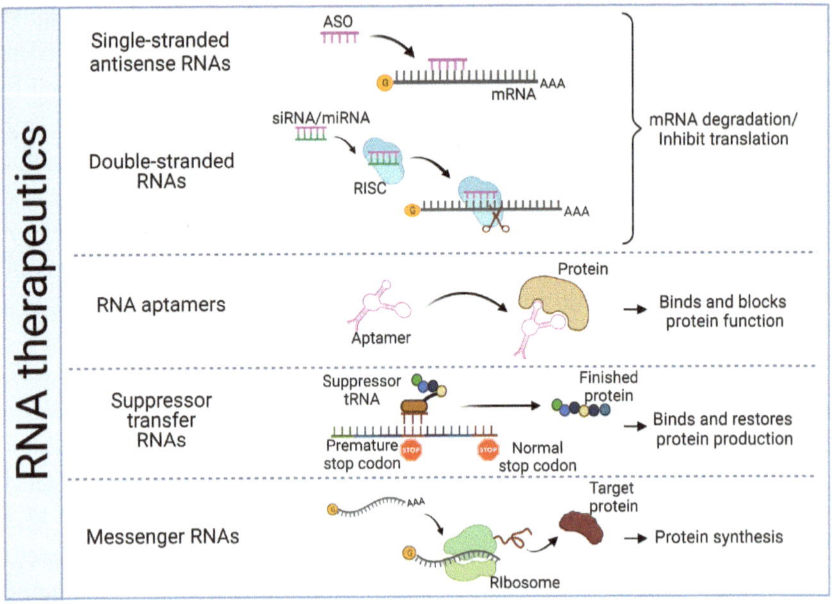

Fig. 1 Schematic illustration of the different RNA-based therapeutic approaches. ASO, antisense oligonucleotide; miRNA, microRNA; mRNA, messenger RNA; RISC, RNA-induced silencing complex; siRNA, small interfering RNA.

diseases (e.g., Cystic fibrosis or Duchenne muscular dystrophy). The fourth group includes RNA sequences with an average length of 1500–2500 base pairs, known as messenger RNAs (mRNAs) that are translated into proteins (Kim, 2020; Kirchner and Ignatova, 2015; Mollocana-Lara et al., 2021).

Therapeutic mRNAs formulated in lipid nanoparticles (LNPs) have a rapid onset of effect, within a few hours after their administration, making them an ideal product to treat acute metabolic decompensations. Non-viral LNPs formulations have been optimized to minimize immunogenicity, allowing single or repeated administrations depending on the disease and the clinical status of the patients. Moreover, LNPs can be designed to achieve liver tropism. Thus, mRNA-based protein replacement strategies may constitute suitable personalized approaches to target hepatic diseases. Recent advances in RNA technology and the urgent need for vaccines caused by the current COVID-19 pandemic have enabled the rapid translation of mRNA therapy to the clinic, paving the way for the therapeutic use of mRNAs in IEM currently lacking effective treatments. Importantly, once the technology has been optimized, it can be universally applied, thus benefiting a broad range of diseases and patients.

1.1 mRNA therapy development

The discovery of mRNA in 1961 was the first milestone in the development of this new emerging therapy (Brenner et al., 1961; Cobb, 2015; Gros et al., 1961), a therapeutic strategy based on the cytoplasmic delivery of an mRNA able to be translated into a protein. The therapeutic capability of mRNA was not explored during the years immediately after its discovery (Sahin et al., 2014) due to the instability and rapid degradation of the molecule and its capacity to induce an immune response. In 1990, Wolff et al. described for the first time the expression of reporter genes in mice skeletal muscle after pure RNA injection (Wolff et al., 1990). One year before, Malone et al. had achieved exogenous mRNA transfection in cultured cells by the use of cationic liposomes (Malone et al., 1989). These events determined the beginning of mRNA therapy, whose applications in vaccines for cancer or infectious and metabolic diseases are likely to have a significant impact in the very near future.

Throughout the development of mRNA therapies certain obstacles had to be overcome. mRNA is an inherently unstable molecule due to the ubiquitous presence of RNases (Houseley and Tollervey, 2009), and it can also induce strong immunogenicity, causing cell toxicity and impaired translation of therapeutic proteins (Damase et al., 2021). Moreover, the intracellular delivery of naked mRNA was challenging (Su et al., 2011) as nucleic acids cannot freely cross cell membranes due to their high electric charge and large molecular weight (Dammes and Peer, 2020; Whitehead et al., 2009).

Significant efforts have been made to optimize mRNA molecules by improving their transport to the cytoplasm of target cells, enhancing their stability and translation capacity, and by dramatically lowering the immune activation in order to increase their therapeutic potential. Most synthetic mRNAs consist of a single-stranded open reading frame (ORF) flanked by untranslated regions (UTR, 5′ and 3′), a 5′ cap and 3′ poly(A) tail sequences (Loomis et al., 2016). These five structural components can be modified to optimize the therapeutic potential and safety of the mRNA molecules (Fig. 2):

(i) The *5′ cap* protects the molecule from exonuclease degradation, initiates mRNA translation, and avoids induction of the innate immune response. This cap can be incorporated by a recombinant viral capping enzyme after the initial synthesis (Martin and Moss, 1975) or by the addition of a synthetic cap analog into the *in vitro* transcription reaction (Malone et al., 1989). Anti-reverse cap analogs (ARCAs) are the most extended type (Mockey et al., 2006; Rabinovich et al., 2006).

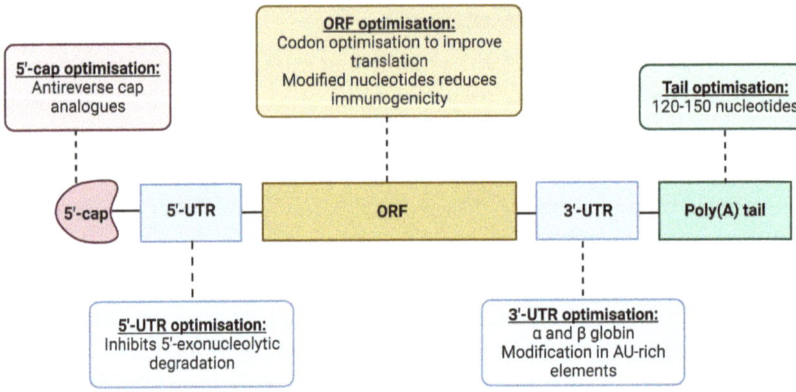

Fig. 2 Structure of an optimized therapeutic mRNA construction. ORF, open reading frame; UTR, untranslated region.

(ii) *Poly A tail* regulates the stability and translational efficiency of the mRNA molecule. Its optimal length is between 120 and 150 nucleotides (Mockey et al., 2006; Sahin et al., 2014).

(iii) The *5′ and 3′ UTR* regions are not translated into proteins. They can be optimized, for example by using the 3′ UTR of the α- and β-globin transcripts, or by removing AU-rich elements, in order to improve the stability and translation efficiency of the target mRNA sequence (Chen and Shyu, 1995; Ross and Sullivan, 1985; Sahin et al., 2014; Wang et al., 1999).

(iv) *ORF* or coding sequence. Its nature also affects translation efficiency. Codon optimization (for example, replacing rare codons with synonymous frequent codons) can lead to accelerated translation rates and increased protein yield (Gustafsson et al., 2004). Furthermore, the use of chemically-modified nucleotides, such as pseudouridine, 2-thiouridine, or 5-methoxyuridine, reduces immunogenicity (Anderson et al., 2010, 2011; Thess et al., 2015). Modified mRNAs decrease their recognition by pathogen-associated molecular receptors, such as Toll-like receptors, minimizing the inflammatory response (Karikó et al., 2005).

Besides the incorporation of modified-nucleosides, purification procedure by high performance liquid chromatography (HPLC) removes DNA template and double-stranded RNA, responsible for innate immune activation (Karikó et al., 2011). HPLC purification process also eliminates truncated RNAs, all enzymes used, and organic and inorganic contaminants improving mRNA translation upon delivery into primary mammalian cells.

One of the major challenges for the success of therapeutic mRNA application in medicine is its intracellular delivery and endosomal escape. For these reasons, different biomaterials such as lipids, polymers, and protein derivatives have been developed as mRNA vehicles. LNPs are the most common delivery system for small molecule drugs, interfering RNA and mRNA (Akinc et al., 2019; Hou et al., 2021; Khurana et al., 2021; Kim et al., 2021; Mollocana-Lara et al., 2021). LNPs are multi-component spherical vesicles (~100 nm) made of amino lipids, phospholipids, cholesterol, and polyethylene glycol (PEG)-conjugated lipids (Khurana et al., 2021; Martini and Guey, 2019; Trepotec et al., 2019). The key component is the amino lipid, whose constitution determines cellular uptake, endosomal escape, and tolerability (Sabnis et al., 2018). Cholesterol enhances particle stability by modulating membrane integrity and stiffness. PEG inhibits interactions with plasma proteins allowing increased LNPs circulation time and escape from phagocytic cells (Dammes and Peer, 2020), while phospholipids provide additional LNPs fusogenicity with the target cell membranes, thus helping endosomal escape and mRNA release into the cytosol (Sabnis et al., 2018). Compared to other GT strategies, one of the advantages of LNPs over viral vectors is their theoretically unlimited packaging size, allowing their application to diseases associated with defects in larger genes. In contrast, transgene size is limited to 4.7 kb in recombinant vectors derived from adeno-associated viruses (Chamberlain et al., 2016).

LNPs protect the mRNA from degradation until it reaches the target cell, where they are internalized by endocytosis mediated by the apolipoprotein E (ApoE) receptor (Akinc et al., 2010). Then, mRNA escapes from the endosome leading to its release into the cytoplasm, where it is translated into functional protein by the ribosome (Patel et al., 2017) (Fig. 3).

2. mRNA therapeutics *vs* other advanced therapies for liver metabolic diseases

Part of the great interest in mRNA-based therapies is due to their advantages over other technologies (Table 1). Some GT products such as plasmid DNA or viral vectors require nuclear localization and their delivery has consequently resulted in low levels of transduction because of the barriers to entering the nucleus (Prieve et al., 2018). In contrast, mRNA only needs to reach the cytoplasm and does not transit to the nucleus, thereby mitigating the risks of insertional mutagenesis. Furthermore, mRNA provides transient protein expression, reducing the risk of unintentional

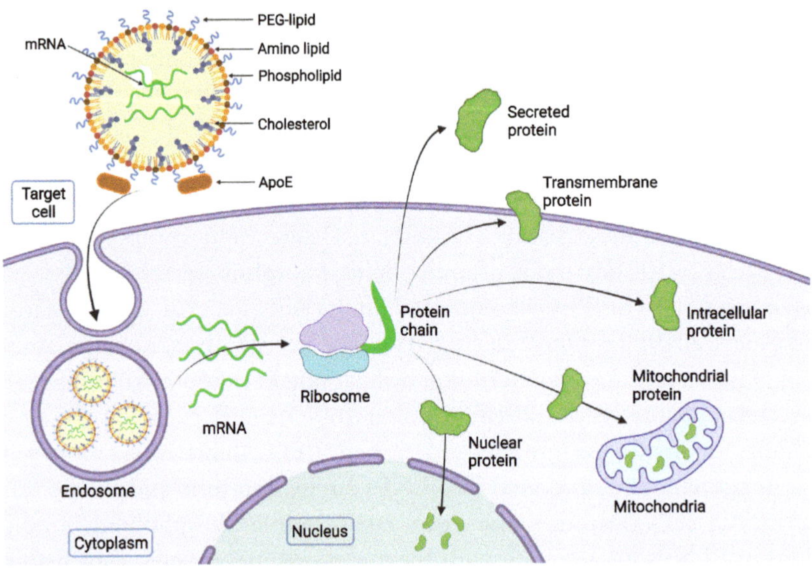

Fig. 3 Lipid nanoparticle-mediated delivery of mRNA. LNPs are taken up by endocytosis, followed by endosomal escape and mRNA release into the cytosol. mRNA is then translated into proteins by ribosomes. Proteins containing a specific signal peptide help to direct them to their specific intracellular location or to secrete them according to their site of action.

overdose due to constitutive gene activation, thus allowing control of pharmacokinetics (Cao et al., 2021; Prieve et al., 2018). Therefore, dosing can be adjusted for each patient from childhood to adulthood, in contrast to viral vectors whose use in pediatric diseases is challenging because hepatocyte proliferation would dilute the presence of the transduced cells (Truong et al., 2019). Another side effect of viral vectors that is avoided by the use of mRNA-LNPs is the immune-mediated hepatitis that may occur due to viral vector infection (Prieve et al., 2018; Raper et al., 2003). In fact, viruses induce such a potent humoral immune response that may preclude their readministration. Moreover, the efficacy of the first dose can also be affected by the presence of pre-existing neutralizing antibodies (Vandamme et al., 2017). In contrast, mRNA-LNPs have proved to be safe and effective in non-human primates after several consecutive administrations (Jiang et al., 2018).

Enzyme replacement therapy (ERT) consists of the direct administration of the defective or absent proteins. ERT is mainly restricted to secreted proteins (Berraondo et al., 2019). Furthermore, most recombinant proteins have difficulty entering the target cells, or do so in very small quantities that

Table 1 Comparison between enzyme replacement therapy, gene therapy, and mRNA therapy for the treatment of rare metabolic disorders.

	Enzyme replacement therapy	Viral vectors gene therapy	mRNA-based therapy
Multiple administration	Yes, daily/weekly	No	Yes, weekly/monthly
Administration route	Subcutaneous, Intramuscular, Intravenous...	Intravenous, Intrathecal, Intramuscular	Subcutaneous, Intramuscular, Intravenous...
Intracellular location	No	Yes, nucleus	Yes
Immunogenicity	Low, after chronic administrations	High, no readministration	Low, modified mRNA (*e.g.*, pseudouridines)
Side effects	Low	High, mutagenesis risk	Low
Beginning of therapeutic action	Hours	3–5 Days	Hours
Relative production cost	High	High	Low
Provide the optimal dosage for each presentation	Yes	No	Yes

are not curative. These facts potentially rule out ERT for those disorders in which the therapeutic protein needs to be internalized to exert its curative effect. Moreover, recombinant proteins may have different biochemical and pharmacological properties from endogenous proteins, such as leader sequence processing, post-translational modifications, or cellular localization signals, which may be critical for protein function (An et al., 2017). Whereas recombinant proteins have a high production cost and need the design of large-scale production processes, the development and manufacturing of mRNA-based drugs is relatively simple and more cost-effective. In addition, exogenous mRNA can encode for any protein sequence and take advantage of the intracellular machinery for the production and the appropriate cellular localization of the protein, whether this be transmembrane, mitochondrial, intracellular, or secreted. Thus, systemic mRNA therapy can restore "undruggable" targets (An et al., 2017; Cao et al., 2021; Kim, 2020). Other limitations of protein-based ERT are nonimmunologic anaphylactic reactions and antibody formation. Conversely, therapeutic mRNA avoids the common complications of recombinant proteins by using the endogenous translational machinery and post-translational mechanisms to produce the therapeutic protein (Prieve et al., 2018).

One of the strengths of mRNA therapy is the ability to rapidly alter the sequence of the mRNA construct for more accurate treatment (Damase et al., 2021). This feature was crucial to accelerate vaccine development during the current COVID-19 pandemic, when two mRNA products were readily available for mass administration: BNT162b2 (Polack et al., 2020) and mRNA-1273 (Baden et al., 2021). Although mRNA vaccines had been studied before for other viral diseases such as influenza, Zika, rabies, and cytomegalovirus (CMV), these two mRNA vaccines were among the first authorized and approved for COVID-19. The efficacy and safety profile shown by mRNA vaccines has increased the confidence in and expectations of mRNA technology as a promising alternative for the treatment of IEM. mRNA would provide the information required to produce the therapeutic protein in the human cells. This protein, like other endogenous proteins, may stay in the body for days to treat an acute presentation. In the case of chronic presentations, a new administration cycle can be applied, and individual treatment decisions could be taken according to the biochemical and clinical status of the patient. These features may define a new era in personalized medicine. Here, we summarize preclinical studies in hepatic IEM that further support the validity of this strategy and which must precede clinical trials.

2.1 Acute intermittent porphyria

Acute intermittent porphyria (AIP, MIM 176000) is caused by a hepatic deficiency of porphobilinogen deaminase (PBGD, EC 2.5.1.61), the third enzyme of the heme synthesis pathway. The disease has an autosomal dominant inheritance, and patients show acute attacks characterized by severe neuropathic abdominal pain, fatigue, insomnia, muscle pain, and in some cases, peripheral neuropathy. The prevalence of AIP is estimated to be 5–10 cases per million, but the occurrence of the genetic defects is much higher in the general population (about 1 in 1700 individuals), implying low penetrance of the disease (Chen et al., 2016; Nordmann et al., 1997).

The clinical manifestations of AIP begin to appear between 15 and 40 years of age, and it is more common among women (5:1 w *vs* m gender ratio). Most patients discover that they have AIP after the first acute attack caused by exposure to precipitating factors including medications and other chemicals (Drugs database: http://www.drugs-porphyria.org), alcohol intake, acute illness, infection, psychological stress, physical exhaustion, fasting and caloric deprivation as well as estrogen and progesterone, steroid hormones regulating reproductive activity in women. Once the precipitating agent is identified and removed, when possible, patients may not experience an attack again. However, most maintain a high accumulation and urinary excretion of the potential neurotoxic heme precursors, δ-aminolevulinic acid (ALA) and porphobilinogen (PBG) for years. The asymptomatic but high excretors do not receive treatment, although a study shows that 46.4% of such patients report the chronic symptoms associated with porphyria, such as abdominal pain, fatigue, muscle pain, and insomnia (Buendía-Martínez et al., 2021). Exposure to heme precursors, mainly ALA, is associated with an increased risk of developing progressive renal failure (Pallet et al., 2015) or hepatocellular carcinoma (Peoc'h et al., 2019), the long-term complications associated with AIP.

Patients can be divided into two groups according to the frequency of the attacks: sporadic (1–3 per year) or frequent (>3 attacks per year). Patients with mild pain and no paresis can be treated with intravenous (i.v.) administration of glucose. Hemin therapy (Hemin arginate, Normosang® in Europe and Lyophilised hematin, Panhematin® in the US, both from Recordati, Milan, Italy) is used in severe attacks. In these serious cases, the standard of care consists of daily i.v. hemin administration with the aim of replenishing the regulatory heme pool in hepatocytes. The decrease in heme precursor excretion typically becomes pronounced on day 3, and pain and nausea frequently resolve by day 4 (Marsden et al., 2015).

Around 5% of symptomatic patients suffer from frequent and disabling acute attacks. The most common therapeutic option for preventing frequent attacks is the off-label prophylactic i.v. administration of hemin. The efficacy of this treatment is very difficult to assess due to significant variability among patients and the low number of patients included in each treatment group. An audit report by The National Acute Porphyria Service in England concluded that prophylactic heme arginate appears to be beneficial in patients with recurrent acute porphyria symptoms (Marsden et al., 2015). However, in a review published by The American Porphyria Foundation in which all the porphyria cases reported between 1976 and 2004 were compiled (71 publications that involve approximately 1000 patients), the main conclusion was that the efficacy of hemin therapy depends on the early i.v. administration of this compound (Anderson et al., 2005).

Repeated administration of hemin causes many unwanted effects and complications, such as iron overload, thrombophlebitis at the peripheral vein infusion site which frequently requires that the drug be administered through a central vein, and transitory renal failure (Kauppinen and Mustajoki, 1992). In addition, repeated administration reduces the therapeutic efficacy of hemin, which is associated with the induction of heme oxygenase (Schmitt et al., 2018), the key enzyme of heme catabolism.

A novel approach to preventing recurrent acute attacks in this subgroup of severely affected patients is a synthetic small interfering RNA (givosiran, Givlaari™, Alnylam Pharmaceuticals) designed to down-regulate the expression of the first and rate-limiting enzyme of heme synthesis in the liver, the ALA synthase 1 (ALAS1, EC 2.3.1.37). Subcutaneous administration of givosiran has shown promising results preventing ALA/PBG accumulation and reducing the frequency of acute attacks (74%) in a phase III trial (NCT03338816). The adverse events associated with givosiran were elevation of serum aminotransferases levels and changes in serum creatinine and estimated Glomerular Filtration Rate, suggesting kidney function issues (Balwani et al., 2020). Downregulation of ALAS1 combined with defective heme synthesis in severely affected patients could lower hepatic heme content and reduce the activity of heme-dependent hepatic proteins causing reduced drug metabolization rates (Bosilkovska et al., 2016) and aggravation of the dysregulation of one-carbon metabolism, resembling classic homocystinuria (Fontanellas et al., 2021; To-Figueras et al., 2021). For severe affected patients that do not respond to current treatments, liver transplantation is the only therapeutic option. Therefore, there is a clear need for a more physiological therapy with fewer adverse effects that can be applicable to all groups of patients, regardless of their kidney function.

Liver-directed GT using recombinant adeno-associated virus (rAAV)-mediated transfer of human PBGD cDNA represents a promising therapeutic option for those patients with AIP suffering from severe and recurrent acute attacks. An over-exposure to the PBGD protein is safe, and partial recovery of hepatic PBGD levels was sufficient to achieve clinical benefit in a mouse model of the disease (Unzu et al., 2011, 2015; Yasuda et al., 2010). However, a clinical trial (NTC020082860) failed to reduce levels of porphyrin precursors in 8 patients with severe disease, probably due to insufficient liver transduction at the doses tested (up to 1.8×10^{13} genome copies/kg) (D'Avola et al., 2016). Although the new generations of AAV vectors are more efficient at transducing the liver, one of the major hurdles of these non-integrative viral vectors is that their action is diluted due to the natural division of hepatocytes (every 4 months in humans), and the fact that they can only be administered once due to the strong immune reaction against the viral capsid at the time of first exposure.

PBGD mRNA-based AIP therapy is likewise a potential candidate for all types of patients with AIP. Given that the therapeutic action occurs in hours and multi-dose administration is possible, it can be used both as a treatment for sporadic acute attacks and as a preventive strategy in patients suffering frequent acute attacks. PBGD mRNA has already been tested in an AIP mouse model characterized by high accumulation of porphyrin precursors, *Alas1* induction, pain, motor disturbances, and hypertension when exposed to a porphyrinogenic drug (Table 2). Systemic administration of a human PBGD mRNA (0.5 mg/kg) proved to be effective in preventing and restoring all these abnormalities more effectively than hemin, the current standard of care (Jiang et al., 2018). Multidose administration in large animals (rabbits and non-human primates) was proved safe and without loss of efficacy after recurrent administration (Jericó et al., 2021; Jiang et al., 2018). These experimental data provide proof-of-concept for efficacy, safety, and translatability of the systemic hPBGD mRNA. The approach holds promise as a singular therapeutic intervention applicable to patients with latent but high porphyrin excretion levels, as well as for sporadic and recurrent presentations. For those cases of recurrent administrations, individual dosage decisions could be made according to the clinical status or excretion levels of urinary porphyrin precursors.

2.2 Phenylketonuria

Phenylketonuria (PKU, MIM 261600) is an autosomal recessive disease of amino acid metabolism caused by a hepatic deficiency in the cytosolic enzyme phenylalanine hydroxylase (PAH, EC 1.14.16.1). A genetic mutation causing reduced activity of the hepatic PAH enzyme, or a deficiency of its cofactor tetrahydrobiopterin (BH4), results in the accumulation of phenylalanine in the blood to potentially toxic levels. This alteration is associated with severe intellectual disability when untreated as well as an eczematous rash, behavioral, psychiatric and movement problems (Blau et al., 2010; van Spronsen et al., 2021). In pregnant patients, phenylalanine levels above the therapeutic range lead to adverse fetal effects (Lichter-Konecki and Vockley, 2019). Phenylketonuria affects about 1 in 12,000 newborns and both genders are affected equally. The onset of symptoms occurs during the first days of life.

Table 2 Current status of the development of new therapeutic strategies for rare metabolic diseases.

Disease	Inheritance	Onset of symptoms	Affected metabolism(enzyme)	Subcellular location	Therapeutic mRNA (reference)		Company	Phase of development
Acute Intermittent Porphyria	Autosomal dominant	15–40 years	Hemo (PBGD)	Cytoplasm	PBGD mRNA		Moderna, Inc	Preclinical
Phenylketonuria	Autosomal recessive	Newborn	Phenylalanine (PAH)	Cytoplasm	PAH mRNA	(mRNA-3283)	Moderna, Inc	Preclinical
Classic Galactosemia	Autosomal recessive	Infancy	Galactose (GALT)	Cytoplasm	hGALT mRNA		Moderna, Inc	Preclinical
Liver arginase deficiency	Autosomal recessive	Childhood	Ciclo Urea (ARG1)	Cytoplasm	hARG1 mRNA		Moderna, Inc	Preclinical
Glycogen storage Disease Type 3	Autosomal recessive	Infancy	Glycogen (AGL)	Cytoplasm	GDE mRNA	(UX053)	Ultragenyx/ Arcturus Therapeutics	Phase 1 (NCT 04574830)
Glycogen storage Disease Type 1a	Autosomal recessive	Infancy	Glucosa (G6Pase)	ER membrane	G6Pase mRNA	(mRNA-3745)	Moderna, Inc	Phase 1 (NCT 05095727)
							Alexion Pharmaceuticals	Preclinical
AAT deficiency	Autosomal recessive	25–50 years	(AAT)	ER	hAAT mRNA		Alexion Pharmaceuticals	Preclinical
							Moderna, Inc	Preclinical
Crigler-Najjar Syndrome Type 1	Autosomal recessive	Infancy	Bilirubin (UGT1A1)	ER	UGT1A1 mRNA		Alexion Pharmaceuticals	Preclinical
							Moderna, Inc	Preclinical

Propionic Acidemia	Autosomal recessive	Infancy	Propionate/Urea cycle (PPC)	Mitochondria	PCCA/PCCB mRNA	(mRNA-3927)	Moderna, Inc	Phase 1/2 (NCT 04159103; NCT 05130437)
Ornithine transcarbamylase deficiency	X-linked	Infancy	Urea cycle (OTC)	Mitochondria	OTC mRNA	(MRT5201)	Translate Bio	Phase 1/2 discontinued (NCT 03767270)
						(PRX-0D2)	CureVac/ Arcturus Therapeutics	Preclinical
						(ARCT-810 / LUNAR-OTC)	CureVac/ Arcturus Therapeutics	Phase 1 (NCT 04442347)
								Phase 2 (EudraCT 2021–001081-38)
							Genevant	Preclinical
							Phase Rx	Preclinical
Methylmalonic Acidemia	Autosomal recessive	Infancy	Propionate (MUT)	Mitochondria	MUT mRNA	(mRNA-3704)	Moderna, Inc	Phase 1/2 (NCT 03810690 discontinued; EudraCT2019-001061-32)
						(mRNA-3705)	Moderna, Inc	Phase 1/2 (NCT 04899310)
Primary hyperoxaluria 1	Autosomal recessive	Childhood	Glyoxylate (AGT)	Peroxisome	AGT mRNA		Alexion Pharmaceuticals	Preclinical
Fabry Disease	X-linked	Childhood/ Teenagehood	Glycolipids (GLA)	Lysosome	Gal A mRNA		Moderna, Inc	Preclinical
							Translate Bio	Preclinical

Although dietary phenylalanine restriction remains the main therapy for PKU (Blau et al., 2010; Lichter-Konecki and Vockley, 2019), two pharmacological treatments are available. Sapropterin dihydrochloride is a synthetic BH4 analog that reduces blood phenylalanine concentrations in a subset of patients with cofactor deficiency. It allows for a dietary tolerance for phenylalanine (Levy et al., 2007) in patients with milder forms of the disease (Bernegger and Blau, 2002; Burton et al., 2007). ERT with phenylalanine ammonia lyase (PAL), an enzyme that degrades phenylalanine through a different pathway than PAH (Levy et al., 2018), is highly effective in most patients, but it induces the formation of anti-drug antibodies, that preclude its readministration and is associated with adverse effects such as skin reactions, arthralgia and, very rarely, anaphylactic responses (Gupta et al., 2018).

In order to develop more effective therapies for all groups of patients, liver-directed GT using viral vector delivery of the functional PAH gene and genome editing have been assayed in animal models of the disease (Harding et al., 2006; Villiger et al., 2018; Yagi et al., 2011). However, only two Phase 1/2, open-label, dose-escalation clinical trials using vectors derived from AAV9 (AAVHSC15; NCT03952156) and AAV8 serotypes (NCT04480567) are currently active to evaluate safety and efficacy in children (>15 y) and adults with PAH deficiency. However, as previously mentioned, the efficacy of the AAV-mediated GT approach would be transient in infants due to rapid hepatocyte proliferation and episomal dilution of the AAV genome vector. In the absence of a selective growth advantage for PAH expressing hepatocytes, only liver-directed GT based on integrating lentiviral vectors or recurrent administration of non-immunogenic vectors could avoid this issue. In this regard, American Gene Technologies, Inc. (Rockville, MD) has publicly announced a liver-directed GT strategy using a proprietary formulation integrating a lentivirus-based vector system (Orphan Drug status granted by the FDA, ref.: DRU2018-6572). On the other hand, Moderna Inc. has included in its pipeline an mRNA-based therapeutic product for PKU (mRNA-3283). However, no preclinical data or any other information regarding the progress of these strategies toward clinical trials have been released to date (https://clinicaltrials.gov/ update, February 2022).

2.3 Classic galactosemia

Type I or Classic galactosemia (CG, MIM 230400) is a rare hereditary disease caused by cytosolic galactose-1-phosphate uridylyltransferase (GALT,

EC 2.7.7.12) deficiency. CG is the most common and the most severe form of galactosemias, a group of disorders in which galactose metabolism is affected. The incidence rate varies according to the population studied but is 1 per 60,000 births for people of European descent. Galactosemia is usually first detected through newborn screening when available.

GALT deficiency prevents the appropriate transformation of uridine diphosphoglucose and galactose-1 phosphate (Gal-1-P) into uridine diphosphogalactose (UDP-Gal) and glucose-1-phosphate, hence, resulting in the accumulation of Gal-1-P (Bosch et al., 2004; Demirbas et al., 2018) in blood and urine. If untreated, CG seriously endangers patients' lives. Neurological deterioration is infrequent and galactose removal from the diet of neonatal infants has shown to be effective in controlling life-threatening manifestations (Welling et al., 2017). However, it is not sufficient to prevent the long-term complications associated with endogenous Gal-1-P accumulation (Schadewaldt et al., 2014), such as brain function impairment with movement disorders, primary ovarian insufficiency, and diminished bone mineral density (Fridovich-Keil et al., 2011; Rubio-Gozalbo et al., 2019).

No clinical trials of gene or molecular therapies for CG have been registered. However, some preclinical studies offer promising results. Rasmussen et al. preclinically assayed AAV therapy efficacy in 3-day-old GALT-null rats that were i.v. administered an AAV9-hGALT vector (with tropism for the liver and central nervous system) in order to increase GALT activity in the liver and brain (Rasmussen et al., 2021). A significant increase in GALT activity in both organs was associated with reduction in GAL-1-P and the accumulation of other galactose metabolites at 2 weeks post-AAV injection.

Systemic or intraperitoneal administration of a single dose of human GALT (hGALT) mRNA soon after birth increased survival in lethal GalT-deficient mice associated with increased protein expression in the liver (Balakrishnan et al., 2020). Although the LNPs formulation only targeted the liver, hGALT mRNA therapy was demonstrated to be effective in reducing symptoms and increasing the growth of GalT-deficient mice (Balakrishnan et al., 2020). Although total elimination of Gal-1-P was not achieved with hGALT mRNA, Yuzyuk et al. suggested that partial removal could improve prognosis and quality of life according to an observational study of a galactose restricted diet in 34 patients (Yuzyuk et al., 2018).

Moreover, four biweekly administrations of 0.5 mg/kg significantly reduced plasma galactose and Gal-1-P in the liver, ovary and brain; and a single high dose (1 mg/kg) of hGALT mRNA increased mouse survival rate by threefold. Remarkably, hGALT mRNA remained effective after consecutive administrations, indicating that the immune system was not activated in GalT-deficient mice (Balakrishnan et al., 2020). Taken together, these observations suggest a significant advantage for mRNA therapy, as the first doses could be applied in newborn children at a concentration to reduce symptomatology and then the concentration and frequency of dosage be adjusted depending on the requirements of the patients.

2.4 Liver arginase deficiency

Arginase deficiency (MIM 207800) is a rare autosomal recessive urea cycle disorder caused by partial or complete loss of cytosolic arginase 1 (ARG1, EC 3.5.3.1) function. It has an estimated incidence of 1:950,000 in the United States. ARG1 hydrolyses arginine to ornithine and urea in the liver. Thus, its deficiency affects the urea cycle, the metabolic pathway in charge of excess nitrogen removal (Sin et al., 2015; Yokoi et al., 2021). Patients with ARG deficiency present hyperargininemia and accumulation of guanidino compounds in plasma, urea, and cerebrospinal fluid. These metabolites are believed to be neurotoxic and associated with the onset of clinical manifestations in late infancy (Deignan et al., 2008, 2010).

Symptoms in patients with arginase deficiency differ from other urea cycle-related diseases. They include seizures, loss of ambulation, microcephaly, intellectual disability, and growth deficiency among others (Carvalho et al., 2012). Prevention of primary manifestations is based on a protein-restricted diet to maintain plasma arginine close to normal values, amino acid supplementation, and nitrogen scavenger drugs to treat hyperammonemia.

A pegylated ARG1 protein for ERT was able to reduce plasma arginine in neonatal and adult Arg1-deficient mice, but failed to reduce hyperammonemia or improve survival after repeated doses. Although pegylated ARG1 therapy was effective in reducing plasma arginine levels, hepatic arginine levels were not reduced because the pegylated enzyme was unable to enter the hepatocytes (Burrage et al., 2015).

Liver transplantation is another therapeutic option to normalize hyperargininemia and halt neurological decline (Sun et al., 1993). This observation supports the hypothesis that therapies aimed at supplementing hepatocytes with the normal version of the ARG1 gene might correct the disease. Confirming this notion, studies in knockout Arg1 mice showed that the administration of an AAV vector encoding the ARG1 gene shortly after birth was able to restore hepatic arginase activity, and to protect against hyperargininemia and guanidino accumulations, leading to normal neurological development, cognitive activity, and behavior (Cantero et al., 2016; Lee et al., 2013). However, given that the first features of arginase deficiency often appear between the ages of one and 3 years old, AAV-based GT might have a decreased therapeutic action due to rapid hepatocyte division during growth.

Systemic administration of a single dose of this human codon-optimized ARG1 (hARG1) mRNA achieved 54% of normal hepatic arginase activity, and protection against plasma and hepatocyte arginine accumulation, a crucial step to avoid guanidino accumulation (Truong et al., 2019). Moreover, multidose administration of mRNA therapy would overcome AAV limitations, because its recurrent injection every 3 days improved survival beyond 77 days in 100% of knockout Arg1 mice without hyperammonemia or weight loss, compared to control animals treated with luciferase mRNA formulated in the same LNPs, all of which persished by day 22. In addition, hARG1 mRNA was found to be safe as no histological or enzymatic evidence of liver injury was detected after repeated injections (Truong et al., 2019).

The experimental data with mRNA technology hold promise as a singular therapeutic intervention applicable to infants with arginase-1 deficiency.

2.5 Glycogen storage diseases

Glycogen storage diseases (GSDs) are a group of rare genetic disorders caused by deficiency of an enzyme involved in glycogen production, degradation, or glucose breakdown, primarily affecting the liver, skeletal muscle, or both (Marion and Paljevic, 2020; Ozen, 2007). Patients with GSDs can present hypoglycemia, exercise intolerance, or abnormal glycogen accumulation (Ozen, 2007; Stone et al., 2021). The overall incidence of GSD is estimated to be between 1 per 20,000–40,000 (Stone et al., 2021). There are several types of GSDs based on the enzyme deficiency and the tissue affected (Hicks et al., 2011). Among them, GSD type Ia, Ib, III and VI showed the clinical features of hypoglycemic liver pathophysiology; and type Ia (GSDIa) and GSDIII currently have ongoing mRNA therapeutic products under development.

2.5.1 Glycogen storage disease type Ia

GSDIa (MIM 232200) is caused by the hepatic deficiency of the glucose-6-phosphatase (G6Pase-α or G6PC, EC 3.1.3.9) enzyme. It has an incidence of 1–2 per 100,000 births and can benefit from prenatal diagnosis

(Marion and Paljevic, 2020; Ozen, 2007). GSD Ia can be differentiated into 4 subtypes depending on the abnormality in the G6Pase system. A homozygous or compound heterozygous mutation in the G6PC gene, encoding the catalytic subunit of the system located inside the ER, causes GSDIa (MIM 232200). G6Pase-α catalyzes the last reaction of glycogenolysis and gluconeogenesis. Its deficiency leads to a lack of glucose and the accumulation of glucose-6-phosphate (G6P).

GSDIa normally first appears in infancy and is characterized by life-threatening hypoglycemia during fasting, lactic acidemia, hypertriglyceridemia, hyperuricemia, hypercholesterolemia and steatosis (Cao et al., 2021; Chou et al., 2002). Glycogen accumulation in the liver and kidneys causes the hepatomegaly and nephromegaly which are the hallmarks of the disease (Chou et al., 2002).

Current treatment is based on strict dietary management, consisting of frequent feedings (every 6–4 h) of uncooked or modified cornstarch and gastric drip-feeding of glucose through the night (mainly in young patients) (Chou et al., 2002; Marion and Paljevic, 2020). However, this does not fully prevent glycogen accumulation and other underlying metabolic abnormalities leading to long-term complications, including hepatocellular adenomas (HCA) in 70–80% of adult patients. HCA transformation into hepatocellular carcinomas (HCCs) occurs in 10% of these patients (Roseman et al., 2018). Currently, liver/kidney transplantation is the only curative option (Marion and Paljevic, 2020).

ERT for GSD Ia is challenging due to the hydrophobic nature of G6Pase-α and its localization in the endoplasmic reticulum (ER) membrane (Cao et al., 2021). Liver-mediated GT by means of AAV vectors or mRNA delivery mediated by LNPs could be a therapeutic option for GSDIa, and proof-of-concepts have been obtained in animal models that reproduce the severe symptoms of the disease. Transfer of a normal copy of cDNA to hepatocytes using an rAAV vector protected against hypoglycemia and prevented the development of HCA in GSDIa mice (Kim et al., 2017; Lee et al., 2015). An AAV8 vector encoding G6Pase-α is currently being evaluated in a Phase 1/2 clinical trial (NCT03517085) sponsored by Ultragenyx Pharmaceutical Inc.

However, recent studies report that adult mice exposed to rAAV gene therapy in the context of chronic liver disease developed HCC with a high frequency (Dalwadi et al., 2021); and therefore, caution and observational follow-up studies are needed given the high prevalence of hepatocellular adenomas and HCCs in patients with GSDIa.

Systemic G6Pase-α mRNA administration also induces a dose-dependent G6Pase-α protein expression in $G6PC^{-/-}$ mouse hepatocytes, leading to reductions in liver G6P, glycogen and triglycerides (hepatic biomarkers of GSD1) and restores normoglycemia in fasting mice (Roseman et al., 2018). The authors also developed an engineered protein variant of human G6Pase-α, which when delivered in a context of mRNA formulated in LNPs has the ability to improve fasting blood glucose levels for up to 12 days post-injection and reduce liver mass rapidly 24h post-injection (Roseman et al., 2018). Cao et al. also reported that a single administration of G6Pase-α mRNA led to significant improvement in blood glucose homeostasis during fasting and a robust decrease in serum triglycerides, correlating with the reduction in hepatic triglycerides levels (Cao et al., 2021). After recurrent administration of the hG6PC mRNA, no anti-drug antibody nor increased serum cytokines (IFNγ, IL-1β, IL-6, and TNFα) were detected in $G6PC^{-/-}$ mice. Of interest, multi-dose administration showed a tendency to reduce the risk of HCA/HCC (Cao et al., 2021).

An open-label dose-escalation study is planning to recruit up to 18 adult and older adult patients with GSDIa in 2022 to evaluate the safety, tolerability, pharmacokinetics, and pharmacodynamics of mRNA-3745 (NCT05095727).

2.5.2 Glycogen storage disease type 3

In GSDIII or Cori disease (MIM 232400), the deficiency of the cytosolic glycogen debranching enzyme results in glycogen accumulation in the liver and muscle (Lim et al., 2020). Typical clinical manifestations include hepatomegaly, hypoglycemia, growth retardation, hyperlipidemia and cardiomyopathy (Dagli et al., 1993; Endo et al., 2006; Lim et al., 2020). There are no approved treatment options for GSDIII, which is estimated to affect more than 10,000 patients in the developed world. As in GSDIa, the current standard of care relies on a strict diet, including frequent doses of cornstarch, to reduce the risk of hypoglycemia and hyperketonemia (Jauze et al., 2019; Ozen, 2007).

The development of an AAV-based GT for GSDIII is a great challenge due to the transgene size that is close to the AAV genome packaging limit, and the complexity of the disease that requires simultaneous liver and muscle expression (Jauze et al., 2019; Vidal et al., 2017). However, of the four subtypes of the disease, 15% of patients are classified as GSDIIIb in which the disease is associated with enzyme deficiency only in the liver (Endo et al., 2006; Hicks et al., 2011).

An investigational mRNA-based therapy encoding glycogen debranching enzyme encapsulated in LNPs (UX053) from Ultragenyx Pharmaceutical Inc/Arcturus Therapeutics) (https://ir.ultragenyx.com/news-releases/news-release-details/ultragenyx-announces-fda-clearance-investigational-new-drug-0, Last access Dec 1st, 2021) has been announced. However, there are no preclinical data currently published, but an interventional Phase 1/2 clinical trial to evaluate the safety, tolerability, and efficacy of UX053 is recruiting 30 adults and older adult patients with GSDIII (NCT04990388).

2.6 AAT deficiency

Alpha-1-antitrypsin deficiency (AATD, MIM 613490) is an autosomal recessive disease caused by an inherited mutation in the SERPINA1 gene resulting in an unfavorably charged Alpha-1-antitrypsin (AAT, KO K03984) protein, also known as SERPINA1 (Serine protease inhibitor, group A, member 1). AATD is characterized by aberrant folding, protein aggregation and accumulation of misfolded SERPINA1 in the ER of hepatocytes leading to low levels and activity of AAT in blood and tissues. It is one of the most common potentially fatal metabolic diseases and affects about 2–5 per 10,000 births in Europe (Michel et al., 2015).

AAT is a secreted protein predominantly synthesized in the liver, whose main function is the regulation and neutralization of the proteolytic effects of neutrophil elastase in the lungs (Connolly et al., 2018; Greene et al., 2016; Pietrangelo et al., 2007). Thus, patients with AATD are predisposed to pulmonary emphysema and liver disease due to toxic SERPINA1 accumulation and polymerization in hepatocytes (Connolly et al., 2018).

Current medical treatment available for AATD with severe pulmonary emphysema is based on aerosol or i.v. administration of ERT consisting of AAT protein derived from the blood of healthy donors. Although this therapy improved the amount of lung AAT, it is costly (Connolly et al., 2018; Karadagi et al., 2020). Liver disease secondary to AATD has no currently approved therapy other than transplantation for patients with advanced disease (Strnad et al., 2020). Emerging therapies such as chemical chaperones, hepatocyte transplantation, and AAV-mediated GT are under clinical investigation (Remih et al., 2021). Although several serotypes have been tested, further improvements in the design or delivery of rAAV-AAT vectors will be required to achieve therapeutic serum AAT concentrations (NCT02168686, NCT00377416 and NCT01054339).

In 2015, Michel et al. showed that SERPINA1 mRNA-transfected cells expressed a highly functional AAT protein and elastase inhibitory activity within the first 24h after *in vitro* delivery of the mRNA (Michel et al., 2015). Connolly et al. have also demonstrated increased AAT protein levels in cell culture media of SERPINA1 mRNA-treated AAT patient's fibroblasts and AAT patient-derived primary hepatocytes. Moreover, systemic administration of SERPINA1 mRNA in both wild-type mice and the NSG-PiZ transgenic mouse model of AAT deficiency results in both SERPINA1 mRNA and protein detection in liver and lungs 48h post-injection (Connolly et al., 2018; Karadagi et al., 2020).

However, AAT therapy is challenging because it involves expressing a secretable protein in individuals with autosomal recessive inheritance, so it can induce an immune response. Furthermore, in addition to administering the appropriate protein, it is necessary to inhibit the endogenous expression of the mutated protein in most hepatocytes given its ability to form toxic intracellular aggregates.

2.7 Crigler-Najjar syndrome type 1

Uridine 5′-diphosphate glucuronyltransferase (UGT1A1, EC 2.4.1.17) is an ER enzyme that is responsible for the glucuronidation of bilirubin in hepatocytes, which allows its secretion *via* the bile ducts by conferring solubility to bilirubin (Canu et al., 2013; Skierka et al., 2013). UGT1A1 is mainly expressed in the liver, although the intestine, kidney, and skin show minor levels of expression (Nakamura et al., 2008). UGT1A1 deficiency causes Crigler-Najjar Syndrome Type 1 (CN1, MIM 218800), an ultra-rare autosomal recessive disease (estimated incidence is 1 in 750,000–1000,000 newborns) in which hydrophobic unconjugated bilirubin is less effectively removed, thus accumulating in the circulation and tissues and being able to pass through the blood brain barrier. Its accumulation in the brain can cause neurological damage such as speech impairment, ataxia, and death in untreated infants (Apgar et al., 2018).

Current treatment is based on phototherapy that converts bilirubin to oxidation products which are easier to eliminate. This therapy is effective in patients up to 10 years old but becomes less effective in patients with more advanced age (Stokowski, 2006). As in other hepatic diseases, orthotopic liver transplantation is a curative treatment option for patients with severe CN1 (Ambrosino et al., 2005). Also, long-term protection in patients with severe CN1 requiring phototherapy is being tested in a clinical trial using an AAV vector expressing the UGT1A1 transgene (NCT03466463).

mRNA therapy for CN1 has been assayed in preclinical studies. A modified mRNA encoding UGT1A1 encapsulated in LNPs was able to restore the UGT1A1 hepatic expression of Gunn rats (animal model of CN1 disease). UGT1A1 mRNA normalized glucuronidation and thus, clearance of bilirubin. Even after multiple doses, it did not show toxic effects (Apgar et al., 2018).

Slow kinetics of bilirubin production enables UGT1A1 mRNA doses to be administered at a low frequency, as normal bilirubin levels are slowly recovered. Importantly, maintenance of reduced bilirubin levels could be achieved without the constant expression of UGT1A1. In fact, Quantitative Systems Pharmacology predicted that a single 0.5 mg/kg dose of UGT1A1 mRNA every 4 weeks should be able to provide a clinically meaningful and sustained reduction of bilirubin levels (Apgar et al., 2018).

Recently, Moderna and the Institute for Life Changing Medicines announced a new collaboration in order to develop an mRNA therapeutic for CN1. Although patients with CN1 show a complete absence of hepatic UGT1A1, induction of an immune response against the transgenic protein is unlikely because UGT1A1 is an intracellular enzyme.

2.8 Propionic acidemia

Propionic acidemia (PA, MIM 606054) is an inborn error of amino acid and odd-chain fatty acid metabolism caused by a defect in the mitochondrial enzyme propionyl-CoA carboxylase (PCC, EC 6.4.1.3) (Pena et al., 2012). PCC is a heterododecamer (750 kD) composed of an equal number of alpha (PCCA, 75 kD) and beta (PCCB, 54 kD) subunits (Jiang et al., 2020; Pena et al., 2012). PA is caused by mutations in either of the two subunit genes. The majority of patients are heterozygotes, but homozygotes for mutations in PCCA or PCCB have also been reported (Pena et al., 2012). Therefore, there are two disease subtypes: PCCA-deficient (Type I) and PCCB-deficient (Type II). As PCCB is intrinsically unstable if not in a complex with PCCA, PCCA-deficient patients are deficient in both protein subunits (Jiang et al., 2020).

PCC enzyme catalyzes the carboxylation of propionyl-CoA to *D*-methylmalonyl-CoA, which is converted to succinyl-CoA that enters the Krebs cycle (Wongkittichote et al., 2017). Thus, its deficiency results in impaired propionate metabolism and abnormal accumulation of propionyl-CoA among other toxic metabolites, leading to hyperammonemia, acidosis, ketonuria, and hematological alterations (Richard et al., 2015;

Shchelochkov et al., 1993). PA is typically diagnosed in early childhood, with 65% of patients being identified in the neonatal period. It affects around 1/100,000 to 1/250,000 individuals worldwide. The initial symptoms include poor feeding, dehydration, vomiting, loss of appetite and weight, weak muscle tone (hypotonia), lack of energy (lethargy), and convulsions, possibly resulting in coma, multi-organ failure, and even death (Haijes et al., 2019). Less commonly, the disease presents after a year or later in life (the intermittent late onset form), with intercurrent metabolic decompensations caused by fasting, fever or infections, failure to thrive and diverse neurological symptoms. Some affected children experience intellectual disability or delayed development (Pena et al., 2012).

PA management is primarily based on a life-long protein-restricted diet to reduce propiogenic substrates and high caloric intake to reverse catabolism (Haijes et al., 2019). Although liver transplantation has shown beneficial outcomes in some patients (Critelli et al., 2018; Rela et al., 2007; Vara et al., 2011), its use continues to be questioned because in addition to its high cost, associated risks and scarcity of donors, it can only correct the enzymatic deficiency in the liver so, the metabolic defect in other organs may maintain an increased production of toxic metabolites (Haijes et al., 2019). Therefore, there is a need for alternative treatment strategies. Current emerging approaches such as GT, enzyme replacement, or mRNA therapy aim to reduce the accumulation of toxic metabolites by improving enzyme activity (Haijes et al., 2019).

GT with adenoviral serotype 5 (Ad5) and AAV8 vectors in knockout *Pcca* mice resulted in low transduction that led to transient rescue from neonatal lethality (Hofherr et al., 2009). Some studies have demonstrated that intrahepatic AAV8 gene transfer can result in a rescue of 64% of the mice and the amelioration of metabolic markers of the disease lasted up to 1.5 years in adult PA hypomorphic mice, indicating a significant increase in PCC enzymatic activity (Chandler et al., 2011; Guenzel et al., 2015). Nevertheless, taking into account that PA typically manifests in childhood, one of the major hurdles of AAV GT is its limited applicability to patients under ~12 years old due to their ongoing organ growth. Moreover, published GT approaches focus on replacing a single subunit of the deficient enzyme complex thus addressing only a subset of patients with PA.

A new therapeutic alternative using a combination of mRNAs encapsulated in LNPs encoding for both human subunits, hPCCA, and hPCCB, producing functional hepatic PCC enzyme has been described (Jiang et al., 2020). Thus, this dual approach is intended to treat all patients with

PA (both subtypes). Proof of concept for dual mRNAs-encoded functional PCC protein with correct mitochondrial localization was obtained in PA mice and fibroblasts from patients with PA. Systemic administration of dual mRNA encapsulated in LNPs induced higher PCC enzyme activity compared to single (PCCA or PCCB) mRNA alone. Preclinical experiments, including 3- and 6-month repeat-dose studies in murine models, demonstrated the long-term therapeutic potential of this strategy. Although the primary disease biomarkers were not normalized to the levels observed in unaffected mice, clinical outcomes were improved, which is consistent with the liver transplantation experience in patients with PA. Dual mRNAs were well-tolerated in these studies with no adverse findings, demonstrating the potential of mRNA technology to chronically administer multiple mRNAs to produce large complex enzymes, with applicability to other genetic disorders (Jiang et al., 2020).

The development of anti-PCCA and/or anti-PCCB antibodies was not evaluated, but the induction of the immune response is unlikely to occur because PCC subunits are expressed intracellularly and trafficked to the mitochondria (Jiang et al., 2020). Nevertheless, anti-PCCA and anti-PCCB antibodies will be assessed in the upcoming Phase 1/2 trial clinical trials testing safety, pharmacokinetics (PK), and pharmacodynamics (PD) of different doses of mRNA-3927 (NCT04159103) and its long-term safety after up to 10 doses administered to participants with PA during the approximate 30-week treatment period (NCT05130437).

2.9 Ornithine transcarbamylase deficiency

Ornithine transcarbamylase deficiency (OTCD, MIM 311250) is a rare X-linked disorder whose prevalence is around 1 in 50,000–80,000 newborns. It is characterized by complete or partial lack of the mitochondrial enzyme ornithine transcarbamylase (OTC, EC 2.1.3.3), which is predominantly expressed in the liver. OTC participates in the urea cycle, and its absence is related to permanent brain damage and death as ammonia cannot be metabolized. Currently, ammonia scavengers and protein-restricted diets are used as treatments although these do not address the origin of the disease.

AAV therapy has shown that increasing hepatic OTC activity in the murine OTCD model could become a therapeutic option for this metabolic disease (Cunningham et al., 2009; Wang et al., 2012). mRNA therapy can only bring about transient expression of OTC protein in comparison with the long-term expression that is offered by AAV therapy, but viral vectors for pediatric patients may soon lose their efficacy due to rapid liver cell division at that age, as previously noted (Cunningham et al., 2009; Yang et al., 2016). Furthermore, these vectors cannot be re-administered due to the activation of the immune system after the first administration. Thus, mRNA expressing OTC is postulated as a good therapeutic option, and different biotechnology companies are developing their own mRNA delivery systems.

Translate Bio Inc. showed that a 0.15 mg/kg i.v. dose of mRNA-OTC (MRT5201) protected against hyperammonemia in an OTCD mouse model for up to 4 weeks. MRT5201 was designed for biweekly i.v. administration, and a Phase 1/2 clinical trial (NCT03767270) was performed, although it was discontinued. LUNAR® based LNPs for OTC (ARCT-810), from Arcturus Therapeutics, protected from weight loss and mortality in OTCD mice even with administration every 3 weeks. A Phase 1 clinical trial (NCT03767270) for the study of ARCT-810 is currently under recruitment. The Genevant Sciences Ltd. mRNA-OTC candidate showed a significant increase in OTCD mice survival after weekly administration and a safe profile in non-human primates after three weekly doses of hOTC-mRNA in preclinical studies. Prieve M. et al. demonstrated that a single mRNA administration produced OTC protein active for up to 96 h and increased mice survival. Furthermore, after 12 weekly injections, there was no loss of OTC activity, suggesting there is no production of neutralizing antibodies (Prieve et al., 2018). Importantly, OTC protein has a half-life of 12 days, so only biweekly or monthly administration would be necessary to control the biochemical and clinical outcomes of the disease. Of note, the frequency of administration in mice may be higher than in humans due to the higher metabolization rate of small animals.

2.10 Methylmalonic acidemia

A complete or partial deficiency of the mitochondrial enzyme methylmalonyl-CoA mutase (MUT, EC 5.4.99.2) results in a metabolic block in the propionate metabolism pathway that causes a pediatric ultra-rare disorder called methylmalonic acidemia (MMA, MIM 251000). MMA is an autosomal recessive inherited disorder with a estimated prevalence of 1 per 90,000 newborns, and characterized by the accumulation of toxic metabolites such as methylmalonic acid. Patients suffer from intermittent life-threatening decompensations, growth retardation, neurological complications, hepatomegaly and chronic renal failure (Fraser and Venditti, 2016).

Currently, management of this disease is based on monitorization of patients, dietary protein restriction, and cofactor supplementation, as no approved therapy exists. ERT is not available for MMA as mitochondrial-specific delivery is a complex challenge (An et al., 2017). AAV therapy has been assayed in preclinical studies. However, new engineered AAV vectors are needed in order to potentiate transduction enhancing MUT expression (Ilyinskii et al., 2021) and to enable repeated doses of the same vector allowing for re-administration as the child grows.

hMUT mRNA administration followed by image analysis demonstrated that the produced MUT protein was subcellularly located in the mitochondria of hepatocytes. This therapy showed hepatic MUT expression for up to 7 days, and protein expression peaked at 16 h with a hepatic concentration 3 times higher than endogenous murine MUT. As acute metabolic decompensation requires immediate treatment, the rapid and high hepatic activity obtained after a single hMUT mRNA administration is essential for the treatment of MMA emergencies (An et al., 2017).

hMUT mRNA administration demonstrated biological efficacy in MMA mice, which increased growth and survival. Additionally, there was no increase in markers of hepatic toxicity, inflammation, or immune response in mice after repeated i.v. dosing. Lack of immune response allowed repeated administrations every few weeks, which has shown sustained pharmacology and durable functional benefit in mice of the MMA model (An et al., 2019).

Despite being a liver-directed therapy, hMUT mRNA decreases methylmalonic acid and 2-methylcitrate in extrahepatic tissues, although no MUT enzyme was produced in those tissues (An et al., 2017). These tissue biomarker reductions were likely due to decreases in circulating toxic metabolites. Likewise, liver transplantation reduces methylmalonic acid in the circulation (An et al., 2019). However, biomarkers were not completely normalized in transplanted patients. This is explained by the production of toxins in other organs such as skeletal muscle, heart, or brain. Remarkably, hMUT mRNA therapy decreased biomarkers to levels observed in transplanted patients, whose metabolic decompensations are reduced (An et al., 2019). For all these reasons, a phase 1/2 clinical trial (NCT04899310) was undertaken by Moderna in order to evaluate the safety profile, pharmacokinetics, and pharmacodynamics of hMUT mRNA (mRNA-3705).

2.11 Primary hyperoxaluria

Primary hyperoxalurias (PHs) are rare inherited disorders caused by enzyme defects in glyoxylate metabolism. There are three types with type 1 (PH1, MIM 259900), an autosomal recessive disorder, being the most severe and responsible for 80% of total PHs, due to peroxisomal alanine glyoxylate aminotransferase (AGT, EC 2.6.1.44) deficiency. The estimation of PH1 prevalence is 1–3 cases per million individuals. AGT deficiency leads to glyoxylate accumulation and, thus, overproduction of oxalate and glycolate. Oxalate is excreted by the kidneys, although it is highly insoluble in urine in the presence of calcium, where it precipitates, provoking calcium oxalate nephrolithiasis and nephrocalcinosis. After kidney failure, the presence of oxalate in plasma and its deposition in bones, kidneys, skin, retina, myocardium, vessel walls, and the central nervous system increases, which is highly associated with mortality. Most patients present PH1 in childhood (Milliner et al., 1993).

In 2020, Lumasiran (Oxlumo™, Alnylam Pharmaceuticals), an RNAi against hepatic glycolate oxidase (EC 1.1.3.15), the enzyme in charge of oxidating glycolate to glyoxylate, received FDA approval for PH1 treatment (Frishberg et al., 2021) (https://www.fda.gov/news-events/press-announcements/fda-approves-first-drug-treat-rare-metabolic-disorder). Currently, Nedosiran, another RNAi from Dicerna Pharmaceuticals Inc., is being evaluated in a phase 3 clinical trial (NCT04042402). Nedosiran targets hepatic lactate dehydrogenase (LDH, EC 1.1.1.27), the enzyme responsible for metabolizing glyoxylate to oxalate (Forbes et al., 2021). Until the approval of these drugs, liver transplantation was the only strategy for metabolic correction.

The expression of the correct version of the mutated gene, instead of blocking metabolic pathways is another option that has been tested for PH1. In fact, an AAV8 and AAV5 encoding for AGT, transduced 80%–90% of hepatocytes achieving almost normal urinary oxalate levels in these PH1 mice (Salido et al., 2011). However, these vectors were not tested in any clinical trial. Although 20% AGT activity can be found in asymptomatics heterozygous for PH1, the mouse model of the disease requires 80%–90% of hepatocytes to express AGT to correct the disease. Thereby, such a high percentage of transduced hepatocytes is difficult to obtain in humans with the available AAV serotypes.

An mRNA encoding AGT developed by Alexion Pharmaceuticals Inc. was assayed as a therapeutic approach in order to replace the deficient peroxisomal enzyme treating the primary cause of PH1. *In vitro* studies in a mouse hepatocyte cell line showed correct localization of active AGT protein in the peroxisome; and *in vivo* studies reported increased hepatic AGT activity for up to 4 days after AGT mRNA administration, followed by a decrease of 40% in urine oxalate levels in a PH1 mouse model (Kukreja et al., 2019). Systemic administration of this mRNA with a 2.5 mg/kg dose led to expression of AGT protein in about 56%–75% of hepatocytes. However, multidose studies showed loss of efficiency after repeated doses, suggested by an unexpected immunogenic response against LNPs. Reducing immunogenicity in this therapeutic product with a better formulation would increase the potential of this drug delivery system (Kukreja et al., 2019).

2.12 Fabry disease

Fabry disease (FD, MIM 301500) is an X-linked lysosomal storage disorder (LSD) caused by a deficiency in the lysosomal enzyme a-galactosidase A (a-Gal A, EC 3.2.1.22) as a result of mutations in the *GLA* gene (Crocker et al., 2021). This enzymatic defect leads to the altered metabolism of glycosphingolipids, such as globotriaosylceramide (Gb3) and its deacylated analog lyso-globotriaosylceramide (lyso-Gb3). Intracellular accumulation of these lipids triggers cellular dysfunction and progressive damage of multiple tissues, including the liver, heart, kidney, spleen, and skin. Clinical manifestations begin during childhood and include acroparesthesia, angiokeratoma, corneal and lenticular opacities, stroke, renal failure, or hypertrophic cardiomyopathy (Brady and Schiffmann, 2000; Crocker et al., 2021; Germain, 2010; Miller et al., 2020).

FD is the second most common LSD with an estimated prevalence of 8–25 per million births (Mehta et al., 2010). However, newborn screening efforts have found a high incidence, indicating that the disease has been underestimated.

Current treatments include ERTs such as Raplagal (Shire Pharmaceuticals) and Fabrazyme (Sanofi Genzyme), as well as a chaperone therapy based on migalastat (Miller et al., 2020) (Galafold-Amicus Therapeutics). None of these therapies are curative, and both are expensive (~$300,000 US/year). Chaperone therapy is only of benefit in patients with certain mutations. ERT is time-consuming (requiring frequent and lengthy infusions) and has other limitations as infusion-associated reactions or immunogenicity of recombinant protein. In addition, this treatment does not completely resolve all symptoms and might not be symptomatically beneficial for all patients

(Crocker et al., 2021; Domm et al., 2021; Miller et al., 2020). Thus, there are alternative treatment modalities in development, including GT, substrate reduction therapy, or messenger RNA therapy.

GTs currently under investigation for FD include *in vivo* and *ex vivo* approaches. For *in vivo* GT, AAV delivery is the most promising strategy. Two therapeutic products (ST-920 and FLT190) based on recombinant AAV vectors have been shown to promote α-gal A transgene expression in the liver of Fabry mice, leading to the reduction of Gb3 and lyso-Gb3 in plasma, liver, spleen, and heart, as well as the kidney (Huston et al., 2019; Jeyakumar et al., 2019; Yasuda et al., 2020; Ziegler et al., 2004). Two clinical trials (NCT04046224 and NCT04040049) using these platforms were initiated in 2019. However, the application of this technique still presents some challenges. Although AAV vectors induce transgene expression for years, it may eventually decline, which in FD would lead to the accumulation of glycosphingolipids, forcing the re-administration of the AAV vector, which would have decreased efficacy because of neutralizing antibodies. The *ex vivo* approach is based on the use of a lentivirus vector to deliver the GLA gene into $CD34^+$ stem cells obtained from the patient and then an autologous transplant is performed. There are two active clinical trials (NCT02800070 and NCT03454893) to assess the efficacy and safety of this therapeutic strategy, whose main disadvantage is that it requires hematopoietic stem/progenitor cell harvesting followed by host myeloablation, and autologous transplant (Domm et al., 2021).

Within the mRNA therapy modality, two recent preclinical studies (DeRosa et al., 2019; Zhu et al., 2019) have evaluated the administration of LNPs-formulated mRNA encoding human a-Gal A (h-a-Gal A) in wild-type mice, Fabry mice and non-human primates. Both studies reported detectable protein levels and activity of h-a-Gal A in serum, liver, spleen, kidney, and heart after a single i.v. administration of h-a-Gal A mRNA and observed a prolonged protein half-life.

Treatment efficacy was determined by monitoring the clearance of Gb3 or lyso-Gb3 after a single i.v. administration of h-a-Gal A mRNA in Fabry mice. mRNA treatment resulted in a decrease of up to 80% in Gb3 and lyso-Gb3 levels in key tissues. Zhu et al. reported that maximum lyso-Gb3 depletion in the liver and spleen was observed 14 days after administration of the dose and this remained through day 42 when re-accumulation began. In the heart and kidney, lyso-Gb3 decreased by 28–35 days and started to re-accumulate after 35 days post-administration. Of note, lyso-Gb3 levels did not return to baseline values at completion of the study (Zhu et al., 2019).

Repeated i.v. administration of h-a-Gal A mRNA in Fabry mice was also evaluated in both studies. The authors reported the restoration of a-Gal A activity in tissues and the reduction of Gb3/lyso-Gb3, without formation of anti-h-a-Gal antibodies, suggesting the absence of immune responses to the treatment. In addition, DeRosa et al. carried out a comparative study between ERT and mRNA therapy. After a single i.v. administration, they observed a greater efficacy of the h-a-Gal A mRNA, resulting in significantly decreased levels of Gb3 and lyso-Gb3. Using multi-dose regimens of eight weekly doses of h-a-Gal A (ERT) or two monthly doses of h-a-Gal A mRNA, equivalent reductions in biomarker clearance were observed, indicating equivalent efficacy for mRNA-based therapy with monthly dosing (DeRosa et al., 2019).

Finally, after weekly administration of h-a-Gal A mRNA in non-human primates sustained expression of functional a-Gal A protein was observed in plasma after each dose, with no immune response to the protein encoded by the mRNA as well as, increased a-Gal A levels in tissues and the absence of markers of liver toxicity after consecutive doses of h-a-Gal A mRNA (Zhu et al., 2019). Taken together, these data provide preclinical proof-of-concept support for systemic delivery of h-a-Gal A mRNA as a therapeutic modality for protein restoration and substrate depletion.

3. Conclusions and perspectives

About 6% of the population is born with small changes distributed in one of the 30,000 genes that we have in our 46 chromosomes. These changes are the origin of specific signs or symptoms that in some individuals evolve into potentially life-threatening conditions. Each disease usually affects a small number of people, and due to their low frequency these diseases are called rare or orphan diseases. Scientific and technological advances in molecular biology represent a revolution for these diseases as they may allow their pathogenesis to be accurately defined, and ideally, molecularly tailored therapeutic solutions to be developed for each one. Rare metabolic diseases are in some cases devastating conditions in which large investments by for-profit organizations are not possible due to the reduced numbers of patients.

Greater efficacy over non-viral vectors and the expectation of long-term expression of the transgene are attractive arguments for undertaking liver-GT strategies using viral vectors. Among these, the rAAV vectors are the most widely used because of their safety and their proven efficacy in a multitude of preclinical trials. However, some clinical trials confirmed that it is still necessary to develop more efficient vectors, and to apply protocols in combination with immunomodulators enabling the repeated administration of the same vector, especially in the case of pediatric patients. However, the low prevalence of such diseases and their heterogeneous presentation are important issues that limit research efforts and hamper expensive AAV development.

Advances in mRNA and LNPs technologies, and the expansion of the indications, will reduce the cost of production and development, making this therapeutic option affordable for a wide variety of rare and ultrarare diseases. Due to the liver targeting properties of the LNPs developed thus far, mRNA technology will be of great interest for those monogenic diseases for which liver transplantation is the only solution in many cases.

Among the advantages of therapeutic mRNAs formulated in LNPs is the rapid onset of their effects, within a few hours of administration, therefore making them ideal products to treat acute metabolic decompensations. Furthermore, unlike AAVs where transgene size is restricted to 4.7 kb, non-viral LNPs formulations are not limited by their mRNA size and can even combine several different mRNAs, as described for propionic acidemia.

These LNPs formulations have been optimized to reduce immunogenicity, allowing repeated administration depending on each disease and the half-life of the expressed protein. In the case of chronic presentations, and especially in infants, readministration and individual dosage decisions could be taken according to the clinical status, as well as the genotypic and biochemical characteristics of the patient. These features are crucial for a new era of personalized and precision medicine.

Acknowledgment

All figures were created with BioRender.com.

Funding

This study was supported by grants from the Spanish Institute of Health Carlos III (FIS) co-financed by European FEDER funds (PI18/00860 to A.F., PI19/01128 to P.B. and PI21/00546 to A.F.) and Spanish Fundación Mutua Madrileña (to A.F.). No funding from Moderna Inc. supports this work.

Competing interest

A.F. reports research funding from Moderna. P.B. reports research funding from Sanofi, Ferring, Hookipa, Moderna, and Bavarian Nordic; and speaker honoraria from BMS, MSD, Novartis, Boehringer Ingelheim, and AstraZeneca. L.J. and P.G.V.M. are employees of Moderna Inc. and may hold equities from the company. The other authors declare no competing interest.

References

Akinc, A., Querbes, W., De, S., et al., 2010. Targeted delivery of RNAi therapeutics with endogenous and exogenous ligand-based mechanisms. Mol. Ther. 18, 1357–1364. https://doi.org/10.1038/mt.2010.85.

Akinc, A., Maier, M.A., Manoharan, M., et al., 2019. The Onpattro story and the clinical translation of nanomedicines containing nucleic acid-based drugs. Nat. Nanotechnol. 14, 1084–1087. https://doi.org/10.1038/s41565-019-0591-y.

Ambrosino, G., Varotto, S., Strom, S.C., et al., 2005. Isolated hepatocyte transplantation for Crigler-Najjar syndrome type 1. Cell Transplant. 14, 151–157. https://doi.org/10.3727/000000005783983250.

An, D., Schneller, J.L., Frassetto, A., et al., 2017. Systemic messenger RNA therapy as a treatment for methylmalonic acidemia. Cell Rep. 21, 3548–3558. https://doi.org/10.1016/j.celrep.2017.11.081.

An, D., Frassetto, A., Jacquinet, E., et al., 2019. Long-term efficacy and safety of mRNA therapy in two murine models of methylmalonic acidemia. EBioMedicine 45, 519–528. https://doi.org/10.1016/j.ebiom.2019.07.003.

Anderson, K.E., Bloomer, J.R., Bonkovsky, H.L., et al., 2005. Recommendations for the diagnosis and treatment of the acute porphyrias. Ann. Intern. Med. 142, 439–450. https://doi.org/10.7326/0003-4819-142-6-200503150-00010.

Anderson, B.R., Muramatsu, H., Nallagatla, S.R., et al., 2010. Incorporation of pseudouridine into mRNA enhances translation by diminishing PKR activation. Nucleic Acids Res. 38, 5884–5892. https://doi.org/10.1093/nar/gkq347.

Anderson, B.R., Muramatsu, H., Jha, B.K., et al., 2011. Nucleoside modifications in RNA limit activation of 2′-5′-oligoadenylate synthetase and increase resistance to cleavage by RNase L. Nucleic Acids Res. 39, 9329–9338. https://doi.org/10.1093/nar/gkr586.

Apgar, J.F., Tang, J.-P., Singh, P., et al., 2018. Quantitative systems pharmacology model of hUGT1A1-modRNA encoding for the UGT1A1 enzyme to treat Crigler-Najjar syndrome type 1. CPT Pharmacometrics Syst. Pharmacol. 7, 404–412. https://doi.org/10.1002/psp4.12301.

Baden, L.R., El Sahly, H.M., Essink, B., et al., 2021. Efficacy and safety of the mRNA-1273 SARS-CoV-2 vaccine. N. Engl. J. Med. 384, 403–416. https://doi.org/10.1056/NEJMoa2035389.

Balakrishnan, B., An, D., Nguyen, V., et al., 2020. Novel mRNA-based therapy reduces toxic galactose metabolites and overcomes galactose sensitivity in a mouse model of classic galactosemia. Mol. Ther. 28, 304–312. https://doi.org/10.1016/j.ymthe.2019.09.018.

Balwani, M., Sardh, E., Ventura, P., et al., 2020. Phase 3 trial of RNAi therapeutic givosiran for acute intermittent porphyria. N. Engl. J. Med. 382, 2289–2301. https://doi.org/10.1056/NEJMoa1913147.

Bernegger, C., Blau, N., 2002. High frequency of tetrahydrobiopterin-responsiveness among hyperphenylalaninemias: a study of 1919 patients observed from 1988 to 2002. Mol. Genet. Metab. 77, 304–313. https://doi.org/10.1016/S1096-7192(02)00171-3.

Berraondo, P., Martini, P.G.V., Avila, M.A., et al., 2019. Messenger RNA therapy for rare genetic metabolic diseases. Gut 68, 1323–1330. https://doi.org/10.1136/gutjnl-2019-318269.

Blau, N., van Spronsen, F.J., Levy, H.L., 2010. Phenylketonuria. Lancet 376, 1417–1427. https://doi.org/10.1016/S0140-6736(10)60961-0.

Bosch, A.M., Grootenhuis, M.A., Bakker, H.D., et al., 2004. Living with classical galactosemia: health-related quality of life consequences. Pediatrics 113, e423–e428. https://doi.org/10.1542/peds.113.5.e423.

Bosilkovska, M., Samer, C., Déglon, J., et al., 2016. Evaluation of mutual drug-drug interaction within Geneva cocktail for cytochrome P450 phenotyping using innovative dried blood sampling method. Basic Clin. Pharmacol. Toxicol. 119, 284–290. https://doi.org/10.1111/bcpt.12586.

Brady, R.O., Schiffmann, R., 2000. Clinical features of and recent advances in therapy for Fabry disease. JAMA 284, 2771–2775. https://doi.org/10.1001/jama.284.21.2771.

Brenner, S., Jacob, F., Meselson, M., 1961. An unstable intermediate carrying information from genes to ribosomes for protein synthesis. Nature 190, 576–581. https://doi.org/10.1038/190576a0.

Buendía-Martínez, J., Barreda-Sánchez, M., Rodríguez-Peña, L., et al., 2021. Health impact of acute intermittent porphyria in latent and non-recurrent attacks patients. Orphanet J. Rare Dis. 16, 106. https://doi.org/10.1186/s13023-021-01742-3.

Burrage, L.C., Sun, Q., Elsea, S.H., et al., 2015. Human recombinant arginase enzyme reduces plasma arginine in mouse models of arginase deficiency. Hum. Mol. Genet. 24, 6417–6427. https://doi.org/10.1093/hmg/ddv352.

Burton, B.K., Grange, D.K., Milanowski, A., et al., 2007. The response of patients with phenylketonuria and elevated serum phenylalanine to treatment with oral sapropterin dihydrochloride (6R-tetrahydrobiopterin): a phase II, multicentre, open-label, screening study. J. Inherit. Metab. Dis. 30, 700–707. https://doi.org/10.1007/s10545-007-0605-z.

Cantero, G., Liu, X.-B., Mervis, R.F., et al., 2016. Rescue of the functional alterations of motor cortical circuits in arginase deficiency by neonatal gene therapy. J. Neurosci. 36, 6680–6690. https://doi.org/10.1523/JNEUROSCI.0897-16.2016.

Canu, G., Minucci, A., Zuppi, C., et al., 2013. Gilbert and Crigler Najjar syndromes: an update of the UDP-glucuronosyltransferase 1A1 (UGT1A1) gene mutation database. Blood Cells Mol. Dis. 50, 273–280. https://doi.org/10.1016/j.bcmd.2013.01.003.

Cao, J., Choi, M., Guadagnin, E., et al., 2021. mRNA therapy restores euglycemia and prevents liver tumors in murine model of glycogen storage disease. Nat. Commun. 12, 3090. https://doi.org/10.1038/s41467-021-23318-2.

Carvalho, D.R., Brum, J.M., Speck-Martins, C.E., et al., 2012. Clinical features and neurologic progression of hyperargininemia. Pediatr. Neurol. 46, 369–374. https://doi.org/10.1016/j.pediatrneurol.2012.03.016.

Chamberlain, K., Riyad, J.M., Weber, T., 2016. Expressing transgenes that exceed the packaging capacity of adeno-associated virus capsids. Hum. Gene Ther. Methods 27, 1–12. https://doi.org/10.1089/hgtb.2015.140.

Chandler, R.J., Chandrasekaran, S., Carrillo-Carrasco, N., et al., 2011. Adeno-associated virus serotype 8 gene transfer rescues a neonatal lethal murine model of propionic acidemia. Hum. Gene Ther. 22, 477–481. https://doi.org/10.1089/hum.2010.164.

Chen, C.Y., Shyu, A.B., 1995. AU-rich elements: characterization and importance in mRNA degradation. Trends Biochem. Sci. 20, 465–470. https://doi.org/10.1016/s0968-0004(00)89102-1.

Chen, B., Solis-Villa, C., Hakenberg, J., et al., 2016. Acute intermittent porphyria: predicted pathogenicity of HMBS variants indicates extremely low penetrance of the autosomal dominant disease. Hum. Mutat. 37, 1215–1222. https://doi.org/10.1002/humu.23067.

Chou, J.Y., Matern, D., Mansfield, B.C., et al., 2002. Type I glycogen storage diseases: disorders of the glucose-6-phosphatase complex. Curr. Mol. Med. 2, 121–143. https://doi.org/10.2174/1566524024605798.

Cobb, M., 2015. Who discovered messenger RNA? Curr. Biol. 25, R526–R532. https://doi.org/10.1016/j.cub.2015.05.032.

Connolly, B., Isaacs, C., Cheng, L., et al., 2018. SERPINA1 mRNA as a treatment for alpha-1 antitrypsin deficiency. J. Nucleic Acids 2018, 8247935. https://doi.org/10.1155/2018/8247935.

Critelli, K., McKiernan, P., Vockley, J., et al., 2018. Liver transplantation for propionic acidemia and methylmalonic acidemia: perioperative management and clinical outcomes. Liver Transpl. 24, 1260–1270. https://doi.org/10.1002/lt.25304.

Crocker, T.F., Brown, L., Lam, N., et al., 2021. Information provision for stroke survivors and their carers. Cochrane Database Syst. Rev. 11, CD001919. https://doi.org/10.1002/14651858.CD001919.pub4.

Cunningham, S.C., Spinoulas, A., Carpenter, K.H., et al., 2009. AAV2/8-mediated correction of OTC deficiency is robust in adult but not neonatal Spf(ash) mice. Mol. Ther. 17, 1340–1346. https://doi.org/10.1038/mt.2009.88.

Dagli, A., Sentner, C.P., Weinstein, D.A., 1993. Glycogen storage disease type III. In: Adam, M.P., Ardinger, H.H., Pagon, R.A., et al. (Eds.), GeneReviews. University of Washington, Seattle (WA).

Dalwadi, D.A., Torrens, L., Abril-Fornaguera, J., et al., 2021. Liver injury increases the incidence of HCC following AAV gene therapy in mice. Mol. Ther. 29, 680–690. https://doi.org/10.1016/j.ymthe.2020.10.018.

Damase, T.R., Sukhovershin, R., Boada, C., et al., 2021. The limitless future of RNA therapeutics. Front. Bioeng. Biotechnol. 9, 628137. https://doi.org/10.3389/fbioe.2021.628137.

Dammes, N., Peer, D., 2020. Paving the road for RNA therapeutics. Trends Pharmacol. Sci. 41, 755. https://doi.org/10.1016/j.tips.2020.08.004.

D'Avola, D., López-Franco, E., Sangro, B., et al., 2016. Phase I open label liver-directed gene therapy clinical trial for acute intermittent porphyria. J. Hepatol. 65, 776–783. https://doi.org/10.1016/j.jhep.2016.05.012.

Deignan, J.L., Marescau, B., Livesay, J.C., et al., 2008. Increased plasma and tissue guanidino compounds in a mouse model of hyperargininemia. Mol. Genet. Metab. 93, 172–178. https://doi.org/10.1016/j.ymgme.2007.09.016.

Deignan, J.L., De Deyn, P.P., Cederbaum, S.D., et al., 2010. Guanidino compound levels in blood, cerebrospinal fluid, and post-mortem brain material of patients with argininemia. Mol. Genet. Metab. 100 (Suppl), S31–S36. https://doi.org/10.1016/j.ymgme.2010.01.012.

Demirbas, D., Coelho, A.I., Rubio-Gozalbo, M.E., et al., 2018. Hereditary galactosemia. Metabolism 83, 188–196. https://doi.org/10.1016/j.metabol.2018.01.025.

DeRosa, F., Smith, L., Shen, Y., et al., 2019. Improved efficacy in a Fabry disease model using a systemic mRNA liver depot system as compared to enzyme replacement therapy. Mol. Ther. 27, 878–889. https://doi.org/10.1016/j.ymthe.2019.03.001.

Domm, J.M., Wootton, S.K., Medin, J.A., et al., 2021. Gene therapy for Fabry disease: progress, challenges, and outlooks on gene-editing. Mol. Genet. Metab. 134, 117–131. https://doi.org/10.1016/j.ymgme.2021.07.006.

Endo, Y., Horinishi, A., Vorgerd, M., et al., 2006. Molecular analysis of the AGL gene: heterogeneity of mutations in patients with glycogen storage disease type III from Germany, Canada, Afghanistan, Iran, and Turkey. J. Hum. Genet. 51, 958–963. https://doi.org/10.1007/s10038-006-0045-x.

Fontanellas, A., Ávila, M.A., Arranz, E., et al., 2021. Acute intermittent porphyria, givosiran, and homocysteine. J. Inherit. Metab. Dis. https://doi.org/10.1002/jimd.12411.

Forbes, T.A., Brown, B.D., Lai, C., 2021. Therapeutic RNA interference: a novel approach to the treatment of primary hyperoxaluria. Br. J. Clin. Pharmacol. https://doi.org/10.1111/bcp.14925.

Fraser, J.L., Venditti, C.P., 2016. Methylmalonic and propionic acidemias: clinical management update. Curr. Opin. Pediatr. 28, 682–693. https://doi.org/10.1097/MOP.0000000000000422.

Fridovich-Keil, J.L., Gubbels, C.S., Spencer, J.B., et al., 2011. Ovarian function in girls and women with GALT-deficiency galactosemia. J. Inherit. Metab. Dis. 34, 357–366. https://doi.org/10.1007/s10545-010-9221-4.

Frishberg, Y., Deschênes, G., Groothoff, J.W., et al., 2021. Phase 1/2 study of lumasiran for treatment of primary hyperoxaluria type 1: a placebo-controlled randomized clinical trial. Clin. J. Am. Soc. Nephrol. 16, 1025–1036. https://doi.org/10.2215/CJN.14730920.
Germain, D.P., 2010. Fabry disease. Orphanet J. Rare Dis. 5, 30. https://doi.org/10.1186/1750-1172-5-30.
Greene, C.M., Marciniak, S.J., Teckman, J., et al., 2016. α1-Antitrypsin deficiency. Nat. Rev. Dis. Prim. 2, 16051. https://doi.org/10.1038/nrdp.2016.51.
Gros, F., Hiatt, H., Gilbert, W., et al., 1961. Unstable ribonucleic acid revealed by pulse labelling of *Escherichia coli*. Nature 190, 581–585. https://doi.org/10.1038/190581a0.
Guenzel, A.J., Collard, R., Kraus, J.P., et al., 2015. Long-term sex-biased correction of circulating propionic acidemia disease markers by adeno-associated virus vectors. Hum. Gene Ther. 26, 153–160. https://doi.org/10.1089/hum.2014.126.
Gupta, S., Lau, K., Harding, C.O., et al., 2018. Association of immune response with efficacy and safety outcomes in adults with phenylketonuria administered pegvaliase in phase 3 clinical trials. EBioMedicine 37, 366–373. https://doi.org/10.1016/j.ebiom.2018.10.038.
Gustafsson, C., Govindarajan, S., Minshull, J., 2004. Codon bias and heterologous protein expression. Trends Biotechnol. 22, 346–353. https://doi.org/10.1016/j.tibtech.2004.04.006.
Haijes, H.A., Jans, J.J.M., Tas, S.Y., et al., 2019. Pathophysiology of propionic and methylmalonic acidemias. Part 1: complications. J. Inherit. Metab. Dis. 42, 730–744. https://doi.org/10.1002/jimd.12129.
Harding, C.O., Gillingham, M.B., Hamman, K., et al., 2006. Complete correction of hyperphenylalaninemia following liver-directed, recombinant AAV2/8 vector-mediated gene therapy in murine phenylketonuria. Gene Ther. 13, 457–462. https://doi.org/10.1038/sj.gt.3302678.
Hicks, J., Wartchow, E., Mierau, G., 2011. Glycogen storage diseases: a brief review and update on clinical features, genetic abnormalities, pathologic features, and treatment. Ultrastruct. Pathol. 35, 183–196. https://doi.org/10.3109/01913123.2011.601404.
Hofherr, S.E., Senac, J.S., Chen, C.Y., et al., 2009. Short-term rescue of neonatal lethality in a mouse model of propionic acidemia by gene therapy. Hum. Gene Ther. 20, 169–180. https://doi.org/10.1089/hum.2008.158.
Hou, X., Zaks, T., Langer, R., et al., 2021. Lipid nanoparticles for mRNA delivery. Nat. Rev. Mater. https://doi.org/10.1038/s41578-021-00358-0.
Houseley, J., Tollervey, D., 2009. The many pathways of RNA degradation. Cell 136, 763–776. https://doi.org/10.1016/j.cell.2009.01.019.
Huston, M.W., Yasuda, M., Pagant, S., et al., 2019. Liver-targeted AAV gene therapy vectors produced by a clinical scale manufacturing process result in high, continuous therapeutic levels of enzyme activity and effective substrate reduction in mouse model of Fabry disease. Mol. Genet. Metab. 126, S77. https://doi.org/10.1016/j.ymgme.2018.12.187.
Ilyinskii, P.O., Michaud, A.M., Rizzo, G.L., et al., 2021. ImmTOR nanoparticles enhance AAV transgene expression after initial and repeat dosing in a mouse model of methylmalonic acidemia. Mol. Ther. Methods Clin. Dev. 22, 279–292. https://doi.org/10.1016/j.omtm.2021.06.015.
Jauze, L., Monteillet, L., Mithieux, G., et al., 2019. Challenges of gene therapy for the treatment of glycogen storage diseases type I and type III. Hum. Gene Ther. 30, 1263–1273. https://doi.org/10.1089/hum.2019.102.
Jericó, D., Córdoba, K.M., Jiang, L., et al., 2021. mRNA-based therapy in a rabbit model of variegate porphyria offers new insights into the pathogenesis of acute attacks. Mol. Ther. Nucleic Acids 25, 207–219. https://doi.org/10.1016/j.omtn.2021.05.010.

Jeyakumar, J., Kia, A., McIntosh, J., et al., 2019. Liver-directed gene therapy corrects Fabry disease in mice. Mol. Genet. Metab. 126, S80. https://doi.org/10.1016/j.ymgme.2018.12.196.

Jiang, L., Berraondo, P., Jericó, D., et al., 2018. Systemic messenger RNA as an etiological treatment for acute intermittent porphyria. Nat. Med. 1. https://doi.org/10.1038/s41591-018-0199-z.

Jiang, L., Park, J.-S., Yin, L., et al., 2020. Dual mRNA therapy restores metabolic function in long-term studies in mice with propionic acidemia. Nat. Commun. 11, 5339. https://doi.org/10.1038/s41467-020-19156-3.

Karadagi, A., Cavedon, A.G., Zemack, H., et al., 2020. Systemic modified messenger RNA for replacement therapy in alpha 1-antitrypsin deficiency. Sci. Rep. 10, 7052. https://doi.org/10.1038/s41598-020-64017-0.

Karikó, K., Buckstein, M., Ni, H., et al., 2005. Suppression of RNA recognition by Toll-like receptors: the impact of nucleoside modification and the evolutionary origin of RNA. Immunity 23, 165–175. https://doi.org/10.1016/j.immuni.2005.06.008.

Karikó, K., Muramatsu, H., Ludwig, J., et al., 2011. Generating the optimal mRNA for therapy: HPLC purification eliminates immune activation and improves translation of nucleoside-modified, protein-encoding mRNA. Nucleic Acids Res. 39. https://doi.org/10.1093/nar/gkr695, e142.

Kauppinen, R., Mustajoki, P., 1992. Prognosis of acute porphyria: occurrence of acute attacks, precipitating factors, and associated diseases. Medicine (Baltimore) 71, 1–13.

Khurana, A., Allawadhi, P., Khurana, I., et al., 2021. Role of nanotechnology behind the success of mRNA vaccines for COVID-19. Nano Today 38. https://doi.org/10.1016/j.nantod.2021.101142, 101142.

Kim, Y.-K., 2020. RNA therapy: current status and future potential. Chonnam Med. J. 56, 87–93. https://doi.org/10.4068/cmj.2020.56.2.87.

Kim, G.-Y., Lee, Y.M., Kwon, J.H., et al., 2017. Glycogen storage disease type Ia mice with less than 2% of normal hepatic glucose-6-phosphatase-α activity restored are at risk of developing hepatic tumors. Mol. Genet. Metab. 120, 229–234. https://doi.org/10.1016/j.ymgme.2017.01.003.

Kim, J., Eygeris, Y., Gupta, M., et al., 2021. Self-assembled mRNA vaccines. Adv. Drug Deliv. Rev. 170, 83–112. https://doi.org/10.1016/j.addr.2020.12.014.

Kirchner, S., Ignatova, Z., 2015. Emerging roles of tRNA in adaptive translation, signalling dynamics and disease. Nat. Rev. Genet. 16, 98–112. https://doi.org/10.1038/nrg3861.

Kukreja, A., Lasaro, M., Cobaugh, C., et al., 2019. Systemic alanine glyoxylate aminotransferase mRNA improves glyoxylate metabolism in a mouse model of primary hyperoxaluria type 1. Nucleic Acid Ther. 29, 104–113. https://doi.org/10.1089/nat.2018.0740.

Lee, E.K., Hu, C., Bhargava, R., et al., 2013. AAV-based gene therapy prevents neuropathology and results in normal cognitive development in the hyperargininemic mouse. Gene Ther. 20, 785–796. https://doi.org/10.1038/gt.2012.99.

Lee, Y.M., Kim, G.-Y., Pan, C.-J., et al., 2015. Minimal hepatic glucose-6-phosphatase-α activity required to sustain survival and prevent hepatocellular adenoma formation in murine glycogen storage disease type Ia. Mol. Genet. Metab. reports 3, 28–32. https://doi.org/10.1016/j.ymgmr.2015.03.001.

Levy, H.L., Milanowski, A., Chakrapani, A., et al., 2007. Efficacy of sapropterin dihydrochloride (tetrahydrobiopterin, 6R-BH4) for reduction of phenylalanine concentration in patients with phenylketonuria: a phase III randomised placebo-controlled study. Lancet 370, 504–510. https://doi.org/10.1016/S0140-6736(07)61234-3.

Levy, H.L., Sarkissian, C.N., Scriver, C.R., 2018. Phenylalanine ammonia lyase (PAL): from discovery to enzyme substitution therapy for phenylketonuria. Mol. Genet. Metab. 124, 223–229. https://doi.org/10.1016/j.ymgme.2018.06.002.

Lichter-Konecki, U., Vockley, J., 2019. Phenylketonuria: current treatments and future developments. Drugs 79, 495–500. https://doi.org/10.1007/s40265-019-01079-z.

Lim, J.-A., Choi, S.J., Gao, F., et al., 2020. A novel gene therapy approach for GSD III using an AAV vector encoding a bacterial glycogen debranching enzyme. Mol. Ther. Methods Clin. Dev. 18, 240–249. https://doi.org/10.1016/j.omtm.2020.05.034.

Loomis, K.H., Kirschman, J.L., Bhosle, S., et al., 2016. Strategies for modulating innate immune activation and protein production of in vitro transcribed mRNAs. J. Mater. Chem. B 4, 1619–1632. https://doi.org/10.1039/c5tb01753j.

Malone, R.W., Felgner, P.L., Verma, I.M., 1989. Cationic liposome-mediated RNA transfection. Proc. Natl. Acad. Sci. U. S. A. 86, 6077–6081. https://doi.org/10.1073/pnas.86.16.6077.

Marion, R.W., Paljevic, E., 2020. The glycogen storage disorders. Pediatr. Rev. 41, 41–44. https://doi.org/10.1542/pir.2018-0146.

Marsden, J.T., Guppy, S., Stein, P., et al., 2015. Audit of the use of regular haem arginate infusions in patients with acute porphyria to prevent recurrent symptoms. JIMD Rep. 22, 57–65. https://doi.org/10.1007/8904_2015_411.

Martin, S.A., Moss, B., 1975. Modification of RNA by mRNA guanylyltransferase and mRNA (guanine-7-)methyltransferase from vaccinia virions. J. Biol. Chem. 250, 9330–9335. https://doi.org/10.1016/S0021-9258(19)40647-9.

Martini, P.G.V., Guey, L.T., 2019. A new era for rare genetic diseases: messenger RNA therapy. Hum. Gene Ther. 30, 1180–1189. https://doi.org/10.1089/hum.2019.090.

Mehta, A., Beck, M., Eyskens, F., et al., 2010. Fabry disease: a review of current management strategies. QJM 103, 641–659. https://doi.org/10.1093/qjmed/hcq117.

Michel, T., Kankura, A., Salinas Medina, M.L., et al., 2015. In vitro evaluation of a novel mRNA-based therapeutic strategy for the treatment of patients suffering from alpha-1-antitrypsin deficiency. Nucleic Acid Ther. 25, 235–244. https://doi.org/10.1089/nat.2015.0537.

Miller, J.J., Kanack, A.J., Dahms, N.M., 2020. Progress in the understanding and treatment of Fabry disease. Biochim. Biophys. acta. Gen. Subj. 1864, 129437. https://doi.org/10.1016/j.bbagen.2019.129437.

Milliner, D.S., Harris, P.C., Cogal, A.G., et al., 1993. Primary hyperoxaluria type 1. In: Adam, M.P., Ardinger, H.H., Pagon, R.A., et al. (Eds.), GeneReviews. University of Washington, Seattle (WA).

Mockey, M., Gonçalves, C., Dupuy, F.P., et al., 2006. mRNA transfection of dendritic cells: synergistic effect of ARCA mRNA capping with poly(A) chains in cis and in trans for a high protein expression level. Biochem. Biophys. Res. Commun. 340, 1062–1068. https://doi.org/10.1016/j.bbrc.2005.12.105.

Mollocana-Lara, E.C., Ni, M., Agathos, S.N., et al., 2021. The infinite possibilities of RNA therapeutics. J. Ind. Microbiol. Biotechnol. https://doi.org/10.1093/jimb/kuab063.

Nakamura, A., Nakajima, M., Yamanaka, H., et al., 2008. Expression of UGT1A and UGT2B mRNA in human normal tissues and various cell lines. Drug Metab. Dispos. 36, 1461–1464. https://doi.org/10.1124/dmd.108.021428.

Nordmann, Y., Puy, H., Da Silva, V., et al., 1997. Acute intermittent porphyria: prevalence of mutations in the porphobilinogen deaminase gene in blood donors in France. J. Intern. Med. 242, 213–217. https://doi.org/10.1046/j.1365-2796.1997.00189.x.

Ozen, H., 2007. Glycogen storage diseases: new perspectives. World J. Gastroenterol. 13, 2541–2553. https://doi.org/10.3748/wjg.v13.i18.2541.

Pallet, N., Mami, I., Schmitt, C., et al., 2015. High prevalence of and potential mechanisms for chronic kidney disease in patients with acute intermittent porphyria. Kidney Int. 88, 386–395. https://doi.org/10.1038/ki.2015.97.

Patel, S., Ashwanikumar, N., Robinson, E., et al., 2017. Boosting intracellular delivery of lipid nanoparticle-encapsulated mRNA. Nano Lett. 17, 5711–5718. https://doi.org/10.1021/acs.nanolett.7b02664.

Pena, L., Franks, J., Chapman, K.A., et al., 2012. Natural history of propionic acidemia. Mol. Genet. Metab. 105, 5–9. https://doi.org/10.1016/j.ymgme.2011.09.022.

Peoc'h, K., Manceau, H., Karim, Z., et al., 2019. Hepatocellular carcinoma in acute hepatic porphyrias: a Damocles Sword. Mol. Genet. Metab. 128, 236–241. https://doi.org/10.1016/j.ymgme.2018.10.001.

Pietrangelo, A., Oude Elferink, R., Prieto, J., et al., 2007. Genetics in liver diseases. J. Hepatol. https://doi.org/10.1016/j.jhep.2007.03.009.

Polack, F.P., Thomas, S.J., Kitchin, N., et al., 2020. Safety and efficacy of the BNT162b2 mRNA Covid-19 vaccine. N. Engl. J. Med. 383, 2603–2615. https://doi.org/10.1056/NEJMoa2034577.

Prieve, M.G., Harvie, P., Monahan, S.D., et al., 2018. Targeted mRNA therapy for ornithine transcarbamylase deficiency. Mol. Ther. 26, 801–813. https://doi.org/10.1016/j.ymthe.2017.12.024.

Rabinovich, P.M., Komarovskaya, M.E., Ye, Z.-J., et al., 2006. Synthetic messenger RNA as a tool for gene therapy. Hum. Gene Ther. 17, 1027–1035. https://doi.org/10.1089/hum.2006.17.1027.

Raper, S.E., Chirmule, N., Lee, F.S., et al., 2003. Fatal systemic inflammatory response syndrome in a ornithine transcarbamylase deficient patient following adenoviral gene transfer. Mol. Genet. Metab. 80, 148–158. https://doi.org/10.1016/j.ymgme.2003.08.016.

Rasmussen, S.A., Daenzer, J.M.I., Fridovich-Keil, J.L., 2021. A pilot study of neonatal GALT gene replacement using AAV9 dramatically lowers galactose metabolites in blood, liver, and brain and minimizes cataracts in GALT-null rat pups. J. Inherit. Metab. Dis. 44, 272–281. https://doi.org/10.1002/jimd.12311.

Rela, M., Battula, N., Madanur, M., et al., 2007. Auxiliary liver transplantation for propionic acidemia: a 10-year follow-up. Am. J. Transplant. 7, 2200–2203. https://doi.org/10.1111/j.1600-6143.2007.01899.x.

Remih, K., Amzou, S., Strnad, P., 2021. Alpha1-antitrypsin deficiency: new therapies on the horizon. Curr. Opin. Pharmacol. 59, 149–156. https://doi.org/10.1016/j.coph.2021.06.001.

Richard, E., Pérez, B., Pérez-Cerdá, C., et al., 2015. Understanding molecular mechanisms in propionic acidemia and investigated therapeutic strategies. Expert Opin. Orphan Drugs 3, 1427–1438. https://doi.org/10.1517/21678707.2015.1092380.

Roseman, D.S., Khan, T., Rajas, F., et al., 2018. G6PC mRNA therapy positively regulates fasting blood glucose and decreases liver abnormalities in a mouse model of glycogen storage disease 1a. Mol. Ther. 26, 814–821. https://doi.org/10.1016/j.ymthe.2018.01.006.

Ross, J., Sullivan, T.D., 1985. Half-lives of beta and gamma globin messenger RNAs and of protein synthetic capacity in cultured human reticulocytes. Blood 66, 1149–1154.

Rubio-Gozalbo, M.E., Haskovic, M., Bosch, A.M., et al., 2019. The natural history of classic galactosemia: lessons from the GalNet registry. Orphanet J. Rare Dis. 14, 86. https://doi.org/10.1186/s13023-019-1047-z.

Sabnis, S., Kumarasinghe, E.S., Salerno, T., et al., 2018. A novel amino lipid series for mRNA delivery: improved endosomal escape and sustained pharmacology and safety in non-human primates. Mol. Ther. 26, 1509–1519. https://doi.org/10.1016/J.YMTHE.2018.03.010.

Sahin, U., Karikó, K., Türeci, Ö., 2014. MRNA-based therapeutics-developing a new class of drugs. Nat. Rev. Drug Discov. 13, 759–780. https://doi.org/10.1038/nrd4278.

Salido, E., Rodriguez-Pena, M., Santana, A., et al., 2011. Phenotypic correction of a mouse model for primary hyperoxaluria with adeno-associated virus gene transfer. Mol. Ther. 19, 870–875. https://doi.org/10.1038/mt.2010.270.

Schadewaldt, P., Kamalanathan, L., Hammen, H.-W., et al., 2014. Endogenous galactose formation in galactose-1-phosphate uridyltransferase deficiency. Arch. Physiol. Biochem. 120, 228–239. https://doi.org/10.3109/13813455.2014.962547.

Schmitt, C., Lenglet, H., Yu, A., et al., 2018. Recurrent attacks of acute hepatic porphyria: major role of the chronic inflammatory response in the liver. J. Intern. Med. 284, 78–91. https://doi.org/10.1111/joim.12750.
Shchelochkov, O.A., Carrillo, N., Venditti, C., 1993. Propionic acidemia. In: Adam, M.P., Ardinger, H.H., Pagon, R.A., et al. (Eds.), GeneReviews. University of Washington, Seattle (WA).
Sin, Y.Y., Baron, G., Schulze, A., et al., 2015. Arginase-1 deficiency. J. Mol. Med. (Berl) 93, 1287–1296. https://doi.org/10.1007/s00109-015-1354-3.
Skierka, J.M., Kotzer, K.E., Lagerstedt, S.A., et al., 2013. UGT1A1 genetic analysis as a diagnostic aid for individuals with unconjugated hyperbilirubinemia. J. Pediatr. 162, 1152. e1–2. https://doi.org/10.1016/j.jpeds.2012.11.042. 1146–52.
Solares, I., Heredia-Mena, C., Castelbón, F.J., et al., 2021. Diagnosis and management of inborn errors of metabolism in adult patients in the emergency department. Diagnostics. https://doi.org/10.3390/diagnostics11112148.
Stokowski, L.A., 2006. Fundamentals of phototherapy for neonatal jaundice. Adv. Neonatal Care 6, 303–312. https://doi.org/10.1016/j.adnc.2006.08.004.
Stone, W.L., Basit, H., Adil, A., 2021. Glycogen Storage Disease. Treasure Island (FL).
Strnad, P., McElvaney, N.G., Lomas, D.A., 2020. Alpha(1)-antitrypsin deficiency. N. Engl. J. Med. 382, 1443–1455. https://doi.org/10.1056/NEJMra1910234.
Su, X., Fricke, J., Kavanagh, D.G., et al., 2011. In vitro and in vivo mRNA delivery using lipid-enveloped pH-responsive polymer nanoparticles. Mol. Pharm. 8, 774–787. https://doi.org/10.1021/mp100390w.
Sun, A., Crombez, E.A., Wong, D., 1993. Arginase deficiency. In: Adam, M.P., Ardinger, H.H., Pagon, R.A., et al. (Eds.), GeneReviews. University of Washington, Seattle (WA).
Thess, A., Grund, S., Mui, B.L., et al., 2015. Sequence-engineered mRNA without chemical nucleoside modifications enables an effective protein therapy in large animals. Mol. Ther. 23, 1456–1464. https://doi.org/10.1038/mt.2015.103.
To-Figueras, J., Wijngaard, R., García-Villoria, J., et al., 2021. Dysregulation of homocysteine homeostasis in acute intermittent porphyria patients receiving heme arginate or givosiran. J. Inherit. Metab. Dis. 44, 961–971. https://doi.org/10.1002/jimd.12391.
Trepotec, Z., Lichtenegger, E., Plank, C., et al., 2019. Delivery of mRNA therapeutics for the treatment of hepatic diseases. Mol. Ther. 27, 794–802. https://doi.org/10.1016/j.ymthe.2018.12.012.
Truong, B., Allegri, G., Liu, X.-B., et al., 2019. Lipid nanoparticle-targeted mRNA therapy as a treatment for the inherited metabolic liver disorder arginase deficiency. Proc. Natl. Acad. Sci. U. S. A. 116, 21150–21159. https://doi.org/10.1073/pnas.1906182116.
Unzu, C., Sampedro, A., Mauleón, I., et al., 2011. Sustained enzymatic correction by rAAV-mediated liver gene therapy protects against induced motor neuropathy in acute porphyria mice. Mol. Ther. 19, 243–250. https://doi.org/10.1038/mt.2010.210.
Unzu, C., Melero, I., Hervás-Stubbs, S., et al., 2015. Helper-dependent adenovirus achieve more efficient and persistent liver transgene expression in non-human primates under immunosuppression. Gene Ther. 22, 856.
van Spronsen, F.J., Blau, N., Harding, C., et al., 2021. Phenylketonuria. Nat. Rev. Dis. Prim. 7, 36. https://doi.org/10.1038/s41572-021-00267-0.
Vandamme, C., Adjali, O., Mingozzi, F., 2017. Unraveling the complex story of immune responses to AAV vectors trial after trial. Hum. Gene Ther. 28, 1061–1074. https://doi.org/10.1089/hum.2017.150.
Vara, R., Turner, C., Mundy, H., et al., 2011. Liver transplantation for propionic acidemia in children. Liver Transpl. 17, 661–667. https://doi.org/10.1002/lt.22279.
Vidal, P., Ronzitti, G., Collaud, F., et al., 2017. TH.O.18—adeno associated vector-based gene therapy strategy for type 3 glycogen storage disease. Neuromuscul. Disord. 27, S246. https://doi.org/10.1016/j.nmd.2017.06.544.

Villiger, L., Grisch-Chan, H.M., Lindsay, H., et al., 2018. Treatment of a metabolic liver disease by in vivo genome base editing in adult mice. Nat. Med. 24, 1519–1525. https://doi.org/10.1038/s41591-018-0209-1.

Wang, Z., Day, N., Trifillis, P., et al., 1999. An mRNA stability complex functions with poly(A)-binding protein to stabilize mRNA in vitro. Mol. Cell. Biol. 19, 4552–4560. https://doi.org/10.1128/MCB.19.7.4552.

Wang, L., Morizono, H., Lin, J., et al., 2012. Preclinical evaluation of a clinical candidate AAV8 vector for ornithine transcarbamylase (OTC) deficiency reveals functional enzyme from each persisting vector genome. Mol. Genet. Metab. 105, 203–211. https://doi.org/10.1016/j.ymgme.2011.10.020.

Welling, L., Bernstein, L.E., Berry, G.T., et al., 2017. International clinical guideline for the management of classical galactosemia: diagnosis, treatment, and follow-up. J. Inherit. Metab. Dis. 40, 171–176. https://doi.org/10.1007/s10545-016-9990-5.

Whitehead, K.A., Langer, R., Anderson, D.G., 2009. Knocking down barriers: advances in siRNA delivery. Nat. Rev. Drug Discov. 8, 129–138. https://doi.org/10.1038/nrd2742.

Wolff, J.A., Malone, R.W., Williams, P., et al., 1990. Direct gene transfer into mouse muscle in vivo. Science 247, 1465–1468. https://doi.org/10.1126/science.1690918.

Wongkittichote, P., Ah Mew, N., Chapman, K.A., 2017. Propionyl-CoA carboxylase—a review. Mol. Genet. Metab. 122, 145–152. https://doi.org/10.1016/j.ymgme.2017.10.002.

Yagi, H., Ogura, T., Mizukami, H., et al., 2011. Complete restoration of phenylalanine oxidation in phenylketonuria mouse by a self-complementary adeno-associated virus vector. J. Gene Med. 13, 114–122. https://doi.org/10.1002/jgm.1543.

Yang, Y., Wang, L., Bell, P., et al., 2016. A dual AAV system enables the Cas9-mediated correction of a metabolic liver disease in newborn mice. Nat. Biotechnol. 34, 334–338. https://doi.org/10.1038/nbt.3469.

Yasuda, M., Bishop, D.F., Fowkes, M., et al., 2010. AAV8-mediated gene therapy prevents induced biochemical attacks of acute intermittent porphyria and improves neuromotor function. Mol. Ther. 18, 17–22. https://doi.org/10.1038/mt.2009.250.

Yasuda, M., Huston, M.W., Pagant, S., et al., 2020. AAV2/6 gene therapy in a murine model of Fabry disease results in supraphysiological enzyme activity and effective substrate reduction. Mol. Ther. Methods Clin. Dev. 18, 607–619. https://doi.org/10.1016/j.omtm.2020.07.002.

Yokoi, K., Nakajima, Y., Yasui, T., et al., 2021. Novel ARG1 variants identified in a patient with arginase 1 deficiency. Hum. Genome Var. 8, 8. https://doi.org/10.1038/s41439-021-00139-9.

Yuzyuk, T., Viau, K., Andrews, A., et al., 2018. Biochemical changes and clinical outcomes in 34 patients with classic galactosemia. J. Inherit. Metab. Dis. 41, 197–208. https://doi.org/10.1007/s10545-018-0136-9.

Zhu, X., Yin, L., Theisen, M., et al., 2019. Systemic mRNA therapy for the treatment of Fabry disease: preclinical studies in wild-type mice, Fabry mouse model, and wild-type non-human primates. Am. J. Hum. Genet. 104, 625–637. https://doi.org/10.1016/j.ajhg.2019.02.003.

Ziegler, R.J., Lonning, S.M., Armentano, D., et al., 2004. AAV2 vector harboring a liver-restricted promoter facilitates sustained expression of therapeutic levels of α-galactosidase a and the induction of immune tolerance in Fabry mice. Mol. Ther. 9, 231–240. https://doi.org/10.1016/j.ymthe.2003.11.015.

CHAPTER THREE

Applications of self-replicating RNA

Kenneth Lundstrom*
PanTherapeutics, Lutry, Switzerland
*Corresponding author: e-mail address: lundstromkenneth@gmail.com

Contents

International Review of Cell and Molecular Biology, Volume 372
ISSN 1937-6448
https://doi.org/10.1016/bs.ircmb.2022.03.006
Copyright © 2022 Elsevier Inc.
All rights reserved.

Abstract

Self-replicating RNA viral vectors have been engineered for both prophylactic and therapeutic applications. Mainly the areas of infectious diseases and cancer have been targeted. Both positive and negative strand RNA viruses have been utilized including alphaviruses, flaviviruses, measles viruses and rhabdoviruses. The high-level of RNA amplification has provided efficient expression of viral surface proteins and tumor antigens. Immunization studies in animal models have elicit robust neutralizing antibody responses. In the context of infectious diseases, immunization with self-replicating RNA viral vectors has provided protection against challenges with lethal doses of pathogens in animal models. Similarly, immunization with vectors expressing tumor antigens has resulted in tumor regression and eradication and protection against tumor challenges in animal models. The transient nature and non-integration of viral RNA into the host genome are ideal features for vaccine development. Moreover, self-replicating RNA viral vectors show great flexibility as they can be applied as recombinant viral particles, RNA replicons or DNA replicon plasmids. Several clinical trials have been conducted especially in the area of cancer immunotherapy.

1. Introduction

Since the advent of recombinant DNA and genetic engineering technologies in the late 1970s both drug discovery and vaccine development have seen major progress. Numerous recombinant expression systems have been engineered to support the development of safer and more efficacious drugs and vaccines. Recent progress in gene therapy and the enormous efforts dedicated to development of antiviral drugs and vaccines against COVID-19 have resulted in unprecedented progress (Lundstrom, 2021).

Gene expression vectors used for gene therapy and vaccine development can be divided into non-viral (Patil et al., 2019) and viral vectors (Chen et al., 2018). In the case of non-viral vectors, utilization of plasmid DNA has been hampered by poor delivery to target cells and also the necessity of transportation of DNA to the cell nucleus has reduced efficacy. DNA delivery has been facilitated by a variety of formulations based on liposomes (Zylberberg et al., 2017) and polymers (Wahane et al., 2020). The utilization of RNA-based delivery vehicles has further accelerated the formulation of liposome nanoparticles due to the sensitivity of degradation of single stranded RNA molecules (Ickenstein and Garidel, 2019). In the context of viral vectors, a variety of expression systems based on adenoviruses (Ad), adeno-associated viruses (AAV), alphaviruses, flaviviruses, herpes simplex viruses (HSV), lentiviruses (LV), retroviruses (RV), measles viruses (MV), rhabdoviruses, Newcastle disease virus (NDV), poxviruses, and picornaviruses have been engineered (Lundstrom, 2018). Many of these viral vectors have been used for antigen production targeting both infectious diseases and various cancers. Moreover, overexpression of toxic, tumor killing, therapeutic, and immunostimulatory genes has been applied for cancer therapy and immunotherapy. However, the focus of this review is on self-replicating RNA. Engineering of self-replicating RNA vectors is described including the delivery options of viral vectors, DNA plasmids or RNA molecules. Furthermore, applications of RNA-based delivery for vaccine development against various infectious diseases and cancers are presented.

2. Self-replicating RNA vectors

Self-replicating RNA vectors are based on RNA viruses accommodating a single-stranded RNA (ssRNA) genome. Their characteristic feature, as indicated by their name, is the capacity of efficient RNA replication in the cytoplasm of infected host cells, leading to the accumulation of approximately 10^6 copies of subgenomic RNA per cell (Frolov et al., 1996).

Alphaviruses and flaviviruses possess a genome of positive polarity, while the genome of MV and rhabdoviruses is of negative polarity. This difference in polarity affects the design of vectors significantly as positive ssRNA molecules can directly act as mRNA for translation in the cytoplasm of host cells, whereas negative ssRNA needs to be copied to a plus strand copy before translation can take place (Fig. 1). The engineering and application of expression vector systems for both positive and negative strand self-replicating RNA viruses are described below.

2.1 Viral vectors

Alphavirus expression systems have been engineered for Semliki Forest virus (SFV) (Liljestrom and Garoff, 1991), Sindbis virus (SIN) (Xiong et al., 1989), and Venezuelan equine encephalitis virus (VEE) (Davis et al., 1989). Typically, the alphavirus expression vector contains the four non-structural protein (nsP1-4) genes and the 26S subgenomic promoter, downstream of which the structural protein genes have been replaced by a foreign gene (Fig. 2A). The structural genes, C-E3-E2-6K-E1, are placed on a helper plasmid for expression of the alphavirus structural proteins in *trans*. Replication-deficient particles can be generated by electroporation of BHK-21 cells with *in vitro* transcribed RNA from both the expression and helper vectors.

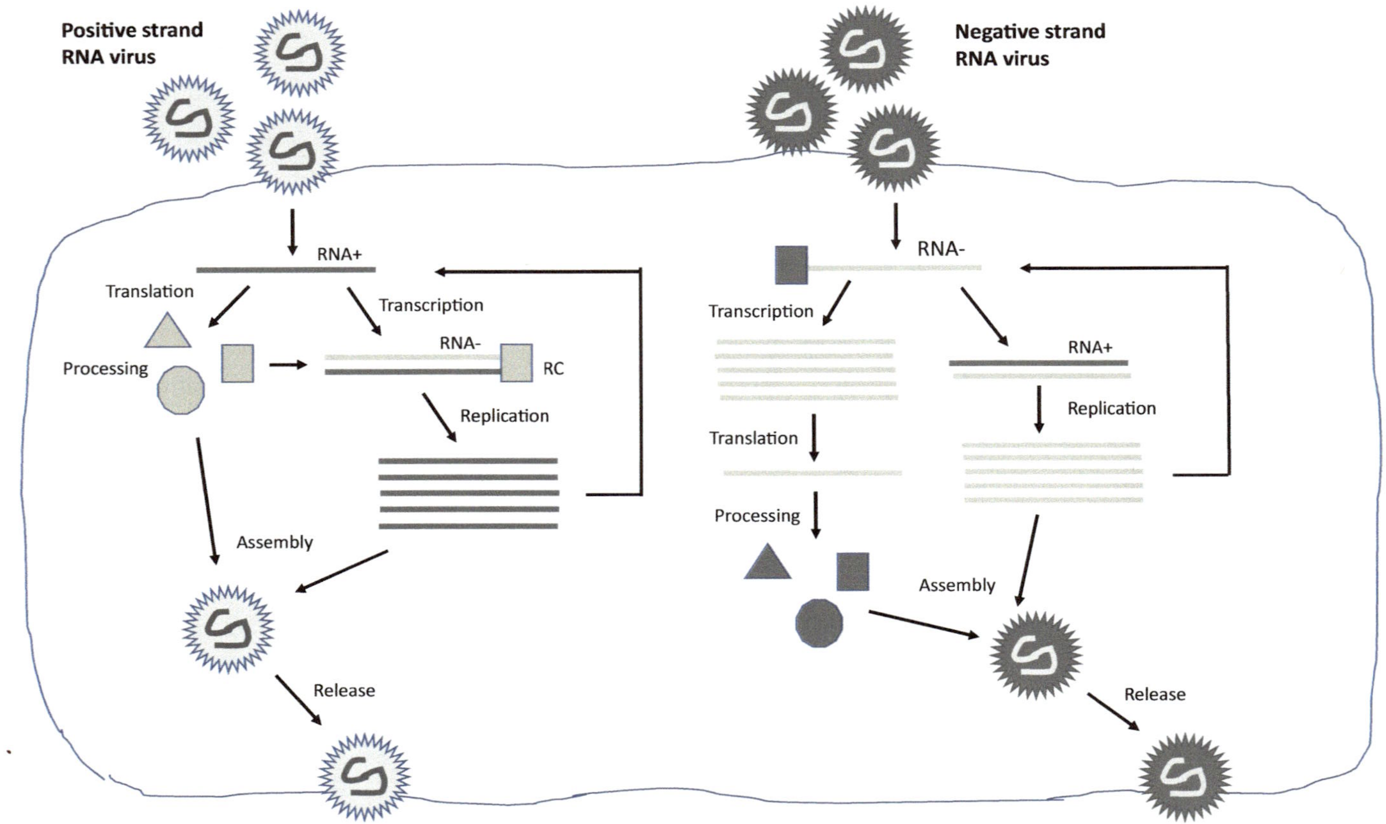

Fig. 1 Schematic presentation of the lifecycles of positive and negative strand self-replicating RNA viruses. Left: steps involved in replication of positive strand RNA virus replication, translation, processing of structural proteins, assembly, and release of mature viral particles. Right: the corresponding steps for negative strand self-replicating RNA viruses. RC, replication complex.

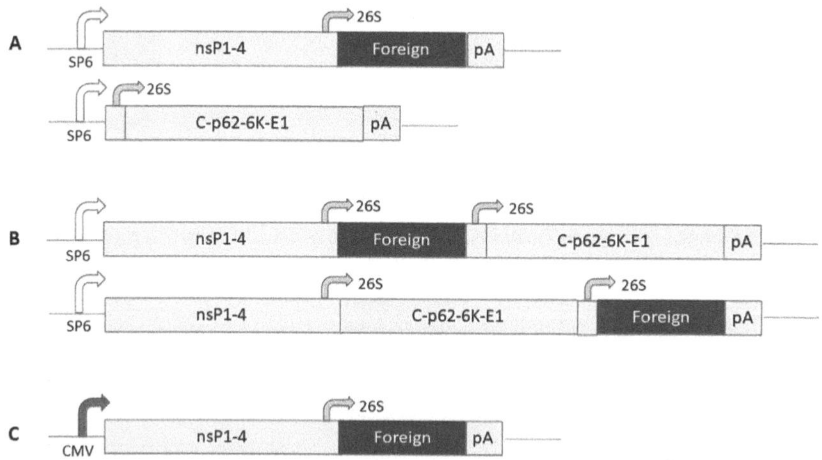

Fig. 2 Schematic presentation of alphavirus expression vector systems. (A) Replication-deficient system, where the foreign gene has been introduced into the expression vector downstream of the non-structural (nsP) genes. A helper vector is required for production of recombinant replication-deficient particles. (B) Replication-proficient system, where the foreign gene is introduced either downstream of the nsP or structural (Capsid-p62-6K-E1) genes and expressed from a second 26S subgenomic promoter. In both cases (A and B) SP6 RNA polymerase promoter (SP6) is applied for *in vitro* transcription of RNA from plasmid DNA templates and electroporated/transfected into baby hamster kidney (BHK-21) cells for production of viral particles. (C) DNA replicon system, where the SP6 promoter is replaced by a cytomegalovirus (CMV) promoter allowing direct transfection of plasmid DNA into mammalian cells.

Due to the localization of the packaging signal in the nsP2 region of SFV and VEE (for SIN in nsP1), only RNA from the expression vector will be packaged into virions providing one round of infection and leading to high expression levels of the foreign gene, but no production of new infectious viral particles. In contrast, replication-proficient particles can be generated by engineering a second 26S subgenomic promoter either upstream or downstream of the structural genes for insertion of foreign genes (Fig. 2B). This approach generates new infectious progeny and expression of the foreign gene. Moreover, replacing the SP6 RNA polymerase promoter with a cytomegalovirus (CMV) promoter has allowed the direct application of DNA-based alphavirus vectors as described below (Fig. 2C).

Among flaviviruses, expression systems have been engineered for Kunjin virus (KUN), where the gene of interest is cloned between the first 60 nucleotides of the core protein (C20) and the last 22 codons of the envelope protein (E22) in frame with the downstream viral polyprotein (Fig. 3)

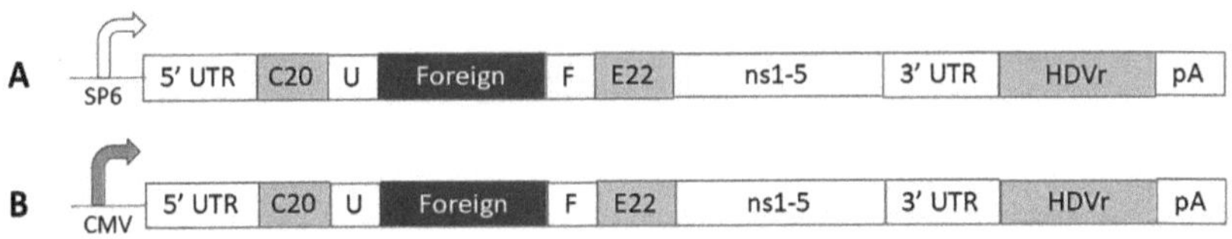

Fig. 3 Schematic presentation of flavivirus expression vector systems. The foreign gene is introduced between the first 60 nucleotides of the core protein (C20) and the last 22 codons of the envelope protein (E22) in frame with the downstream viral polyprotein. (A) RNA-based system using the SP6 RNA polymerase promoter. (B) DNA-based system using the CMV promoter. 5′UTR, 5′ untranslated region; C20, first 20 amino acids of KUN C protein; E22, last 22 amino acids of KUN E protein; U, mouse ubiquitin sequence; F, Foot-and-mouth disease virus 2A autoprotease; E22, last 22 amino acids of KUN E22 protein; 3′ UTR, 3′ untranslated region; HDVr, Hepatitis delta virus ribozyme; pA, polyadenylation signal.

(Pijlman et al., 2006). As the heterologous protein is expressed as part of a larger polyprotein, introduction of the FMDV-2A protease sequence into the vector allows removal of KUN proteins from the recombinant protein of interest (De Felipe, 2004). A KUN packaging system based on the tetKUNCprMEC cell line has been engineered to facilitate recombinant KUN particle production (Khromykh et al., 1998). Additionally, expression systems have been engineered for other flaviviruses such as West Nile virus (WNV) (Scholle et al., 2004), yellow fever virus (YFV) (Jones et al., 2005a, b), Dengue virus (DENV) (Pang et al., 2001) and tick-borne encephalitis virus (TBEV) (Hayasaka et al., 2004).

In the case of negative strand RNA viruses, MV strains have been subjected to the engineering of packaging systems for rescuing of replicating MV vectors from DNA expression constructs (Radecke et al., 1995). It has been achieved by transfection of HEK293 cells stably expressing the MV-N, MV-P and T7 polymerase with a DNA vector coding for the full-length MV genome, the foreign gene and T7 polymerase-driven MV-L (Mühlebach, 2017). Another approach comprises the expression of the MV-N, MV-P and MV genome from transfected HEK293 cells, which can be superinfected with a replication deficient vaccinia virus (MVA) vector expressing T7 polymerase (Schneider et al., 1997). Additionally, expression of the MV genome and the RNA-dependent RNA polymerase (RdRp) from a CMV promoter has allowed MV rescue (Martin et al., 2006). Expression of recombinant proteins can be achieved by insertion of the foreign gene between the P and the matrix (M) genes or alternatively between the hemagglutinin (HA) and the large protein (L) genes in the MV vectors (Fig. 4A).

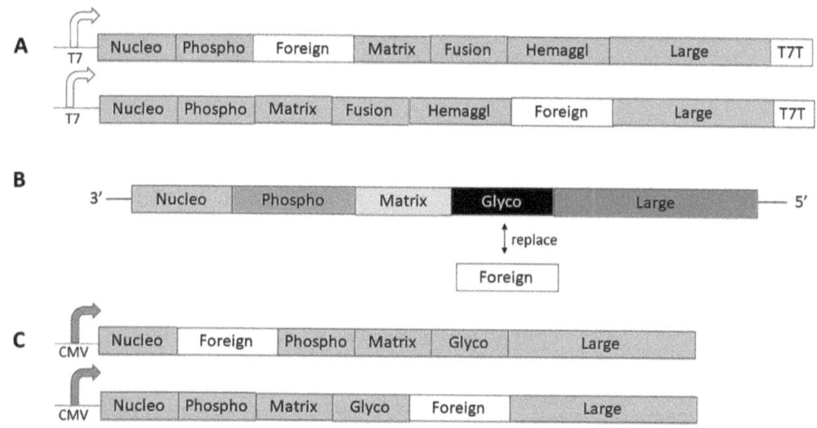

Fig. 4 (A) Measles virus system, where the foreign gene can be introduced either between the phosphoprotein and the matrix protein or the hemagglutinin and the large protein. T7, T7 RNA polymerase promoter, T7T, T7 terminator. (B) VSV system, where the foreign gene replaces the glycoprotein gene. (C) RABV system, where the foreign gene can be introduced either between the nucleoprotein and the phosphoprotein or the glycoprotein and the large protein. CMV, cytomegalovirus promoter.

In the case of rhabdoviruses, expression systems for rabies virus (RABV), vesicular stomatitis virus (VSV), and more recently Maraba virus have been engineered. For VSV, two approaches have been taken. VSV pseudoviruses have been engineered by replacing the VSV glycoprotein (G) with another envelope protein (Fig. 4B) (Li et al., 2018). Recombinant VSV particles can be engineered by insertion of the foreign gene between the G and L genes. Mammalian cells infected with pseudotyped VSV particles allow recombinant protein expression but no new virus production (Schnell et al., 1996). Application of reverse genetics can generate VSV-G complemented recombinant virus capable of producing infectious progeny virus in producer cells (Tani et al., 2012). In the case of RABV, expression vectors have been engineered where the foreign gene is inserted between the nucleoprotein and phosphoprotein in the RABV genome (Osakada and Callaway, 2013) (Fig. 4C). Additionally, Maraba virus has demonstrated oncolytic activity and expression vectors have been engineered, where the foreign gene is inserted between the G and L genes of the Maraba virus genome (Pol et al., 2014).

2.2 RNA vectors

An alternative to delivery of recombinant viral particles is to use RNA. Synthetic RNA for preventive and therapeutic applications has shown promise (Burnett and Rossi, 2012). However, replicon RNA from self-replicating RNA viruses can provide substantial advantages. The self-amplification of replicon RNA in the host cell cytoplasm (Frolov et al., 1996) provides substantially enhanced RNA levels compared to conventional mRNA, which has allowed to apply significantly lower doses in immunization experiments. For instance, protection against influenza virus challenges in mice was achieved with administration of only 1.25 μg of a VEE RNA replicon expressing the influenza virus HA gene compared to 80 μg of synthetic RNA (Vogel et al., 2018). Another positive feature is that self-replicating RNA viruses are unable to reverse transcribe RNA into DNA, which eliminates the possibility of their integration into the host genome. The single stranded nature of the RNA makes it sensitive to degradation within 5–7 days and renders the expression transient, which is ideal for cancer therapy and vaccine development. However, the susceptibility to degradation has a negative impact on RNA delivery. This problem has been addressed by encapsulation of RNA in liposome nanoparticles (LNPs), which has improved both RNA stability and delivery (Raouane et al., 2012). Another issue relates to the transfection efficiency of RNA. Although methods for improved delivery have been developed, RNA transfection is still inferior compared to viral infection. In this context, it was demonstrated that delivery of a self-replicating RNA-based vaccine encapsulated in LNPs improved immunogenicity significantly compared to unformulated RNA (Geall et al., 2012).

2.3 DNA vectors

Self-replicating RNA viral vectors can also be used as plasmid DNA vehicles. In this case, the expression of the non-structural viral genes is achieved by replacing the SP6 or T7 RNA polymerase promoter with a CMV promoter (Fig. 2C) (DiCiommo and Bremner, 1998). DNA-based vectors contain the same RNA replicon present in recombinant viral particles and have the advantage of RNA amplification in the cytoplasm of transfected cells. However, the drawback is that DNA-based vectors must first be delivered to the cell nucleus, where mRNA is transcribed and transported to the cytoplasm for translation. Although DNA is more resistant to degradation compared to RNA the inefficiency of nuclear delivery has required some engineering of better DNA transfection and nuclear uptake methods.

3. Self-replicating RNA for infectious diseases

Infectious diseases have been commonly targeted for vaccine development by introduction of surface proteins or epitopes into expression vectors for immunization studies in animal models. In this context, self-replicating RNA viral vectors have been frequently used for vaccine development as described through numerous examples and summarized in Table 1.

3.1 Alphaviruses

Although frequently used as expression vectors some alphaviruses have caused epidemics such as Chikungunya virus (CHIKV) in Africa (Kelvin, 2011) and Polynesia (Jansen, 2013) and VEE in South America (Weaver et al., 1996). A modified VSV vector where the VSV G protein envelope was replaced by the CHIKV envelope polyprotein (E3-E2-6K-E1) and additionally expressing the Zika virus (ZIKV) membrane-envelope glycoproteins (ME) has been evaluated in mice (Chattopadhyay et al., 2018). A single immunization with 1×10^7 VSV particles elicited neutralizing antibodies and provided protection of A129 mice. Another CHIKV vaccine candidate was engineered based on a recombinant live-attenuated MV strain expressing CHIKV capsid and envelope proteins resulting in CHIKV-like particles (Brandler et al., 2013). A single immunization with the MV-CHIKV vector elicited high CHIKV antibody titers and protected mice challenged with lethal doses of CHIKV. Moreover, the MV-CHIKV vaccine protected macaques from viremia (Rossi et al., 2019). The MV-CHIKV vaccine candidate has been subjected to the first-in-man, randomized, double-blind phase I clinical trial, where healthy volunteers received a low dose (1.5×10^4 $TCID_{50}$), a medium dose (7.5×10^4 $TCID_{50}$) or a high dose (3.0×10^5 $TCID_{50}$) (Ramsauer et al., 2015). The seroconversion rates after a single dose were 44%, 92%, and 90% after a single vaccination with low, medium, and high doses, respectively. A second immunization generated 100% seroconversion for all doses. Furthermore, in a double-blind, randomized, placebo-controlled phase II trial, healthy volunteers were intramuscularly injected with 5×10^4 $TCID_{50}$ or 5×10^5 $TCID_{50}$ MV-CHIKV vaccine, which elicited neutralizing CHIKV antibodies at both doses (Reisinger et al., 2018). Depending on the dose and administration schedule, the seroconversion rates ranged from 50% to 95.9% and no serious adverse events were reported. In the context of VEE, western equine encephalitis virus (WEE), and eastern equine encephalitis

Table 1 Examples of preclinical and clinical vaccine development against infectious diseases.

Disease	Expression system	Target	Response	Ref.
Alphaviruses				
CHIKF	VSV-CHIKV Env	E3-E2-6K-E1	Protection in mice	Chattopadhyay et al. (2018)
CHIKF	MV-CHIKV env + Z-ME	E3-E2-6K-E1	Protection in mice	Brandler et al. (2013)
CHIKF	MV-CHIKV env + Z-ME	E3-E2-6K-E1	Protection in macaques	Rossi et al. (2019)
CHIKF	MV-CHIKV env	E3-E2-6K-E1	Phase I: 100% seroconversion	Ramsauer et al. (2015)
CHIKF	MV-CHIKV env	E3-E2-6K-E1	Phase II: 50-95.9% seroconversion	Reisinger et al. (2018)
EEE	EEE-E2-E3 mutant	E3-E2-6K-E1	Protection in mice, primates	Reed et al. (2014)
WEE	WEE-E2-E3 mutant	E3-E2-6K-E1	Partial protection in primates	Reed et al. (2014)
VEE	VEE-E2-E3 mutant	E3-E2-6K-E1	Protection in mice, primates	Reed et al. (2014)
VEE	VEE-LAV RNA-CNE	VEE TC-83	Protection in mice	Samsa et al. (2019)
VEE	VEE-IAV RNA-CNE	VEE ΔTC-83	Partial protection in mice	Samsa et al. (2019)
VEE	VEE V4020 DNA	VEE TC-83 C	Protection in mice	Tretyakova et al. (2019)
VEE	VEE DNA	E3-E2-6K-E1	Phase I: nAbs after IM-EP, ID-EP	Hannaman et al. (2016)
Arenaviruses				
LHF	VSV-LASFV-GPC	GPC	Protection in guinea pigs & macaques	Safronetz et al. (2015)
LHF	MV-LASV-GPC-NP	GPC, NP	Protection in macaques	Mateo et al. (2019)
LHF	MV-LASV-GPC-NP	GPC, NP	Long-lasting protection in macaques	Mateo et al. (2021)
LHF	MV-LASV-GPC-NP	GPC, NP	Phase I: study in progress	NCT04055454
LHF	YFV-LASV-GPC	GPC	Protection in guinea pigs	Bredenbeek et al. (2006)

LHF	YFV-LASV-GP1/GP2	GP1/GP2	Protection in guinea pigs	Jiang et al. (2011)
LHF	VEE-LASV-GPC/NP	GPC/NP	Protection in guinea pigs	Pushko et al. (2001)
LHF	VEE-LASV-GPCwt/ΔGPfib	GPCwt/ΔGPfib	Immunogenicity, protection in mice	Wang et al. (2018)
AHF	VEE-JUNV/MACV-GPC	GPC	Protection against JUNV	Johnson et al. (2020)
BHF	VEE-JUNV/MACV-GPC	GPC	Protection against MACV	Johnson et al. (2020)
Filoviruses				
EVD	KUN-EBOV-GP/D637L	GP/D637L	Protection in guinea pigs	Reynard et al. (2011)
EVD	KUN-EBOV-GP/D637L	GP/D637L	Protection in 3 of 4 primates	Pyankov et al. (2015)
EVD	VSV-EBOV-GP	GP	Protection in primates	Marzi et al. (2015)
EVD	ΔVSV-ZEBOV-GP	GP	Protection in mice	Jones et al. (2007)
EVD	ΔVSV-ZEBOV-GP	GP	Protection in primates	Jones et al. (2005a, 2005b)
MHF	ΔVSV-MARV-GP	GP	Protection in primates	Jones et al. (2005a, 2005b)
MHF	VEE-MARV-GP	GP	Protection in primates	Herbert et al. (2013)
EVD	VEE-SUDV-GP	GP	Protection in primates	Herbert et al. (2013)
EVD	VEE-SUDV-GP	GP	Partial protection against EBOV	Herbert et al. (2013)
EVD	VEE-SUDV-GP + EBOV-GP	GP	Protection against SUDV & EBOV	Herbert et al. (2013)
EVD	VSV-ZEBOV	GP	Phase III: 100% vaccine efficacy	Henao-Restrepo et al. (2015)
EVD	VSV-ZEBOV	GP	Phase III: vaccine efficacy confirmed	Henao-Restrepo et al. (2017)
EVD	VSV-ZEBOV	GP	Ervebo approved by the FDA, EMA	Maxmen (2017)

Continued

Table 1 Examples of preclinical and clinical vaccine development against infectious diseases.—cont'd

Disease	Expression system	Target	Response	Ref.
Flaviviruses				
DF	VEE-DENV-prME	prME	Protection in mice	Khalil et al. (2014)
DF	VEE-DENV-prME + DNA	prME	+DNA > Superior protection	Chen et al. (2007)
DF	MV-DENV-ED3	ED3	Partial protection in mice	Hu et al. (2016)
DF	Tetravalent YFV-DENV	prME	Safety, tolerability in macaques	Ravel et al. (2017)
DF	Tetravalent YFV-DENV	prME	Phase 2b: good safety, 30.2% efficacy	Sabchareon et al. (2012)
DF	Tetravalent YFV-DENV	prME	Phase III: 56.5% efficacy	Capeding et al. (2014)
DF	Tetravalent YFV-DENV	prME	Phae III: 42.3–77.7% efficacy	Villar et al. (2015)
ZVD	VEE-ZIKV NLC RNA	prME	Protection in mice with 10 ng RNA	Erasmus et al. (2018)
ZVD	VSV-ZIKV-prME	prME	Protection of mice	Chattopadhyay et al. (2018)
ZVD	MV-ZIKV-E	E	Protection in mice	Kurup et al. (2022)
ZVD	MV-ZIKV-E/ZIKV-NS1	E, NS1	ZIKV clearance from reproductive tract	Kurup et al. (2022)
ZVD	MV-ZIKA	E	Phase I: competed, no results available	NCT02996890
ZVD	MV-ZIKA-RSP	RSP, E	Phase I: study in process	NCT04033068
LIV	SFV-LIV prME RNA	LIV prME	Protection of 70% of mice	Fleeton et al. (2001)
Hepatotropic viruses				
HB	MV-HBsAg	HBsAg	Partial protection in rhesus monkeys	Del Valle et al. (2007)
HB	SFV-MHB-VLVs	MHB	Immunogenicity, protection in mice	Reynolds et al. (2015)
HB	SFV-HBcAg-VLVs	HBcAg	No immunogenicity or protection	Reynolds et al. (2015)

Lentiviruses				
AIDS/HIV	MV-HIV gp160	HIV gp160	Cellular and humoral response in mice	Lorin et al. (2004)
AIDS/HIV	VSV-HIV-1 Env	HIV-1 Env	Immune response in mice	Rabinovich et al. (2014)
AIDS/HIV	VSV-HIV-1 Env	HIV-1 Env	67% protection in primates	Parks et al. (2013)
AIDS/HIV	SFV-HIV-Env	HIV Env	Superior to DNA and Env protein	Brand et al. (1998)
AIDS/HIV	SIN-HIV-1 Env	HIV-1 Env	Prime boost with Env protein	Center et al. (2013)
AIDS/HIV	SFV-HIV Env RNA	HIV Env	Immune response in mice 12H2 mAb	Giraud et al. (1999)
AIDS/HIV	SFV-HIV particles/RNA	HIV-Env/ Gag-PolRT	Superior immune response of particles compared to RNA	Ajbani et al. (2017)
AIDS/HIV	VEE-HIV gp160 DNA	HIV Envgp160	10–100-fold less DNA per dose	Ljungberg et al. (2007)
AIDS/HIV	SFV-HIV Env/Gag-Pol-Nef	HIV Env/GPN	Efficient priming with MVA boost	Knudsen et al. (2015)
AIDS/HIV	VEE-HIV gp140 RNA-CNE	HIV gp140	Superior to particles and protein	Bogers et al. (2015)
AIDS/HIV	VEE-HIV Gag	HIV Gag	Phase I: vaccine stability problems, termination of study	Wecker et al. (2020)
Influenza viruses				
IFVA	MV-AIK-C-HA	IFA HA	Protection in cotton rats	Ito et al. (2020)
IFVA	VSVΔG-H5N1	HA, NA	Protection in mice	Ryder et al. (2015)
IFVA	VSV-HAfl	HAfl (H5N1)	Protection in mice	Furuyama et al. (2020)

Continued

Table 1 Examples of preclinical and clinical vaccine development against infectious diseases.—cont'd

Disease	Expression system	Target	Response	Ref.
IFVA	VEE-HA	HA	Protection in chickens	Schultz-Cherry et al. (2000)
IFVA	SFV-HA RNA	HA	Protection in 90% of mice	Fleeton et al. (2001)
IFVA	VEE-HA RNA	HA	64-fold less RNA for protection	Vogel et al. (2018)
IFVA	SFV-HA taRNA	HA	Protection in mice with 50 ng of RNA	Beissert et al. (2020)
IFVA	CSFV-HA/NP VRPs/PEI RNA	HA/NP	Strong immunogenicity in pigs	Démoulins et al. (2021)
IFVA	VEE-HA (AVX502)	HA	Phase I/II: Immunogenicity in 80–86%	NCT00440362
IFVA	VEE-HA (AVX502)	HA	Phase I/II: Immunogenicity in 50%	NCT00706732
Orthopneumoviruses				
RSV	SFV-RSV F RNA	RSV F	32-fold reduced lung RSV titers	Fleeton et al. (2001)
RSV	VEE-RSV F LNP RNA	RSV F	Superior to RNA or particles in rats	Geall et al. (2012)
Coronaviruses				
SARS	VEE-SARS-CoV S	SARS-CoV S	Protection in mice	Sheahan et al. (2011)
MERS	VSVΔG-MERS-CoV S	MERS-CoV S	nAb and T-cell responses in macaques	Liu et al. (2018)
PED	BVDV-PEDV S	PEDV S	BVDV and PEDV Ab responses in mice	Tao et al. (2020)
COVID	MV-SARS-CoV-2 S	SARS-CoV-2 S	Th1-biased Ab and T-cell responses	Hörner et al. (2020)

COVID	MV-SARS-CoV-2 S (TMV-083)	SARS-CoV-2 S	Phase I: Weak response, termination	NCT04497298
COVID	VSV-SARS-CoV-2 S	SARS-CoV-2 S	Immunogenicity, protection in mice	Case et al. (2020)
COVID	VSV-SARS-CoV-2 S (V590)	SARS-CoV-2 S	Phase I: Weak response, termination	NCT04569786
COVID	VSVΔG-SARS-CoV-2 S	SARS-CoV-2 S	Protection in hamsters	Yahalom-Ronen et al. (2020)
COVID	VSVΔG-SARS-CoV-2 S	SARS-CoV-2 S	Phase I/II: Study in progress	NCT04608305
COVID	VSVΔG-SARS-CoV-2 S	SARS-CoV-2 S	Phase IIb/III: Study approved	NCT04990466
COVID	VEE-SARS-CoV-2 S LNP-RNA	SARS-CoV-2 S	Strong Ab responses in mice	McKay et al. (2020)
COVID	VEE-LIONrepRNA-CoV2S	SARS-CoV-2 S	nAbs, T-cell resp. in mice and macaques	Erasmus et al. (2020)
COVID	LNP-nCoVsaRNA	SARS-CoV-2 S	Phase I: Safe, dose-related response	Pollock et al. (2021)
COVID	VEE LUNAR RNA	SARS-CoV-2 S	Protection in mice	De Alwis et al. (2021)
COVID	SIN-SARS-CoV-2 S + αOX40	SARS-CoV-2 S	Protection in mice	Scaglione et al. (2021)
COVID	LNP-saRNA + ChAdOx1	SARS-CoV-2 S	Heterologous superior to homologous	Spencer et al. (2021)
Bacterial				
Anthrax	SIN-PA	*B. anthracis* PA	Some protection in mice	Thomas et al. (2009)
Anthrax	SFV-PA VRPs, DNA	*B. anthracis* PA	Superior to conventional DNA, protection in mice	Wang et al. (2015)
Brucellosis	SFV-IF3	*B. abortus* IF3	Protection in mice	Cabrera et al. (2009)

Continued

Table 1 Examples of preclinical and clinical vaccine development against infectious diseases.—cont'd

Disease	Expression system	Target	Response	Ref.
Parasitic				
Malaria	SFV-Pf332 VRPs, DNA	*P. falciparum* Pf332	Th1-biased immune responses in mice	Andersson et al. (2001)
Malaria	SIN-SYVPSAEQI-CS	*P. yoelii* CS epi	Protection against malaria in mice	Tsuji et al. (1998)
Leishmaniasis	SFV-PpSP15-LmSTI1	PpSP15-LmSTI1	Superior replicon RNA responses	Savar et al. (2021)

AHF, Argentine hemorrhagic fever; BHF, Bolivian hemorrhagic fever; BVDV, bovine viral diarrhea virus; CHIKF, Chikungunya fever; CHIKV, Chikungunya virus; CNE, cationic nanoemulsion; CS, circumsporozoite; DENV, Dengue virus; DF, Dengue fever; EBOV, Ebola virus; ED3, DENV envelope protein domain III; EVD, Ebola virus disease; GPC. Glycoprotein; GPN, Gag-Pol-Nef; HB, Hepatitis B; HBcAg, Hepatitis B core antigen; HBsAg, Hepatitis B surface antigen; IAV, irreversibly attenuated VEE; ID-EP, intradermal administration + electroporation; IFVA, influenza virus A; IM-EP. Intramuscular injection + electroporation; JUNV, Junin virus; LASV, Lassa virus; LAV, live-attenuated VEE; LHF, Lassa hemorrhagic fever; LIV, louping ill virus; LNPs, lipid nanoparticles; mAb, monoclonal antibody; MACV, Machupo virus; MARV, Marburg virus; MHB, HBV middle surface envelope glycoprotein; MHF, Marburg hemorrhagic fever; MV, measles virus; MVA, Modified Vaccinia Ankara virus; nAbs, neutralizing antibodies; NP, nucleoprotein; PED, porcine epidemic diarrhea; PEDV, procine epidemic diarrhea virus; prME, premembrane-envelope protein of DENV; RSV, respiratory syncytial virus; saRNA, self-amplifying RNA; SFV. Semliki Forest virus; SIN, Sindbis virus; taRNA, trans-amplifying RNA; VEE, Venezuelan equine encephalitis virus; VLVs, virus-like vesicles; VRPs, viral replicon particles; YFV, yellow fever virus; ZIKV, Zika virus; Z-ME. ZIKV membrane-envelope polyprotein; ZVD, Zika virus disease.

virus (EEE), replication-deficient expression vectors have been engineered by the removal of the furin cleavage site between the E2 and E3 envelope proteins (Reed et al., 2014). Immunization of mice elicited strong neutralizing antibody responses and protection for both the combination of VEE, WEE and EEE particles and individual alphavirus particles. In macaques, robust antibody responses and protection were seen after immunization with the combined particles and the respective alphaviruses against VEE and EEE. In contrast, immunization with WEE particles did not provide strong protection against WEE challenges. In another approach, two VEE vaccine formulations have been designed based on live-attenuated VEE (LAV) or irreversibly attenuated VEE (IAV) combined with synthetic nanoemulsion (CNE) (Samsa et al., 2019). Both vaccine candidates are based on the TC-83 VEE genome although in the latter case the capsid gene has been removed, which produces a single cycle of RNA amplification but generates no virus progeny. The LAV-CNE vaccine candidate elicited strong immune responses and provided 100% protection against challenges with VEE in mice. Although the IAV-CNE induced immune responses the protection in mice was only partial. DNA based VEE vaccine candidates have been developed by engineering the V4020 vector based on the attenuated TC-83 strain, where the capsid gene is expressed from a second subgenomic promoter inserted downstream of the envelope protein genes (Tretyakova et al., 2019). A single dose of V4020 DNA showed no adverse reactions in BALB/c mice and elicited high neutralizing antibody titers. Moreover, immunized mice were protected against challenges with lethal doses of VEE. The VEE V4020 DNA vaccine was also evaluated in cynomolgus macaques (Tretyakova et al., 2020). Intramuscular immunization elicited robust neutralizing antibody responses and protected macaques after aerosol challenges with VEE. In a clinical setting, a VEE DNA vaccine candidate expressing the E3-E2-6K-E1 genes was subjected to a phase I clinical trial to assess the safety, reactogenicity, tolerability and immunogenicity of intramuscular (IM-EP) or intradermal (ID-EP) administration combined with electroporation (Hannaman et al., 2016. In the case of IM-EP injections doses of 0.5 or 2.0 mg of VEE DNA were administered and for ID-EP 0.08 or 0.3 mg of VEE DNA was applied. The vaccination caused no vaccine- or device-related serious adverse events and both routes were generally accepted. Neutralizing antibodies were detected in all individuals after both low and high dose of IM-EP administration. For ID-EP delivery, seven out of eight individuals (87.5%) receiving the high dose elicited neutralizing antibodies, whereas five out of eight (62.5%) showed it for the low dose.

3.2 Arenaviruses

Arenaviruses have also been targeted for vaccine development, especially Lassa virus (LASV) but also Junin virus (JUNV) and Machupo virus (MACV). In this context, a recombinant VSV vector expressing the Josiah strain LASV glycoprotein (GPC) was evaluated in two animal models (Safronetz et al., 2015). Complete protection against challenges with LASV isolates from Liberia, Mali and Nigeria was achieved in guinea pigs and macaques after immunization with 1×10^6 and 6×10^7 pfu of VSV-GPC, respectively. In the case of MV-based LASV vaccines, a single immunization of macaques with 6×10^6 pfu of MV-GPC provided protection (Mateo et al., 2019). However, the superior vaccine candidate was the MV vector expressing both LASV GPC and nucleoprotein (NP). In another study, MV-based expression of LASV GPC and NP was used for immunization of cynomolgus macaques, which provided protection against challenges with lethal doses of LASV (Mateo et al., 2021). Moreover, when another cohort of primates were challenged 1 year after immunization with the vaccine candidate, it was demonstrated that a long-lasting immune response was elicited as well as protection against diverse LASV strains. The MV-LASV vaccine candidate was subjected to two doses in a randomized, placebo-controlled, dose-finding, phase I trial in healthy volunteers to evaluate the safety, tolerability, and immunogenicity (NCT04055454). The replication-competent flavivirus YFV has also been applied for LASV vaccine development by introduction of the LASV GPC gene between the envelope (E) and non-structural protein 1 (NS1) genes (Bredenbeek et al., 2006). Although the YFV-LASV-GPC replicated poorly in guinea pigs, LASV-specific antibodies were elicited, and immunized animals were protected against LASV challenges. Moreover, introduction of the GP1 and GP2 subunits of the LASV-GPC enhanced the stability of recombinant viruses (Jiang et al., 2011). Immunization of mice elicited specific $CD8^+$ T-cell responses and vaccination of guinea pigs with YFV-LASV-GP1 and GP2 provided protection against challenges with LASV. Importantly, although good protection was obtained in rodents, prime-boost vaccination with the YFV-LASV-GPC vaccine candidate failed in marmosets demonstrating that success seen in rodents is not automatically reproducible in primates (Lukashevich and Pushko, 2016). Alphaviruses have also been used as vectors for LASV vaccines. Expression of LASV GPC or NP from an attenuated VEE strain resulted in protection of guinea pigs from LASV challenges (Pushko et al., 2001). In another approach, bicistronic VEE replicon particles were applied for the expression of wild-type LASV GPC (GPCwt) and a modified non-cleavable version with a C-terminal deletion (ΔGPfib) simultaneously from two subgenomic 26S promoters (Wang et al., 2018). The LASV- ΔGPfib was expressed as trimers, which accumulated in the endoplasmic reticulum (ER), inducing apoptosis and promotion of antigen cross-priming. Immunization of mice elicited robust immune responses and provided protection against challenges with LASV. In the context of JUNV and MACV, replication-restricted VEE particles based on the TC83 strain were engineered for co-expression of JUNV and MACV GPCs (Johnson et al., 2020). Immunization of guinea pigs with VEE-JUNV/MACV-GPC elicited robust humoral immune response and provided complete protection against challenges with both JUNV and MACV.

3.3 Filoviruses

Filoviruses, especially Ebola virus (EBOV), due to numerous outbreaks (Coltart et al., 2017), have been targeted for vaccine development. For example, the EBOV glycoprotein (GP) mutant D637L, showing superior cleavability and shedding compared to wild-type EBOV GP, was inserted into the KUN vector (Reynard et al., 2011). Evaluation in guinea pigs demonstrated dose-dependent protection against challenges with guinea pig-adapted EBOV. Moreover, administration of two subcutaneous doses of 1×10^9 KUN-GP/D637L virus-like particles (VLPs) protected three out of four non-human primates from EBOV challenges (Pyankov et al., 2015). VSV-based vectors have shown great promise related to EBOV vaccines. In this context, macaques were immunized with 5×10^7 VSV-EBOV-GP particles and challenged with the EBOV-Makona strain (Marzi et al. 2015). Complete protection was observed when the immunization took place 7 days prior to the challenge. However, if the VSV-EBOV-GP vaccine candidate was administered 3 days before the challenge, the protection was only partial. Moreover, ΔVSV vectors have been engineered, where the VSV G protein has been replaced by the EBOV Zaire strain (ZEBOV), the Sudan virus (SUDV) (SEBOV), a member of the EBOV family, and Marburg virus (MARV) GP, respectively (Geisbert and Feldmann, 2011). BALB/c mice immunized with ΔVSV-ZEBOV GP showed complete protection against ZEBOV challenges 28 days after immunization (Jones et al., 2007). Moreover, a single immunization of primates with ΔVSV-ZEBOV GP resulted in protection against ZEBOV challenges (Jones et al., 2005a, b). However, no cross-protection was established as back-challenges of ZEBOV-resistant primates with SEBOV did not provide protection. In the case of MARV, a single administration of ΔVSV-MARV-GP provided protection against challenges with MARV in cynomolgus macaques (Jones et al., 2005a, b). VEE vectors have also been applied for the expression of MARV-GP. Immunization with 1×10^7 pfu of VEE-MARV-GP particles protected macaques against challenges with MARV (Herbert et al., 2013). Similarly, VEE particles expressing SUDV-GP were intramuscularly administered to cynomolgus macaques resulting in protection against SUDV challenges (Herbert et al., 2013). Additionally, immunization of macaques with VEE-SUDV-GP provided partial protection against EBOV challenges, whereas co-administration of VEE-SUDV-GP and VEE-EBOV-GP resulted in protection against both SUDV and EBOV challenges (Herbert et al., 2013). In the context of clinical trials, two phase III clinical trials have been conducted. In the first trial on 7651 individuals, 4123 were immediately vaccinated with the VSV-ZEBOV vaccine candidate and 3528 received a delayed vaccination after 21 days in July 2015 in Guinea, west Africa (Henao-Restrepo et al., 2015). Based on interim results, 43 cases of serious adverse events were reported, of which one febrile episode was suggested to be causally associated with the vaccination. This case was properly resolved without consequences. No cases of EVD were found in the group of immediately vaccinated individuals and only 16 cases were detected in the group of delayed vaccination, suggesting a vaccine efficacy of 100%. In the other phase III trial conducted in Guinea and Sierra Leone in 2015, the vaccine efficacy was confirmed with no EVD cases among neither the 2119 immediately vaccinated subjects nor in the 2041 individuals receiving a 21-day delayed administration of the VSV-ZEBOV vaccine (Henao-Restrepo et al., 2017). A total of 80 cases of serious adverse events were recorded, of which one case of febrile reaction and another case of anaphylaxis were indicated to be associated with the vaccination. Both the FDA and the EMA have approved the VSV-ZEBOV vaccine (Maxmen, 2017). In a follow-up systematic data review on 17,600 adults and 234 children receiving the VSV-ZEBOV vaccine in phase I-III studies, a safety, immunogenicity and risk-benefit analysis indicated a good risk-benefit balance in adults, children, pregnant and lactating women, and HIV-infected individuals (Bache et al., 2020).

3.4 Flaviviruses

Flaviviruses have become important vaccine targets due to recent outbreaks of DENV (Guo et al., 2017) and ZIKV (Pielnaa et al., 2020). In the context of DENV, expression of the ectodomain of the DENV envelope protein from VEE vectors showed protective immunity in immunized mice (Khalil et al., 2014). In another approach, VEE replicon particles expressing the DENV type 1 premembrane and envelope proteins (prME) was evaluated in cynomolgus macaques (Chen et al., 2007). Animals were immunized with three doses of VEE-DENV-prME (VVV) and compared to three doses of a conventional plasmid DNA expressing the DENV-prME, and two doses of DNA followed by a third dose of VEE-DENV-prME (DDV). The highest DENV-specific IgG antibody responses and virus-neutralizing antibody titers were obtained after immunization with the DDV regimen. Macaques immunized with the DDV regimen showed complete protection against DENV challenges, while the DDD and VVV regimens managed to provide only reduced viremia. Application of MV vectors expressing the DENV envelope protein domain III (ED3) induced strong immune responses and provided partial protection against challenges with DENV in mice (Hu et al., 2016). In another approach, a chimeric live-attenuated flavivirus YFV-DENV tetravalent vaccine based on the YFV 17D backbone expressing four DENV prME serotypes was engineered (Guy et al., 2010). Subcutaneous administration of the YFV-DENV tetravalent vaccine was well tolerated and showed no toxicology in cynomolgus macaques (Ravel et al., 2017). The YFV-DENV tetravalent vaccine has been submitted to several clinical trials (Torresi et al. 2017). For example, the efficacy of YFV-DENV was evaluated in a proof-of-concept phase IIb trial in Thailand (Sabchareon et al., 2012). The vaccine was well tolerated showing no safety issues and the vaccine efficacy was 30.2%. A large phase III trial in South-east Asia showed a good safety profile and the primary endpoint was achieved with a 56.5% vaccine efficacy (Capeding et al., 2014). Another phase III trial was conducted in a total of 20,869 healthy children in Latin America (Villar et al., 2015). The vaccination safety was good, and the efficacy ranged from 42.3% to 77.7% depending on the DENV serotype. The vaccine efficacy against hospitalization for Dengue fever (DF) was 80.3%. In the context of ZIKV, a VEE replicon RNA expressing the optimized ZIKV prME was introduced into nanocarrier lipids (NLCs) to enhance RNA delivery (Erasmus et al., 2018). The strategy involved a two-vialed approach, which allowed manufacturing and storage of highly stable NLCs to which RNA can be added prior to immunization. This approach allowed to establish complete protection of C57BL/6 mice against ZIKV after immunization with a single dose of only 10 ng. VSV has also been applied for ZIKV vaccine development. Immunization of mice with

VSV-ZIKV-prME elicited strong ZIKV-specific antibody responses and protected mice against ZIKV challenges (Chattopadhyay et al., 2018). In another approach, immunization with a live MV-based vector expressing the ZIKV E gene (MV-E2) protected mice from challenges with the non-lethal ZIKV Asian strain PRVABC59 and the lethal African strain MR766 (Kurup et al., 2022). Despite the 100% survival of challenged mice, the viral clearance in the brain and reproductive tract was incomplete. It was addressed by co-administration of MV-ZIKV-NS1, which contributed to the clearance of ZIKV from the female reproductive tract providing fetal protection (Kurup et al., 2022). The MV-ZIKA (V186-001) vaccine candidate was subjected to a double-blinded, randomized, placebo-controlled, dose-finding, phase I trial in 48 healthy volunteers (NCT02996890). Currently, no results have been published. A novel vaccine formulation MV-ZIKA-RSP (V187-001) has been subjected to another phase I trial to evaluate safety and immunogenicity in healthy volunteers (NCT04033068).

3.5 Hepatocytic viruses

Hepatocytic viruses such as hepatitis B virus (HBV) have been the target of vaccine development resulting in approval of the prophylactic vaccines Engerix and Recombivax HB for HBV (Roldão et al., 2010). Self-replicating RNA virus vectors have also been applied, including MV vectors (Del Valle et al., 2007). Hepatitis B surface antigen (HBsAg) was expressed at different levels from three MV vectors, which all secreted HBsAg subviral particles and induced humoral responses in mice, however, with significantly different HBsAg antibody levels. Rhesus monkeys were immunized with the two MV vectors providing the highest HBsAg antibody levels showing low or undetectable viral loads. Only the MV vector showing the highest HBsAg antibody levels was able to partially protect monkeys from challenges with the pathogenic MV Davis 87 strain. In another approach, a vaccine platform for generating virus-like vesicles (VLVs) was established by introducing SFV RNA replicons into a VSV G-protein based envelope (Reynolds et al., 2015). Additionally, the HBV middle surface envelope glycoprotein (MHB) and the HBV core antigen (HBcAg) were introduced into SFV RNA replicons, which were evaluated in mice immunized with VLVs. A single immunization with VLV-MHB induced functional $CD8^+$ T-cell responses and resulted in protection against HBV challenges. In contrast, immunization with VLV-HBcAg neither elicited $CD8^+$ T-cell responses nor provided protection against HBV in mice.

3.6 Lentiviruses

Development of vaccines against the AIDS/HIV epidemic has obviously caught plenty of attention. For example, a live-attenuated MV vector with a HIV gp160 envelope was administered to mice, inducing strong cellular and humoral responses (Lorin et al., 2004). VSV-based vaccines have also been developed for HIV (Racine et al., 2017). An essential question has been the identification of the envelope glycoprotein (Env) which provides the best immunogen. The strategy for VSV-based expression has been to select an Env maximizing both antibodies for antibody-dependent cell-mediated cytotoxicity (ADCC) and for broad neutralization/inhibition of most HIV-1 strains (McCoy et al., 2016; Kong et al., 2016). Several approaches have been exercised for VSV-based HIV vaccines. For example, the pseudotyped VSV carrying HIV-1 Env spikes elicited anti-Env binding antibodies and cell-mediated immune responses in immunized mice (Rabinovich et al., 2014). Moreover, VSV-HIV-1 Env immunization provided 67% protection in non-human primates (Parks et al., 2013). Alphavirus recombinant particle-based HIV vaccines expressing HIV Env have proven to elicit superior antibody titers compared to plasmid DNA and recombinant Env-based vaccine candidates (Brand et al., 1998). Also, SIN replicon particles expressing the HIV-1 Env gene were assessed in a mouse model (Center et al., 2013). Env-specific antibody responses were detected after a boost immunization with Env protein. It was also demonstrated that SIN-HIV-1 Env priming elicited superior humoral responses in comparison to immunization with Env protein alone. In another approach, intramuscular injection of SFV-HIV Env replicon RNA elicited Env-specific immune responses in four out of five immunized mice and generated the 12H2 monoclonal antibody against gp41 (Giraud et al., 1999). Immunization with SFV-based recombinant particles and replicon RNA was compared for an SFV vector carrying the Indian HIV-1C Env-Gag-PolRT in mice (Ajbani et al., 2017). Significant HIV-specific immune responses were obtained for both individual and combination prime-boost regimens. The immunogenicity of vaccination with SFV particles was superior to RNA immunizations. In another approach, the HIV gp160 gene was expressed from a VEE DNA replicon, which was subjected to immunization of mice (Ljungberg et al., 2007). Compared to a conventional DNA vector, the VEE DNA replicon significantly enhanced the humoral immune responses. Moreover, the same immune responses seen for the conventional DNA immunization could be achieved with 10–100-fold lower doses of

DNA replicons. The DNA-launched VEE replicon also served efficiently in a prime-boost vaccination with VEE particles. In another study on DNA replicons, the HIV Env and the Gag-Pol-Nef fusion proteins were expressed from an SFV DNA vector (Knudsen et al., 2015). Especially, using the SFV DNA vector for prime vaccination followed by a heterologous boost with Modified Vaccinia Ankara (MVA) virus also expressing Env and the Gag-Pol-Nef fusion proteins elicited strong immune responses in BALB/c mice. Using as little as 0.2 μg of SFV DNA was sufficient for efficient priming. In the case of replicon RNA delivery, the stability of VEE-HIV-1 gp140 RNA was improved by cationic CNE formulations, which in comparative studies demonstrated that the immune responses in rhesus macaques were stronger after intramuscular administration of 50 μg of VEE RNA-CNE than after immunization with VEE particles or HIV gp140 protein combined with the MF59 adjuvant (Bogers et al., 2015). In the context of clinical trials, the safety and immunogenicity of replication-deficient VEE replicon particles expressing HIV-1 Gag (AVX101) were investigated in a double-blind, randomized, placebo-controlled trial in healthy volunteers in the US and South Africa, but the study was terminated because of instability issues with the vaccine (Wecker et al., 2020).

3.7 Influenza virus and orthopneumoviruses

Vaccine development against influenza viruses has also been conducted using self-replicating RNA viruses. For example, MV vectors expressing the HA gene from the influenza virus A/Sapporo/107/2013 (H1N1pdm) strain induced strong immune responses in immunized cotton rats and showed protection against challenges with lethal doses of influenza A virus (Ito et al., 2020). In the context of VSV, a pseudotyped VSV vector expressing the HA gene from the highly pathogenic avian influenza virus (HPAIV) A/Vietnam/1203/04 (VN1203) strain and the neuraminidase (NA) gene of the mouse adapted H1N1 influenza virus A/Puerto Rico/8/34 (PR8) provided protection in mice against challenges with H5N1 virus after a single immunization (Ryder et al., 2015). In another study the VSV-based H5N1 influenza virus vector was engineered as a platform for a fast-acting pan-H5 influenza virus vaccine involving multiple antigen designs for an optimized vaccine (Furuyama et al., 2020). A single dose of VSV expressing the full-length HA (HAfl) gene provided 100% protection against the homologous H5 clade 1 virus in mice. Moreover, cross-protective H5-specific antibodies and protection against challenges with various H5 clade 2 viruses were achieved after a single immunization of mice. In the context of alphaviruses, 1×10^7 VEE particles expressing the HA gene from the Hong Kong Influenza A isolate (A/HK/156/97) were administered to chickens (Schultz-Cherry et al., 2000). The immunization induced HA-specific antibody responses and protected the animals from challenges with influenza virus. In the case of RNA-based delivery, SFV replicon RNA expression of the influenza HA gene was evaluated in BALB/c mice (Fleeton et al., 2001). Moreover, immunization with SFV replicon RNA expressing the louping ill virus (LIV) (flavivirus) membrane-envelope protein (prME), and the respiratory syncytial virus (RSV) fusion envelope protein (F) was conducted. Immunization of mice with 10 μg SFV-HA RNA and SFV-LIV prME resulted in 90% and 70% protection, respectively. The RSV lung titers were reduced by 32% in mice immunized with SFV-RSV F RNA. In another study on RSV F, VEE-RSV F RNA was encapsulated in lipid nanoparticles (LNPs), which significantly enhanced the immunogenicity (Geall et al., 2012). In comparison to naked RNA, 0.1 μg VEE-RSV F LNP-RNA was significantly more immunogenic than 1 μg of naked RNA. The F-specific IgG response to 1 μg of VEE-RSV F LNP-RNA was equivalent to administration of 1×10^6 IU of VEE-RSV F particles. A single immunization with doses as low as 0.01 μg established seroconversion in mice in comparison to the required two immunizations of 1 μg of naked RNA. Protection against challenges with RSV was conducted in a cotton rat intranasal RSV challenge model, comparing VEE-RSV F as LNP-RNA, naked RNA, recombinant VEE particles, and an RSV F subunit vaccine (Prince et al., 1985). Although all vaccine candidates provided protection against nasal RSV challenges, the LNP-RNA elicited approximately 8-fold higher F-specific IgG titers and 10-fold higher neutralization titers for the same dose of naked RNA. Moreover, the LNP-RNA vaccine candidate was statistically superior to naked RNA and equivalent to immunization with 5×10^6 IU of VEE-RSV F particles.

An indication of the efficacy of self-replicating replicon RNA for vaccination was confirmed in a comparative study, where only 1.25 μg of VEE-HA RNA provided protection against influenza virus challenges compared to 80 μg of synthetic mRNA (Vogel et al., 2018). Furthermore, an interesting approach is the engineering of a bipartite SFV-based system using trans-amplifying RNA (taRNA) (Beissert et al., 2020). In this context, the gene of interest is provided as a trans-replicon from a self-amplifying SFV RNA vector from which the nsP1-4 genes have been deleted. Either a standard self-replicating RNA or an optimized non-replicating RNA (nrRNA) is engineered to provide the replicase activity. The nrRNA was demonstrated to deliver 10–100-fold higher transgene expression, which most likely could be attributed to its higher translation efficacy and lack of interference with cellular translation (Beissert et al., 2020). Application of the taRNA system for SFV-HA RNA, immunization of mice with doses as low as 50 ng, induced neutralizing antibodies and provided protection against challenges with influenza virus. In another approach, the 23-residue external domain of the influenza virus M2 protein (M2e) was inserted as an N-terminal extension of the E2 membrane protein in a SIN vector (Krishnavajhala et al., 2018). Intranasal immunization of mice with SIN-E2S1-M2e particles provided protection against virulent influenza A virus challenges. The Classic Swine Fever virus (CSFV), a member of flaviviruses, has also been engineered as a vaccine vector targeting influenza virus (Démoulins et al., 2021). Both CSFV particles and polyethylenimine (PEI) replicon RNA complexes expressing the influenza virus nucleoprotein (NP) and HA genes have been used for immunization studies in pigs. Superior HA- and NP-specific humoral immune responses were obtained after immunization with CSFV-HA/NP particles although CSFV-HA/NP PEI-RNA also generated strong immune responses. In contrast, naked CSFV-HA/NP RNA elicited only weak immunogenicity. In the context of clinical trials using self-replicating RNA viruses, two phase I/II trials have been conducted for VEE particles expressing the influenza virus HA gene (Sebastian and Lambe, 2018). In a double-blind Phase I/II study in 18–40-year-old healthy volunteers. 80% and 86% of vaccinated individuals generated robust antibody titers after one and two immunizations, respectively (NCT00440362). The antigen T-cell response remained elevated for more than 4 months. In another phase I/II trial in volunteers aged 65 years or older, the immunogenicity of the VEE-HA (AVX-502) vaccine was poorer with only 50% of vaccinated individuals showing a 4-fold increase in antibody titers even after two immunizations (NCT00706732).

3.8 Coronaviruses

Vaccine development against coronaviruses has certainly seen a major boost due to the current COVID-19 pandemic. In addition to the nucleic acid- and adenovirus-based COVID-19 vaccines receiving Emergency Use Approval (EUA) (Lythgoe and Middleton 2021) and approval of the BNT162b2 mRNA by the FDA in August 2021 (Lamb, 2021), self-replicating RNA virus vectors have been investigated as delivery vehicles for SARS-CoV-2 and other coronaviruses. The prior SARS and Middle East respiratory syndrome (MERS) epidemics had already started vaccine development against coronaviruses. For example, VEE particles expressing the SARS-CoV Spike (S) protein were administered to mice, which were protected against challenges with SARS-CoV (Sheahan et al., 2011). Similarly for MERS a vaccine candidate has been designed applying the VSVΔG vector, where the VSV G protein was replaced by the MERS-CoV S protein (Liu et al., 2018). Intranasal immunization of rhesus macaques with VSVΔG-MERS S particles induced robust neutralizing antibody and T-cell responses. Related to animal coronaviruses, the bovine viral diarrhea virus (BVDV), belonging to flaviviruses, has been engineered for the expression of a fragment of the spike antigen of the porcine epidemic diarrhea virus (PEDV) (Tao et al., 2020). The chimeric BVDV-PEDV S vector elicited BVDV- and PEDV-specific antibody responses in immunized BALB/c mice.

In the context of SARS-CoV-2, several self-replicating RNA virus vectors have been applied for vaccine development. Expression of SARS-CoV-2 S from an MV vector showed strong Th1-biased antibody and T-cell responses in immunized mice (Hörner et al., 2020). Furthermore, the MV-SARS-CoV-2 S vaccine candidate has been subjected to a phase I clinical trial under the name TMV-083 (NCT04497298). However, the weak immune responses detected in vaccinated volunteers led to an early termination of the trial. Expression of the SARS-CoV-2 S protein from a replication-competent VSV vector elicited neutralizing antibody responses in immunized mice (Case et al., 2020). Furthermore, immunized mice were protected against SARS-CoV-2-related pathogenesis. The VSV-SARS-CoV-2 vaccine candidate named V590 was subjected to a phase I clinical trial in healthy volunteers (NCT04569786). Despite showing good safety and tolerability profiles, the immune responses were weaker than those detected in recovered COVID-19 patients, resulting in the discontinuation of the study. The replication-competent VSVΔG vector has also been applied for the expression of the SARS-CoV-2 S protein (Yahalom-Ronen et al., 2020). Strong neutralizing antibody responses and protection against SARS-CoV-2 were achieved in Syrian golden hamsters after a single immunization with 5×10^6 VSVΔG-SARS-CoV-2 S particles. A phase I/II study is in progress for 18–55-year-old healthy volunteers receiving a single escalating dose of 5×10^5, 5×10^6, and 5×10^7 pfu of VSVΔG-SARS-CoV-2 S particles, respectively (NCT04608305). In the second part, phase II, the study will be expanded to include elderly subjects receiving either one or two injections of low, medium, or high doses. Furthermore, a phase

IIb/III trial has been approved with the purpose to document the non-inferiority of VSVΔG-SARS-CoV-2 S with an already approved COVID-19 vaccine (NCT04990466).

Self-replicating RNA viruses have also been evaluated as RNA-based COVID-19 vaccine candidates. The efficacy of RNA-based delivery of RNA replicons has been hampered by the sensitivity of RNA degradation, which has led to the engineering of various LNP formulations. For example, VEE replicon RNA expressing the prefusion-stabilized SARS-CoV-2 S gene encapsulated in LNPs, showed robust and dose-dependent SARS-CoV-2 S-specific antibody responses in intramuscularly immunized BALB/c mice (McKay et al., 2020). The antibody titers were superior to those detected in convalescent COVID-19 patients. In another approach, a VEE replicon RNA vaccine candidate expressing the SARS-CoV-2 S protein (repRNA-CoV2S) was encapsulated in lipid inorganic nanoparticles (LIONs) (Erasmus et al., 2020). A single intramuscular immunization with the LION-repRNA-CoV2S vaccine elicited robust Th1-biased immune responses in mice. A single administration or a prime-boost regimen in pig-tail macaques resulted in modest T-cell and robust antibody responses, which lasted for at least 70 days and neutralized SARS-CoV-2. A phase I first-in-human dose-ranging study with the vaccine candidate named LNP-nCoV saRNA was conducted in healthy 18–45-year-old volunteers with no previous history of COVID-19 (Pollock et al., 2021). In the open-label dose-escalation study, two intramuscular injections at six doses from 0.1 to 10 μg were given. The vaccination was well tolerated, and no serious adverse events occurred. The seroconversion was related to the dose, ranging from 8% to 61% (ELISA) and 46% to 87% (immunoblot), respectively. Patient sera demonstrated neutralization of SARS-CoV-2 in 15–48% of the subjects depending on the dose level. Although immunogenicity was achieved with low dose levels, optimization of humoral responses is desired to improve the vaccine efficacy. Another approach of self-replicating RNA virus-based vaccines relates to the VEE-based RNA (STARR)-based LUNAR-COV19 vaccine (De Alwis et al., 2021). The encapsulated VEE replicon RNA contains the full-length SARS-CoV-2 S protein. Immunization of BALB/c mice with single doses as low as 2 and 10 μg of the vaccine candidate induced strong neutralizing antibody responses and provided protection against SARS-CoV-2 challenges in transgenic humanized ACE2 mice. The SIN vector has also been engineered for the expression of the SARS-CoV-2 S protein (Scaglione et al., 2021). In combination with the OX40 immunostimulatory antibody (αOX40)

SIN-SARS-CoV-2 S particles elicited long-lasting neutralizing antibody and T-cell responses in immunized mice. Moreover, the immunization provided strong immune protection against challenges with SARS-CoV-2 in humanized transgenic ACE2 mice. Finally, a heterologous vaccination regimen was evaluated in mice for the self-amplifying RNA vaccine candidate LNP-nCoVsaRNA and the ChAdOx1 nCoV-19 vaccine (Spencer et al., 2021). Antibody responses were stronger after a two-dose heterologous immunization than after a single-dose administration. Moreover, neutralizing antibody titers were similar or higher for heterologous vaccination compared to homologous prime-boost immunizations with ChAdOx1 nCov-19. Most importantly, superior responses were seen for the heterologous approach due the domination of cytotoxic T-cells and Th1 $CD4^+$ T-cells.

3.9 Bacterial and parasitic targets

Not only viruses have been targets for vaccine development applying self-replicating RNA viruses. Bacterial infections and toxins have also been addressed. In this context, the protective antigen (PA) of *Bacillus anthracis* has been expressed from SIN vectors (Thomas et al., 2009). PA-specific IgG and neutralizing antibody responses were detected in Swiss Webster mice immunized with SIN-PA particles. Moreover, intramuscular immunization offered some protection against challenges with a lethal Ames strain and synergized protection was obtained with ciprofloxacin treatment. In another study, SFV particles and SFV replicon DNA were compared to a conventional DNA plasmid for the expression of the *B. anthracis* PA (Wang et al., 2015). Both the SFV particles and replicon DNA elicited stronger PA-specific immune responses in immunized mice compared to conventional plasmid DNA. Moreover, potent protection against challenges with the *B. anthracis* strain A16R was achieved. In another approach, BALB/c mice were immunized with SFV particles expressing the *Brucella abortus* translation initiation factor 3 (IF3), which induced IF3-specific IgM antibody and T-cell proliferative responses (Cabrera et al., 2009). Moreover, the immunization led to protection of mice against *B. abortus* challenges.

Parasites have also been targeted for the development of self-replicating RNA-based vaccines. For example, BALB/c mice were immunized with SFV particles and DNA replicons expressing the *Plasmodium falciparum* Pf332 antigen in comparison to conventional plasmid DNA (Andersson et al., 2001). Surprisingly, the highest immune responses were detected for immunization with the conventional DNA plasmid. However, plasmid

DNA immunizations elicited Th2 type-biased immune responses, whereas the SFV particle responses were biased toward Th1 type responses. Additionally, SIN vectors have been applied for vaccine development against malaria. In this context, the *Plasmodium yoelii* circumsporozoite (CS) protein class I major histocompatibility complex-restricted-9-mer epitope SYVPSAEQI was introduced into a SIN vector named SIN-Mal (Tsuji et al., 1998). SIN-Mal particles were administered subcutaneously into BALB/c mice, which elicited strong epitope-specific $CD8^+$ T-cell responses and provided protection against malaria. In addition to malaria, leishmaniasis has also been targeted. In this context, a fusion protein of the *Plebotomus papatasi* SP15 (PpSP15) and the *Leishmania major* stress inducible protein 1 (LmSTI1) was introduced into the SFV vector (Savar et al., 2021). The expression of PpSP15-LmSTI1 was compared for SFV replicon RNA and DNA with a conventional DNA plasmid. SFV replicon RNA-based expression was significantly higher than levels seen from replicon DNA- and conventional DNA plasmid-based expression. However, these findings need to be confirmed in preclinical animal models and clinical trials.

4. Self-replicating RNA for cancer therapy

Self-replicating RNA virus vectors have also found a large number of applications in cancer therapy not only for overexpression of tumor-associated antigens (TAAs) for vaccine development, but also immunotherapy based on expression of immunostimulatory genes. Moreover, delivery of cytotoxic and anti-tumor genes presents means of killing tumor cells. Virotherapy based on oncolytic viruses specifically replicating in and killing of tumor cells while not affecting normal tissue has shown great promise (Fukuhara et al., 2016). Additionally, induction of apoptosis, especially by alphaviruses is a feature to take advantage of in cancer therapy (Olkkonen et al., 1993). Examples of applications of self-replicating RNA viruses in cancer therapy are described below and summarized in Table 2.

4.1 Brain cancer

Self-replicating RNA viruses have frequently been evaluated for therapeutic interventions of brain tumors, especially glioblastomas. For example, SFV particles expressing endostatin (SFV-Endostatin) showed enhanced inhibition of tumor growth and reduced intratumoral vascularization compared to SFV-LacZ and a retrovirus vector in a B16 mouse brain tumor model

Table 2 Examples of preclinical and clinical cancer therapy.

Cancer	tExpression system	Target	Response	Ref.
Brain				
GBM	SFV-Endostatin	Endostatin	Superior to retrovirus, SFV-LacZ	Yamanaka et al. (2001)
GBM	SFV-IL-18 in DCs	IL-18 + IL-12	Anti-tumor, protective immunity	Yamanaka et al. (2003)
RG2	SFV-IL-12	IL-12	Dose-dependent tumor reduction	Roche et al. (2010)
B16	SIN-gp100/IL-12 DNA	gp100, IL-12	Synergistic therapeutic effect	Yamanaka and Xanthopoulos (2005)
GBM	SFV VA-GFP	Oncolytic SFV	Oncolytic activity, prolonged survival	Heikkilä et al. (2010)
CT-2A	SFV4-miRT124	Oncolytic SFV	Less neurotropism, tumor inhibition	Martikainen et al. (2015)
GBM	ZIKV	Oncolytic ZIKV	Potential treatment of GBM	Su and Balasubramaniam (2019)
GBM	VSVΔG-CHIKV Env	CHIKV Env	Infection, killing of tumor cells	Zhang et al. (2018)
ICM	VSVΔG-CHIKV Env	CHIKV Env	Tumor targeting	Zhang et al. (2018)
G62	Semi-repl. VSV	VSVΔG/VSVΔL	Long-term tumor regression	Muik et al. (2012)
GL261	GLESS-FAST-VSV	FAST	Significantly improved survival	Jiang et al. (2021)
RG5	GLESS-FAST-VSV	FAST	Significantly improved survival	Jiang et al. (2021)
GBM	GSC-oMV	GFP, CEA, NIS	Prolonged survival in mice	Allen et al. (2013)

Continued

Table 2 Examples of preclinical and clinical cancer therapy.—cont'd

Cancer	tExpression system	Target	Response	Ref.
Breast				
A2L2/HER2	SIN-HER2/neu DNA	HER2/neu	Inhibition of tumor growth in mice	Wang et al. (2005b)
A2L2/HER2	SIN-HER2/neu DNA + Ad-neu	HER2/neu	Significantly prolonged survival	Wang et al. (2005b)
A2L2/HER2	SIN-HER2/neu DNA	HER2/neu	80% less DNA required	Lachman et al. (2001)
A2L2/HER2	VEE-HER2-ECD/TM	HER2-ECD/TM	Complete protection against tumors	Wang et al. (2005a)
HER2	VEE-HER2-ECD/TM	HER2-ECD/TM	Phase I: PR and SD	NCT03632941
HER2	VEE-HER2-ECD/TM	HER2-ECD/TM	Phase II: Study in progress	Crosby et al. (2019)
4T1	SFV-IL-12 +LVR01	IL-12 +LVR01	Prolonged survival in 90% of mice	Kramer et al. (2015)
TNBC	M1	oM1 + Dox	Superior tumor activity with Dox	Zhang et al. (2021)
4T1	VSV(M51R)-LacZ	oVSV	Targeting metastases, better survival	Ebert et al. (2005)
TNBC	VSV(M51R)/TNBC ICV	VSV ICV	Enhanced NK and $CD8^+$ T-cell activity	Niavarani et al. (2020)
MCF7	MV Edm	oMV	No radio- or chemoresistance	Yang et al. (2020)
MCF7	rMV-SLAMblind	SLAMblind	Superior oncolytic activity	Sugiyama et al. (2013)
Cervical				
HPV	VEE-HPV16-E7	HPV16-E7	Protection against tumor challenges	Velders et al. (2001)
HPV	SIN AR339	oSIN	Strong tumor regression	Unno et al. (2005)
HPV	SFVenh-HPV E6/E7	HPV E6-E7	Eradication of established tumors	Daemen et al. (2003)

HPV	SFV-sHELP-E7SH	HPV-E6-E7	Tumor regression, protection in mice	Ip et al. (2015)
HPV	SFV HPV E-6-E7 DNA	HPV E6-E7	85% Tumor-free mice with 0.05 μg DNA	Van de Wall et al. (2018)
HPV	SFVenh-HPV-E6/E7(Vvax001)	HPV E6-E7	Phase I: safe, immunogenicity	Komdeur et al. (2021)
HPV	MV-HPV-L1	HPV-L1	Robust humoral immune response	Cantarella et al. (2009)
HPV	MV-HPV-L1 +rec L1	HPV-L1	Neutralizing HPV-specific Abs	Gupta et al. (2016)
CRPV	VSV-CRPV E1/E2/E6/E7	CRPV-E1-E7	Tumor volume reduced 96.9%	Brandsma et al. (2007)
HPV	VSV-HPV E7	HPV E7	Reduced tumor volume	Liao et al. (2008)
Colon				
MC38	SFV-IL-12/SFVeng-IL-12	IL-12	Complete tumor regression	Rodriguez-Madoz et al. (2005)
MC38-CEA	VEE-IL-12/VEE-CEA	IL-12, CEA	Combination: superior immunogenicity	Osada et al. (2012)
CT26	SFV-VEGFR-2	VEGFR-2	Inhibition of tumors and metastases	Lyons et al. (2007)
CT26	SFV-LacZ RNA	LacZ	Tumor regression, prolonged survival	Ying et al. (1999)
CRC	VEE-CEA	CEA	Phase I: long-term survival	Crosby et al. (2020)
CT26	KUN-GM-CSF	GM-CSF	Tumor regression, cure of 50% of mice	Hoang-Le et al. (2009)
MC38-CEA	MV-GM-CSF	GM-CSF	Complete remission in one third of mice	Grossardt et al. (2013)
PC	Maraba MG1-IL-12	IL-12	Reduced tumor burden, better survival	Alkayyal et al. (2017)
CRC	VSV-NDV/F(L289A)	oVSV-NDV/F	Long-term survival in rats	Yamaki et al. (2013)

Continued

Table 2 Examples of preclinical and clinical cancer therapy.—cont'd

Cancer	tExpression system	Target	Response	Ref.
Lung				
H358a	SFV-EGFP	EGFP	Complete tumor regression in 3/7 mice	Murphy et al. (2000)
A549	SFV VA-GFP	Oncolytic SFV	Superior survival compared to Ad	Määttä et al. (2008)
CT26.CL25	SIN-LacZ	Oncolytic SIN	Tumor regression, prolonged survival	Granot et al. (2014)(
LLC	MV-Hu191	Oncolytic MV	Prolonged survival in mice	Zhao et al. (2013)
Adenocarcinoma	MV Schwarz	Oncolytic MV	Inhibition of tumor growth	Boisgerault et al. (2013)
A549, H2009	MV Edmonston CEA	CEA	Tumor regression in mice	Patel et al. (2014)
A549, H2009	VSV-IFNβ	IFNβ	Tumor growth inhibition	Patel et al. (2015)
LM2	VSV-IFNβ	IFNβ	Tumor regression, cure of 30% mice	Patel et al. (2015)
Melanoma				
B16	VEE-TRP-2	TRP-2	Prolonged survival in mice	Avogadri et al. (2010)
B16	VEE-TRP2	TRP-2 +GITR	Complete tumor regression 90% of mice	Avogadri et al. (2014)
B16	VEE-TRP-2	TRP-2 + CTLA-4	Complete tumor regression 50% of mice	Avogadri et al. (2014)
B16	SFV-VEGFR-2/IL-12 DNA + SFV-Survivin/β-hCG Ag DNA	VEGFR-2/IL12 + Surviving/β-hCG	Tumor growth inhibition, prolonged survival, superior combination	Yin et al. (2015)
Mel III-IV	LSFV-IL-12	IL-12	Phase I: good safety. 10-fold IL-12	Ren et al. (2003)
B16-OVA	KUN-GM-CSF	GM-CSF	Tumor regression, cure of >50% of mice	Hoang-Le et al. (2009)
B16-OVA	YFV-SIINFEKL	SIINFEKL epit.	Protection in mice	McAllister et al. (2000)
Mel Z	MV L-16	oMV	Inhibition of tumor growth	Ammour et al. (2020)
B16-OVA	VSV-LCMP	oVSV	Prolonged survival in mice	Kimpel et al. (2018)

Ovarian				
OMC	SIN AR339	oSIN	Suppression of ascites formation in mice	Unno et al. (2005)
ES2	SIN-IL-12	IL-12 + irinotecan	Long-term survival in mice	Granot and Meruelo (2012)
MOSEC	SFV-OVA	OVA + VV-OVA	Enhanced anti-tumor activity in mice	Zhang et al. (2010)
A2780	VSV-LCMV-GP	LCMV-GP + Rux	Enhanced tumor regression in mice	Dold et al. (2016)
SKOV3.ip.1	MV-αFR	scFv-αFR	Reduced tumors, prolonged survival	Hasegawa et al. (2006a)
SKOV3.ip.1	MV.CEA/MV-NIS	CEA + NIS	Superior therapeutic effect	Hasegawa et al. (2006b)
RROC	MV-CEA	CEA	Phase I: Safe, SD in 66% of patients	Galanis et al. (2010)
Pancreatic				
LAPC	M1	oM1 + IRE	Tumor inhibition, prolonged survival	Sun et al. (2021)
MePC	VEE-CEA	CEA	Phase I. Prolonged overall survival	Morse et al. (2010)
PADC	VSV	oVSV	Superior to Sendai virus and RSV	Murphy et al. (2012)
PADC	VSV-GFP, VSV-ΔM51-GFP	oVSV + Gem	Superior oncolytic activity with Gem	Hastle et al. (2013)
KLM1	MV SLAMBlind	SLAMBlind	Reduced tumor growth in mice	Awano et al. (2016)
Capan-2	MV SLAMBlind	SLAMBlind	Reduced tumor growth in mice	Awano et al. (2016)

Continued

Table 2 Examples of preclinical and clinical cancer therapy.—cont'd

Cancer	tExpression system	Target	Response	Ref.
Prostate				
TRAMP	VEE-PSMA	PSMA	Th-1 biased, CTL activity, IgG2a/2b	Durso et al. (2007)
TRAMP-C	VEE-mSTEAP	mSTEAP + DNA	Immune response, tumor protection	Garcia-Hernandez et al. (2007)
TRAMP	VEE-PSCA	PSCA	>12 months survival in 90% of mice	Garcia-Hernandez et al. (2008)
CRPC	VEE-PSMA	PSMA	Phase I. Safe, but weak immunogenicity	Slovin et al. (2013)
PC-3	MV-CEA	CEA	Tumor regression, prolonged survival	Msaouel et al. (2009)
PC-3	oMV + MuV	oMV + MuV	Superior combination of oMV and MuV	Son et al. (2018)
DU145, PC-3	VSV-ΔM51-GFP	oVSV	Apoptosis, killing of tumor cells	Zhao et al. (2014)
DU145	VSV-ΔM51-GFP	oVSV	Prolonged survival in mice	Zhao et al. (2014)
Du145, 22Rv1	VSV-LCMV-GP	LCMV-GP	Long-term remission in mice	Urbiola et al. (2018)

CEA, carcinoembryonic antigen; CHIKV, Chikungunya virus; CRC, colorectal cancer; CRPC, castration resistant metastatic prostate cancer; CTLA-4, CTL antigen-4 monoclonal antibody; Dox, doxorubicin; GBM, glioblastoma multiforme; GITR, glucocorticoid-induced tumor necrosis factor receptor monoclonal antibody; GM-CSF, granular macrophage-colony stimulating factor; ICM, intracranial melanoma; ICV, infected cell vaccine; IFNβ, interferon-β; IRE, irreversible electroporation; KUN, Kunjin virus; LAPC, locally advanced pancreatic cancer; LLC, Lewis lung carcinoma; LSFV, liposome encapsulated SFV; MIII-IV, stage III-IV melanoma; MePC, metastatic prostate cancer; MOSEC, murine ovarian surface epithelial carcinoma; MV, measles virus; MuV, mumps virus; mSTEAP, mouse six-transmembrane epithelial antigen of the prostate; OMC, ovarian metastatic carcinoma; PC, peritoneal carcinomatosis; PSCA, prostate stem cell antigen; PSMA, prostate-specific membrane antigen; RROC, refractory recurrent ovarian cancer; Rux, ruxolitinib; SFV. Semliki Forest virus; SIN, Sindbis virus; TNBC, triple-negative breast cancer; TRAMP, transgenic adenocarcinoma of the prostate; TRP-2, tyrosine-related protein-2; VEE, Venezuelan equine encephalitis virus; VEGFR-2, vascular endothelial growth factor receptor-2; VV, vaccinia virus; YFV, yellow fever virus; ZIKV, Zika virus.

(Yamanaka et al., 2001). In another approach, dendritic cells (DCs) transduced with SFV-IL-18 particles were administered together with IL-12 protein to mice implanted with B16 brain tumors (Yamanaka et al., 2003). The combination therapy was more efficient than systemic administration of IL-12 alone related to Th1 responses, anti-tumor activity, and protective immunity. Dose-dependent therapeutic effects were demonstrated for SFV-IL-12 particles, where 5×10^7 SFV particles reduced the tumor volume by 70% compared to 87% with a dose of 5×10^8 particles in a syngeneic RG2 rat glioma model (Roche et al., 2010). In the context of DNA replicons, expression of the human gp100 and mouse IL-18 from SIN replicons was evaluated in mice implanted with B16-gp100 brain tumors (Yamanaka and Xanthopoulos, 2005). Intramuscular co-administration of SIN-gp100 and SIN-IL-18 DNA replicons resulted in both therapeutic activity and prolonged survival in mice. Replication-proficient alphaviruses have also been evaluated for brain tumors. Intravenous administration of oncolytic SFV VA-GFP particles to BALB/c mice with subcutaneous orthotopic tumors completely inhibited stable expression of firefly luciferase (Luc) (Heikkilä et al., 2010). The treatment also resulted in long-term survival in 16 out of 17 immunized mice. However, the use of replication-proficient particles for brain cancers has raised some concerns due to their neurotropism (Griffin 2016). In one approach it was demonstrated that intravenous administration of SFV-Luc at a dose of 6×10^9 particles/mL resulted in a broad distribution of Luc in the brain (Vasilevska et al., 2012). However, a lower dose of 2×10^8 particles/mL generated predominantly tumor targeting. In another approach, neuron-specific micro-RNA (miRT124) sequences introduced into the replication-proficient SFV4 vector resulted in significant inhibition of tumor growth and prolonged survival of immunized C57BL/6 mice carrying CT-2A orthotopic gliomas (Martikainen et al., 2015). Among flaviviruses, oncolytic ZIKV vectors have demonstrated their potential in cancer therapy (Su and Balasubramaniam, 2019). One concern of using ZIKV vectors has been the large percentage of the population showing immunity against ZIKV, which will reduce the therapeutic efficacy. Therefore, ZIKV vector-based therapy might only be used for glioblastoma multiforme (GBM) patients tested negative for ZIKV. In the context of VSV, a chimeric vector with the CHIKV E3-E2-6K-E2 replacing the VSV G protein, selectively infected tumor cells (Zhang et al., 2018). Immunization of mice with implanted tumors resulted in prolonged survival for more than 100 days. Additionally, immunization of mice with intracranial melanoma xenografts showed tumor killing with only minor effect on normal cells. Semi-replication-competent VSV systems have also been engineered by applying two trans-complementing, propagation-deficient VSV vectors (Muik et al., 2012). The VSV polymerase proteins P- and L-based deletion mutants VSVΔP and VSVΔL, co-propagated *in vitro* and *in vivo* without recombination to replication competent VSV, efficiently lysed human glioblastoma cells and subcutaneous mouse xenografts resulting in long-term tumor regression *in vivo*. Moreover, injection of the VSVΔG/VSVΔL combination into the brain did not cause neurotoxicity. In another approach, a G protein-less VSV vector containing the fusion-associated small transmembrane (FAST) gene was evaluated in F344 rats and C57BL/6 mice implanted with RG2 and GL261 tumors, respectively (Jiang et al., 2021). Intratumoral injections of GLESS-FAST-VSV was well tolerated and significantly prolonged the survival compared to control animals. Also, MV vectors have been applied for brain tumors. Expression of GFP, carcinoembryonic antigen (CEA) and sodium iodide symporter (NIS) from an oncolytic MV vector resulted in replication and cytotoxic effects in glioma stem cells (GSC) (Allen et al., 2013). Moreover, implantation of GSCs transduced with MV-GFP into the right caudate nucleus provided a significant prolongation of survival in nude mice.

4.2 Breast cancer

In the context of breast cancer, alphavirus vectors have been applied as DNA replicons, recombinant viral particles, and oncolytic viruses. Tumor growth inhibition by SIN DNA replicons expressing the HER2/neu gene was observed when the immunization of mice took place prior to challenges with A2L2 breast cancer cells (Wang et al., 2005b). However, the inhibition was inefficient if the immunization took place after the tumor challenge. In a prime-boost strategy with SIN-HER2/neu DNA replicons followed by adenovirus-based neu expression the survival of mice was significantly prolonged (Wang et al., 2005b). Interestingly, the superiority of DNA replicons to conventional DNA plasmids was demonstrated by the requirement of 80% less SIN-HER2/neu DNA to elicit strong antibody responses and protection against challenges in BALB/c mice (Lachman et al., 2001). Several applications of VEE-based breast cancer therapy have been conducted. For example, VEE particles expressing the extracellular domain (ECD) and transmembrane (TM) domains of HER2 administered to mammary tissue or intravenously, inhibited growth of HER2/neu tumors in mice (Wang et al., 2005a). Mice immunized with VEE-HER2-ECD/TM particles were completely protected against tumor challenges. Furthermore, VEE-HER2-ECD/TM particles showed good tolerance, partial response (PR), and stable disease (SD) in stage IV HER2 overexpressing breast cancer patients in a phase I clinical trial (NCT03632941). Moreover, a phase II trial with VEE-HER2-ECD/TM particles in combination with pembrolizumab in HER2-positive breast cancer patients has been initiated (Crosby et al., 2019). Combination therapy of SFV-IL-12 particles and the *Salmonella typhimurium* aroC strain (LVR01) was evaluated in mice implanted with 4T1 mammary tumors (Kramer et al., 2015). Immunization with 2×10^8 SFV-IL-12 particles and LVR01 provided complete inhibition of lethal lung metastasis formation and long-term survival in 90% of BALB/c mice. Among oncolytic alphaviruses, the M1 virus was subjected to the treatment of triple-negative breast cancer (TNBC) (Zhang et al., 2021). Co-administration of M1 particles and doxorubicin enhanced the oncolytic effect of M1 and significantly reduced tumor growth in mice. Related to VSV, pseudotyped VSV vectors replacing the VSV G protein with SIN surface glycoprotein (Sindbis-ZZ) containing the Fc-binding domain of *Staphylococcus aureus* protein A, severely reduced native binding function showing 50-fold titer increase when conjugated with the 4D5 mouse mAb directed against the HER2/neu receptor (Bergman et al., 2003). The VSV-Sindbis-ZZ particles showed 36-fold higher titers on human MCF7/H2 breast cancer cells. In the context of metastatic breast cancer, an oncolytic VSV vector with the M51R mutation in the matrix (M) protein gene expressing the LacZ gene, was intravenously administered to BALB/c mice with syngeneic 4T1 tumors (Ebert et al., 2005). After repeated tail vein injections, VSV(M51R)-LacZ showed infection of multiple breast cancer lesions in the lungs, no cytotoxicity, and significantly prolonged survival of mice. In an infected cell vaccine (ICV) approach, oncolytic VSV(M51R) particles were *ex vivo* transduced into TNBC cells, which enhanced natural killer and $CD8^+$ T-cell functionality as a promising approach for breast cancer (Niavarani et al., 2020). In the case of MV, combining doxorubicin and ionizing radiation with the MV Edmonston strain showed a synergistic therapeutic effect and overcame the radio- or chemotherapy resistance observed for breast cancer (Yang et al., 2020). In another approach the rMV-SLAMblind vector was engineered to prevent using the signaling lymphocyte activation molecule (SLAM) (Sugiyama et al., 2013). The rMV-SLAMBlind vector demonstrated efficient infection of breast cancer cells and superior oncolytic activity in primates.

4.3 Cervical cancer

Cervical cancer has been the target for several studies with self-replicating RNA viruses. For example, VEE particles expressing the human papilloma virus (HPV) E7 protein induced strong $CD8^+$ T-cell responses and provided protection against tumor challenges in immunized C57BL/6 mice (Velders et al., 2001). The oncolytic SIN AR339 strain induced cytotoxic and apoptotic responses in human HeLaS3 and C33A cervical cancer cells (Unno et al., 2005). Moreover, SIN AR339 immunization resulted in significant regression of established cervical tumors. SFV particles and DNA replicons have also been frequently used for cervical cancer. For example, the SFVenh vector carrying the translation enhancer signal from the SFV capsid gene was used for the expression of the HPV E6-E7 fusion protein (Daemen et al., 2003). C57BL/6 mice immunized with SFVenh-HPV E6-E7 showed long-term CTL activity and complete eradication of established tumors. The SFV-sHELP-E7SH vector was engineered to contain helper T-cell epitopes and an endoplasmic reticulum (ER) targeting signal, fused to the HPV E6 and E7 proteins for immunization studies in C57BL/6JOlaHsd mice (Ip et al., 2015). Immunization with 1×10^5 SFV-sHELP-E7SH particles generated substantial tumor regression and protected mice from tumor challenges. DNA replicon-based intradermal immunization with SFV-HPV E6-E7 combined with electroporation has been compared to conventional plasmid DNA immunizations (Van de Wall et al., 2018). In contrast to conventional DNA, which failed to prevent tumor growth in C57BL/6 mice, immunization with SFV HPV E6-E7 DNA replicons resulted in 85% tumor-free mice at 200-fold lower doses, equivalent to 0.05 μg of DNA. In the first-in-human clinical phase I trial, 12 patients with cervical intraepithelial neoplasia were immunized with 5×10^5, 5×10^6, 5×10^7, or 2.5×10^8 SFVenh-HPV E6/E7 particles (Vvax001 vaccine candidate) (Komdeur et al., 2021), which provided safe administration and resulted in HPV-specific immunogenicity in all 12 patients. In the context of MV, immunization of transgenic mice with MV expressing the HPV-16 L1 capsid protein elicited robust humoral immune responses (Cantarella et al., 2009). MV-HPV-16 L1 particles have also been compared to HPV-16 L1 and HPV-18 L1 recombinant protein vaccines produced in *Pichia pastoris* yeast (Gupta et al., 2016). Although both MV- and recombinant protein-based vaccines induced HPV-specific total IgG and neutralizing antibodies, a prime-boosting regimen was superior. In the case of VSV, rabbits have been immunized with VSV vectors expressing cottontail rabbit papillomavirus (CRPV) E1, E2, E7 and E7 proteins (Brandsma et al., 2007). Significant reduction in papilloma tumors was detected, especially for VSV-E7, which reduced the tumor volume by 96.9% and eradicated the disease. Moreover, syngeneic TC-1 tumor-bearing C57BL/6 mice were immunized with 5×10^6 VSV-HPV E7 particles, which resulted in HPV16 E7-specific T-cell responses and anti-tumor activity and a 10-fold decrease in tumor volume (Liao et al., 2008).

4.4 Colon cancer

Colon cancer has been evaluated with SFV-IL-12 and SFVenh-IL-12 particles in the poorly immunogenic MC38 colon adenocarcinoma mouse model (Rodriguez-Madoz et al., 2005). Although expression from the SFVenh generated 8-fold higher IL-12 levels, administration of either 1×10^8 SFV-IL-12 or SFVenh-IL-12 particles resulted in complete tumor regression and prolonged tumor-free survival of MC38 tumor-bearing mice. The SFVenh-IL-12 vector generated better anti-tumor responses than the SFV-IL-12 vector and was also more efficient than a first-generation adenovirus vector in tumor elimination. Moreover, C57BL/6 mice with implanted MC38-CEA-2 tumors were immunized with VEE-IL-12 and VEE-CEA particles (Osada et al., 2012). In comparison to administration of recombinant IL-12 protein, VEE-IL-12 particles provided superior immune responses. Combination therapy with VEE-IL-12 and VEE-CEA particles generated stronger tumor activity and prolonged survival time compared to immunization with either VEE-IL-12 or VEE-CEA particles alone. SFV particles expressing the vascular endothelial growth factor receptor-2 (VEGFR-2) have been administered to mice with CT26 colon tumors, which resulted in a significant inhibition of tumor growth and prevented metastatic spread (Lyons et al., 2007). However, co-distribution of SFV-VEGFR-2 and SFV-IL-12 particles did neither provide strong antibody responses nor anti-tumor activity. In contrast, the combination of SFV-VEGFR-2 and SFV-IL-4 particles enhanced VEGFR-2-specific antibody titers and prolonged survival of mice. The classic example of immunization with alphavirus RNA replicons is illustrated by a single intramuscular immunization of CT26 tumor-bearing mice with 0.1 μg of SFV-LacZ RNA replicons (Ying et al., 1999). Immunized mice elicited strong immune responses and were protected against tumor challenges and the survival was extended in mice with pre-existing tumors. VEE-CEA particles have been subjected to a phase I clinical trial in patients with stages III and IV colorectal cancer (CRC) (Crosby et al., 2020). Antigen-specific immune responses and long-term survival of stage III and IV patients were observed. Flaviviruses have also been evaluated for colon cancer therapy. Noncytopathic KUN particles expressing the granulocyte macrophage colony stimulating factor (GM-CSF) were administered intratumorally into BALB/c mice with CT26 colon xenografts (Hoang-Le et al., 2009). Immunization induced $CD8^+$ T-cell responses, tumor regression, and 50% of mice were cured. Also, oncolytic MV vectors have been utilized for GM-CSF expression showing adaptive immune responses in mice with MC38cea tumors (Grossardt et al., 2013). Tumor progression was delayed, and the survival time extended after intratumoral injection of MV-GM-CSF particles. Moreover, complete remission was achieved in one third of immunized mice. In another approach, an infected cell vaccine (ICV) consisting of tumor cells *ex vivo* infected with the oncolytic Maraba MG1 virus expressing IL-12 was evaluated in a colon cancer model of peritoneal carcinomatosis (PC) (Alkayyal et al., 2017). Intraperitoneal administration promoted the migration of activated natural killer (NK) cells, reduced tumor burden, and provided better survival of mice. Complete radiologic responses and 100% long-term survival was achieved in mice with bulky larger than 8mm tumors. Oncolytic VSV expressing a mutant NDV fusion protein (VSV-NDV/F(L289A)) was evaluated in multiple CRC lung metastases and multifocal CRC liver metastases in immune competent rats (Yamaki et al., 2013). Immunization of syngeneic F344/DuCrl rats with 4×10^6 pfu of VSV-NDV/F(L289A) significantly enhanced survival in a liver metastasis model, resulting in long-term survival of more than 100 days in four out of seven rats. In contrast, although the survival was significantly prolonged in the lung metastasis model, none of the immunized rats showed long-term survival.

4.5 Lung cancer

Lung cancer has also been targeted by self-replicating RNA viruses. For example, transduction of SFV-EGFP particles resulted in growth inhibition and killing of human H358a non-small cell lung cancer (NSCLC) cells and H358a spheroids (Murphy et al., 2000). Furthermore, intratumoral injection of SFV-EGFP into nu/nu mice with implanted H358a xenografts showed complete tumor regression in three out of seven mice. The oncolytic SFV VA strain was compared to a conditionally replicating adenovirus (Ad-Delta24TK-GFP) in mice with implanted A549 lung adenocarcinomas

(Määttä et al., 2008). Administration of SFV VA-GFP particles resulted in better survival in nude mice compared to Ad-Delta24TK-GFP injections. In the context of SIN vectors, mice implanted with CT26.CL25 lung tumors showed complete tumor regression and prolonged survival after immunization with SIN-LacZ particles (Granot et al., 2014). The oncolytic MV Hu-191 strain was administered to C57BL/6 mice with implanted Lewis lung carcinomas (LLC), which resulted in efficient suppression of tumor growth and a significant prolongation of survival (Zhao et al., 2013). Moreover, the oncolytic MV Schwarz strain inhibited the growth of established lung tumors in nude mice with implanted xenografts (Boisgerault et al., 2013). In another approach, immunization with the MV Edmonston strain expressing the CEA gene showed regression of A549 and H2009 tumors in nude mice (Patel et al., 2014). Similarly, VSV particles expressing interferon-β (VSV-IFNβ) showed significant tumor growth inhibition in A549 and H2009 tumor-bearing mice after intratumoral immunization (Patel et al., 2015). Moreover, intratumoral administration provided prolonged survival and cure in 30% of mice carrying syngeneic LM2 lung tumors (Patel et al., 2015).

4.6 Melanoma

Self-replicating RNA virus vectors have been evaluated for melanoma therapy. Expression of the tyrosine-related protein-2 (TRP-2) from VEE particles resulted in humoral responses, anti-tumor activity, and prolonged survival in a B16 mouse melanoma model (Avogadri et al., 2010). Synergistic effects were obtained when VEE-TRP-2 administration was combined with antagonist anti-CTL antigen-4 (CTLA-4) or agonist anti-glucocorticoid-induced tumor necrosis factor receptor (GITR) monoclonal antibodies (mAbs), resulting in tumor regression in 50% and 90% of mice, respectively (Avogadri et al., 2014). In a combination therapy approach with two SFV DNA replicon vectors expressing VEGFR-2 and IL-12 from one plasmid and survivin and β-hCG antigen from another plasmid, inhibition of tumor growth and prolongation of survival in mice with implanted B16 melanoma tumors were detected (Yin et al., 2015). The anti-tumor efficacy was superior for the combination therapy than either SFV DNA replicon alone. Stage III and IV metastasizing melanoma and renal cell carcinoma patients have been subjected to a phase I clinical trial with liposome encapsulated SFV particles expressing IL-12 (LSFV-IL-12 (Ren et al., 2003). Intravenous administration of 1×10^8 or 1×10^9 replication-deficient

LSFV-IL-12 particles caused no major toxicity in any of the 18 patients, only temporary and mild inflammatory reactions such as itching, slight fever and flu-like symptoms after receiving the higher dose. Ten-fold higher IL-12 concentrations compared to baseline, which lasted for 3–4 days, were detected in the peripheral blood. Re-administration of LSFV-IL-12 did not cause any reaction against SFV, liposomes or cytotoxic responses. In the context of flaviviruses, KUN-GM-CSF particles were administered to C57BL/6 mice with B16-OVA melanoma xenografts, which provided substantial tumor regression and cure of more than half of the immunized mice (Hoang-Le et al., 2009). Another flavivirus, YFV, has been engineered for the expression of the chicken ovalbumin CTL epitope SIINFEKL, which protected immunized C57BL/6 mice from challenges with B16-OVA and B16FO melanoma cells (McAllister et al., 2000). The oncolytic MV Leningrad-16 (L-16) strain has been evaluated in the Mel Z mouse melanoma model (Ammour et al., 2020). Subcutaneous administration of MV L16 resulted in efficient inhibition of tumor growth in BALB/c nu/nu mice. The pseudotyped VSV vector, where the lymphocytic choriomeningitis virus (LCMV) glycoprotein replaced the VSV-G protein, was evaluated in subcutaneous A375 xenograft and B16-OVA syngeneic mouse tumor models (Kimpel et al., 2018). The treatment resulted in tumor regression, reduction in size of lung metastasis and prolonged survival of mice.

4.7 Ovarian cancer

Ovarian cancer has also been the target for self-replicating RNA viruses. For example, the oncolytic SIN AR339 strain induced apoptosis and killed efficiently HOC-1, HAC-2, and OMC-3 ovarian cancer cell lines (Unno et al., 2005). Intraperitoneal administration of SIN AR339 in an ovarian metastatic cancer model, resulted in tumor targeting and suppression of ascites formation. Combination therapy of SIN-IL-12 and irinotecan, a CPT-11 topoisomerase inhibitor, resulted in long-term survival of SCID mice implanted with highly aggressive human ES2 ovarian tumors (Granot and Meruelo, 2012). Immunization with SFV particles has been combined with vaccinia virus (VV) administration, where immunization with SFV-OVA was followed by vaccination with VV-OVA in C57BL/6 mice grafted with murine ovarian surface epithelial carcinoma (MOSEC) (Zhang et al., 2010). The prime-boost regimen elicited OVA-specific $CD8^+$ T-cell responses and enhanced the anti-tumor activity compared to immunization with either SFV or VV alone. The previously described pseudotyped

VSV-LCMV-GP vector effectively killed the ovarian cancer cell lines A2780, HOC7, and SKOV6 (Dold et al., 2016). Tumor regression was observed in the A2780 ovarian mouse tumor model, which was enhanced by combination therapy with ruxolitinib. In the case of MV, a tumor targeting MV vector carrying the single-chain antibody (scFv) sequence for the alpha-folate receptor (αFR) was administered to mice with ovarian SKOV3ip.1 xenografts (Hasegawa et al., 2006a). Intratumoral administration of MV-αFR reduced tumor volumes and prolonged overall survival. The SKOV3.ip.1 ovarian mouse model has also been used for the evaluation of MV-CEA and MV-NIS administration individually or together (Hasegawa et al., 2006b). Clearly, the combination therapy resulted in better efficacy than either MV-CEA or MV-NIS alone. Intraperitoneal administration of MV-CEA in a phase I trial in refractory recurrent ovarian cancer (RROC) was well tolerated demonstrating dose-dependent activity in patients who had been heavily pre-treated with other therapeutics (Galanis et al., 2010). SD was achieved in 14 out of 21 patients.

4.8 Pancreatic cancer

The poor prognosis of pancreatic cancer has encouraged the evaluation of therapeutic intervention of self-replicating RNA viral vectors. For example, the oncolytic M1 alphavirus has been combined with irreversible electroporation (IRE), which triggered synergistic enhanced apoptosis in pancreatic cancer cells (Sun et al., 2021). Moreover, the M1-IRE combination therapy resulted in inhibition of tumor proliferation and extended survival in immunocompetent mice carrying orthotopic pancreatic tumors. VEE-CEA particles have been demonstrated to elicit high levels of T-cell responses in preclinical studies. In a phase I trial in patients with metastatic pancreatic cancer, intramuscular administration of VEE-CEA induced clinically relevant T-cell and antibody responses, cellular cytotoxicity against tumor cells and prolonged overall survival (Morse et al., 2010). The oncolytic activity of VSV vectors has been compared to Sendai virus and respiratory syncytial virus (RSV) in aggressive pancreatic ductal adenocarcinoma (PDAC) cell lines (Murphy et al., 2012). Although heterogeneity occurs among PDAC cell lines, VSV showed superior oncolytic activity compared to both Sendai virus and RSV. In another PADC-related study, wildtype VSV, VSV-GFP and the oncolytic VSV-ΔM51-GFP were evaluated for oncolytic activity in five PDAC cell lines (Hastle et al., 2013). Clearly, VSV-ΔM51-GFP showed the highest oncolytic activity, which could be further enhanced by co-administration of gemcitabine. Significant suppression was observed in SCID mice with implanted KLM1 and Capan-2 pancreatic tumor xenografts after immunization with MV-SLAMBlind (Awano et al., 2016).

4.9 Prostate cancer

Prostate cancer has been frequently targeted by expression of various prostate-specific antigens from VEE vectors. For example, VEE particles expressing the prostate-specific membrane antigen (PSMA) induced robust T- and B-cell responses characterized by Th-1 cytokines, CTL activity, and IgG2a/IgG2b antibodies in the transgenic adenocarcinoma of the prostate (TRAMP) mouse model (Durso et al., 2007). Moreover, mice immunized with VEE particles expressing the mouse six-transmembrane epithelial antigen of the prostate (mSTEAP) received a gene gun-based booster vaccination with gold-coated conventional pcDBA-3-mSTEAP plasmid DNA, which induced specific immune responses against a recently defined mSTEAP epitope and substantially protected immunized mice against challenges with tumor cells (Garcia-Hernandez et al., 2007). Expression of the prostate stem cell antigen (PSCA) from VEE particles in TRAMP mice resulted in long-term survival lasting for at least 12 months in 90% of mice (Garcia-Hernandez et al., 2008). The previously mentioned VEE-PSMA particles have been subjected to a phase I clinical study in patients with castration resistant metastatic prostate cancer (CRPC), which showed good safety and tolerance (Slovin et al., 2013). Due to the weak PSMA-specific immune response, expression and dose optimization are still required. Intratumoral immunization with MV-CEA significantly delayed tumor growth and extended survival in a prostate PC-2 mouse model (Msaouel et al., 2009). Combination therapy with oncolytic MV and mumps virus (MuV) showed better anti-tumor activity and longer survival in the PC-3 prostate cancer model compared to immunizations with either MV or MuV alone (Son et al., 2018). Moreover, the modified VSV-ΔM51-GFP vector induced apoptosis and killed efficiently human DU145 and PC-3 cells and eradicated malignant cells leaving normal cells unaffected in immunized nude mice (Zhao et al., 2014). Intratumoral and intravenous immunization of mice carrying DU145 and 22Rv1 prostate tumors with the oncolytic VSV-LCMV-GP resulted in long-term remission of subcutaneous tumors and bone metastases (Urbiola et al., 2018).

5. Conclusions

Self-replicating RNA viruses have demonstrated efficacy as vaccine vectors for both prophylactic and therapeutic applications in various animal models and in selected clinical trials for both infectious diseases (Table 1) and cancers (Table 2). Robust immune responses, and protection against challenges with pathogens or tumor cells have been achieved. Although vaccines have not yet been authorized for use for many indications, the approval of the VSV-based EBOV vaccine Ervebo has been of great encouragement for future endeavors. Naturally, the success in recent COVID-19 vaccine development has also favorably accelerated the application of COVID-19 vaccines based on self-replicating RNA virus particles or RNA molecules. Additionally, the promising results from the phase I trial on SFV-based HPV vaccines should attract further interest in applications of self-replicating RNA viruses.

There are several features of self-replicating RNA viruses that make them attractive. The self-amplification of RNA viruses provides extreme levels of RNA for transient expression of the gene of interest without any risk of integration into the host genome, which is indeed preferable for vaccine development. The high level of expression means that 100–1000-fold lower concentrations of RNA or DNA is needed to achieve the same immune response observed for synthetic mRNA or conventional plasmid DNA vectors. Self-replicating RNA viruses further provide flexibility as viral particles can be engineered as replication-deficient, replication-proficient, and oncolytic viral particles. Moreover, naked or LNP encapsulated RNA molecules as well as DNA replicons can be used.

Although self-replicating RNA viruses have presented their attractive and promising features, there is no denying in certain shortcomings and problems related to their use. For example, viral particle production has been hampered by the lack of adequate and efficient packaging cell lines, especially for alphaviruses. The RNA delivery has been compromised by the sensitivity of degradation of single-stranded RNA. The application of viral vectors has always to be viewed in the light of safety. As for any other vaccine approach, the success achieved in preclinical models has often not been directly compatible in clinical applications, which has required additional vector engineering and optimization studies related to delivery, dosing and immunization regimens. In any case, self-replicating RNA viruses represent an attractive alternative to other types of vaccine development. In comparison to other non-viral and viral vector systems, self-replicating RNA viruses have certainly established themselves as serious candidates for vaccine development.

References

Ajbani, S.P., Velhal, S.M., Kadam, R.B., Patel, V.V., Lundstrom, K., Bandivdekar, A.H., 2017. Immunogenicity of virus-like Semliki Forest virus replicon particles expressing Indian HIV-1C gag, env and pol RT genes. Immunol. Lett. 190, 221–232.

Alkayyal, A.A., Tai, L.-H., Michael Kennedy, M.A., Tanese de Souza, C., Zhang, J., Lefebvre, C., Sahi, S., Ananth, A.A., Mahmoud, A.B., Makrigiannis, A.P., Cron, G.O., Macdonald, B., Marginean, E.C., Stojdl, D.F., Bell, J.C., Auer, R.C., 2017. NK-cell recruitment is necessary for eradication of peritoneal carcinomatosis with an IL12-expressing Maraba virus cellular vaccine. Cancer Immunol. Res. 5, 211–221.

Allen, C., Opyrchal, M., Aderca, I., Schroeder, M.A., Sarkaria, J.N., Domingo, E., Federspiel, M.J., Galanis, E., 2013. Oncolytic measles virus strains have a significant anti-tumor activity against glioma stem cells. Gene Ther. 2, 444–449.

Ammour, Y., Ryabaya, O., Shchetinina, Y., Prokofeva, E., Gavrilova, M., Khochenkov, D., Vorobyev, D., Faizuloev, E., Shohin, I., Zverev, V.V., Svitich, O., Nasdkina, T., 2020. The susceptibility of human melanoma cells to infection with the Leningrad-16 vaccine strain of measles virus. Viruses 12, 173.

Andersson, C., Vasconcelos, N.M., Sievertzon, M., Haddad, D., Liljeqvist, S., Berglund, P., Liljestrom, P., Ahlborg, N., Ståhl, S., Berzins, K., 2001. Comparative immunization study using RNA and DNA constructs encoding a part of the Plasmodium falciparum antigen Pf332. Scand. J. Immunol. 54 (1-2), 117–124.

Avogadri, F., Merghoub, T., Maughan, M.F., Hirschhorn-Cymerman, D., Morris, J., Ritter, E., 2010. Alphavirus replicon particles expressing TRP-2 provide potent therapeutic effect on melanoma through activation of humoral and cellular immunity. PLoS One 5, e12670.

Avogadri, F., Zappasodi, R., Yang, A., Budhu, S., Malandro, N., Hisrchhorn-Cymerman, D., 2014. Combination of alphavirus replicon particle-based vaccination with immunomodulatory antibodies: therapeutic activity in the B16 melanoma mouse model and immune correlates. Cancer Immunol. Res. 2, 448–458.

Awano, M., Fujiyki, T., Shoji, K., Amagai, Y., Murakami, Y., Furukawa, Y., Sato, H., Yoneda, M., Kai, C., 2016. Measles virus selectively blind to signaling lymphocyte activity molecule has oncolytic efficacy against nectin-4 expressing pancreatic cells. Cancer Sci. 107, 1647–1652.

Bache, B.E., Grobusch, M.P., Agnandji, S.T., 2020. Safety, immunogenicity and risk-benefit analysis of rVSV-ΔG-ZEBOV-GP (V920) Ebola vaccine in Phase I-III clinical trials across regions. Future Microbiol. 15. https://doi.org/10.2217/fmb-2019-0237.

Beissert, T., Perkovic, M., Vogel, A., Erbar, S., Walzer, K.C., Hempel, T., Brill, S., Haefner, E., Becker, R., Türeci, Ö., Sahin, U., 2020. A trans-amplifying RNA vaccine strategy for induction of potent protective immunity. Mol. Ther. 28 (1), 119–128.

Bergman, I., Whitaker-Dowling, P., Gao, Y., Griffin, J.A., Watkins, S.C., 2003. Vesicular stomatitis virus expressing a chimeric Sindbis glycoprotein containing an Fc antibody binding domain targets to Her2/neu overexpressing breast cancer cells. Virology 316, 337–347.

Bogers, W.M., Oostermeijer, H., Mooij, P., Koopman, G., Verschoor, E.J., Davis, D., Ulmer, J.B., Brito, L.A., Cu, Y., Bannerjee, K., Otten, G.R., Burke, B., Dey, A., Heeney, J.L., Shen, X., Tomaras, G.D., Labranche, C., Montefiori, D.C.,

Liao, H.-X., Haynes, B., Geall, A.J., Barnett, S.W., 2015. Potent immune responses in rhesus macaques induced by nonviral delivery of self-amplifying RNA vaccine expressing HIV type 1 envelope with a cationic emulsion. J. Infect. Dis. 211, 947–955.
Boisgerault, N., Guillerme, J.B., Pouliquen, D., Mesel-Lemoine, M., Achard, C., Combredet, C., Fonteneau, J.-F., Tangy, F., Grégoire, M., 2013. Natural oncolytic activity of live-attenuated measles virus against human lung and colorectal adenocarcinomas. Biomed. Res. Int. 2013, 387362.
Brand, D., Lemiale, F., Turbica, I., Buzelay, L., Brunet, S., Barin, F., 1998. Comparative analysis of humoral immune responses to HIV type 1 envelope glycoproteins in mice immunized with a DNA vaccine, recombinant Semliki Forest virus RNA, or recombinant Semliki Forest virus particles. AIDS Res. Hum. Retrovir. 14, 1369–1377.
Brandler, S., Ruffié, C., Combredet, C., Brault, J.-B., Najburg, V., Prevost, M.-C., Habel, A., Tauber, E., Desprès, P., Tangy, F., 2013. A recombinant measles vaccine expressing chikungunya virus-like particles is strongly immunogenic and protects mice from lethal challenge with chikungunya virus. Vaccine 31 (36), 3718–3725.
Brandsma, J.L., Shylankevich, M., Su, Y., Roberts, A., Rose, J.K., Zelterman, D., Buonocore, L., 2007. Vesicular stomatitis virus-based therapeutic vaccination targeted to the E1, E2, E6, and E7 proteins of cottontail rabbit papillomavirus. J. Virol. 81, 5749–5758.
Bredenbeek, P.J., Molenkamp, R., Spaan, W.J.M., Deubel, V., Marianneu, P., Salvato, M.S., Moshkoff, D., Zapata, J., Tikhonov, I., Patterson, J., Carrion, R., Ticer, A., Brasky, K., Lukashevich, I.S., 2006. A recombinant Yellow Fever 17D expressing Lassa virus glycoproteins. Virology 345 (2), 299–304.
Burnett, J.C., Rossi, J.J., 2012. RNA-based therapeutics: current progress and future prospects. Chem. Biol. 19, 60–71.
Cabrera, A., Saez, D., Cespedes, S., Andrews, E., Onate, A., 2009. Vaccination with recombinant Semliki Forest virus particles expressing translation initiation factor 3 of Brucella abortus induces protective immunity in BALB/c mice. Immunobiology 214 (6), 467–474.
Cantarella, G., Liniger, M., Zuniga, A., Schiller, J.T., Billeter, M., Naim, H.Y., Glueck, R., 2009. Recombinant measles virus-HPV vaccine candidates for prevention of cervical carcinoma. Vaccine 27, 3386–3390.
Capeding, M.R., Tran, N.H., Hadinegoro, S.R., Ismail, H.I., Chotpitayasunondh, T., Chua, M.N., Luong, C.Q., Rusmil, K., Wirawan, D.N., Nallusamy, R., Pitisuttithum, P., Thisyakorn, U., Yoon, I.K., van der Vliet, D., Langevin, E., Laot, T., Hutagalung, Y., Frago, C., Boaz, M., Wartel, T.A., Tornieporth, N.G., Saville, M., Bouckenooghe, A., CYD14 Study Group, 2014. Clinical efficacy and safety of a novel tetravalent dengue vaccine in healthy children in Asia: a phase 3, randomised, observer-masked, placebo-controlled trial. Lancet 384 (9951), 1358–1365.
Case, J.B., Rothlauf, P.W., Chen, R.E., Kafai, N.M., Fox, J.M., Smith, B.K., Shrihari, S., McCune, B.T., Harvey, I.B., Keeler, S.P., Bloyet, L.-M., Zhao, H., Ma, M., Adams, L.J., Winkler, E.S., Holtzman, M.J., Fremont, D.H., Whealan, S.P.J., Diamond, M.S., 2020. Replication-competent vesicular stomatitis virus vaccine vector protects against SARS-CoV-2-mediated pathogenesis in mice. Cell Host Microbe 28 (3), 465–474.
Center, R.J., Miller, A., Wheatley, A.K., Campbell, S.M., Siebentritt, C., Purcell, D.F., 2013. Utility of the Sindbis replicon system as an Env-targeted HIV vaccine. Vaccine 31 (18), 2260–2266.
Chattopadhyay, A., Aquilar, P.V., Bopp, N.E., Yarovinsky, T.O., Weaver, S.C., Rose, J.K., 2018. A recombinant virus vaccine that protects both against Chikungunya and Zika virus infections. Vaccine 36, 3894–3900.
Chen, L., Ewing, D., Subramanian, H., Block, K., Rayner, J., Alterson, K.D., Sedegah, M., Hayes, C., Porter, K., Raviprakash, K., 2007. A heterologous DNA prime-Venezuelan

equine encephalitis virus replicon particle boost dengue vaccine regimen affords complete protection from virus challenge in cynomolgus macaques. J. Virol. 81 (21), 11634–11639.

Chen, Y.H., Keiser, M.S., Davidson, B.L., 2018. Viral vectors for gene transfer. Curr. Protoc. Mouse Biol. 8 (4), e58.

Coltart, C.E., Lindsey, B., Ghinai, I., Johnson, A.M., Heymann, D.L., 2017. The Ebola outbreak 2013-2016: old lessons for new epidemics. Philos. Trans. R. Soc. Lond. Ser. B Biol. Sci. 372 (1721), 20160297.

Crosby, E.J., Gwin, W., Blackwell, K., Marcom, P.K., Chang, S., Maecker, H.T., Broadwater, G., Hyslop, T., Kim, S., Rogatko, A., Lubkov, V., Snyder, J.C., Osada, T, Hobeika, A.C., Morse, M.A., Lyerly, H.K., Hartman, Z.C., 2019. Vaccine-induced memory CD8+ T cells provide clinical benefit in HER2 expressing breast cancer: a mouse to human translational study. Clin. Cancer Res. 25 (9), 2723–2736.

Crosby, E.J., Hobeika, A.C., Niedzweicki, D., Rushing, C., Hsu, D., Berglund, P., Smith, J., Osada, T., Gwin Iii, W.R., Hartman, Z.C., Morse, M.A., Lyerly, H.K., 2020. Long-term survival of patients with stage III colon cancer treated with VRP-CEA(6D), an alphavirus vector that increases the CD8+ effector memory T cell to Treg ration. J. Immunother. Cancer 8, e001662.

Daemen, T., Riezebos-Brilman, A., Bungener, L., Regts, J., Dontje, B., Wilschut, J., 2003. Eradication of established HPV16-transformed tumours after immunisation with recombinant Semliki Forest virus expressing a fusion protein of E6 and E7. Vaccine 21, 1082–1088.

Davis, N.L., Willis, L.V., Smith, J.F., Johnston, R.F., 1989. In vitro synthesis of infectious Venezuelan equine encephalitis virus RNA from a cDNA clone: analysis of a viable deletion mutant. Virology 171, 189–204.

De Alwis, R., Gan, E.S., Chen, S., Leong, Y.S., Tan, H.C., Zhang, S.L., Yau, C., Low, J.G.H., Kalimuddin, S., Matsuda, D., Allen, E.C., Hartman, P., Park, K.J., Alayyoubi, M., Bhaskaran, H., Dukanovic, A., Bao, Y., Clemente, B., Vega, J., Roberts, S., Gonzalez, J.A., Sablad, M., Yelin, R., Taylor, W., Tachikawa, K., Parker, S., Karmali, P., Davis, J., Sullivan, B.M., Sullivan, S.M., Hughes, S.G., Chivukula, P., Ooi, E.E., 2021. A single dose of self-transcribing and replicating RNA-based SARS-CoV-2 vaccine produces protective adaptive immunity in mice. Mol. Ther. 29, 1970–1983.

De Felipe, F., 2004. Skipping the co-expression problem: the new 2A "CHYSEL" technology. Genet. Vaccines Ther. 2 (1), 13.

Del Valle, J.R., Devaux, P., Hodge, G., Wegner, N.J., McChesney, M.B., Cattaneo, R., 2007. A vectored measles virus induces hepatitis B surface antigen antibodies while protecting macaques against virus challenge. J. Virol. 81 (19), 10597–10605.

Démoulins, T., Ruggli, N., Gerber, M., Thomann-Harwood, L.J., Ebensen, T., Schulze, K., Guzman, C.A., McCullough, K.C., 2021. Self-amplifying pestivirus replicon RNA encoding influenza virus nucleoprotein and hemagglutinin promote humoral and cellular immune responses in pigs. Front. Immunol. 11, 622385.

DiCiommo, P.D., Bremner, R., 1998. Rapid, high level protein production using DNA-based Semliki Forest virus vectors. J. Biol. Chem. 273 (29), 18060–18066.

Dold, C., Rodriguez Urbiola, C., Wollmann, G., Egerer, L., Muik, A., Bellmann, L., Fiegl, H., Marth, C., Kimpel, J., Von Laer, D., 2016. Application of interferon modulators to overcome partial resistance to ovarian cancers to VSV-GP oncolytic viral therapy. Mol. Ther. Oncolytics 3, 16021.

Durso, R.J., Andjelic, S., Gardner, J.P., Margitich, D.J., Donovan, G.P., Arrigale, R.R., Wang, X., Maughan, M.F., Talarico, T.L., Olmsted, R.A., Heston, W.D., Maddon, P.J., Olson, W.C., 2007. A novel alphavirus vaccine encoding prostate-specific membrane antigen elicits potent cellular and humoral immune responses. Clin. Cancer Res. 13, 3999–4008.

Ebert, O., Harbaran, S., Shinozaki, K., Woo, S.L., 2005. Systemic therapy of experimental breast cancer metastases by mutant vesicular stomatitis virus in immune-competent mice. Cancer Gene Ther. 12, 350–358.

Erasmus, J.H., Khandhar, A.P., Guderian, J., Granger, B., Archer, J., Archer, M., Cage, E., Fuerte-Stone, J., Larson, E., Lin, S., Kramer, R., Coler, R.N., Fox, C.B., Stinchcomb, D.T., Reed, S.G., Van Hoeven, N., 2018. A nanostructured lipid carrier for delivery of a replicating viral RNA provides single, low-dose protection against Zika. Mol. Ther. 26 (10), 2507–2522.

Erasmus, J.H., Khandhar, A.P., O'Connor, M.A., Walls, A., Heymann, E., Murapa, P., Archer, J., Leventhal, S., Fuller, J.T., Heydenburg Fuller, D., 2020. An Alphavirus-derived replicon RNA vaccine induces SARS-CoV-2 neutralizing antibody and T cell responses in mice and nonhuman primates. Sci. Transl. Med. 12, eabc9396.

Fleeton, M.N., Chen, M., Berglund, P., Rhodes, G., Parker, S.E., Murphy, M., Atkins, G.J., Liljestrom, P., 2001. Self-replicative RNA vaccines elicit protection against influenza A virus, respiratory syncytial virus, and a tickborne encephalitis virus. J. Infect. Dis. 183, 1395–1398.

Frolov, I., Hoffman, T.A., Pragal, B.M., Dryga, S.A., Huang, H., Schlesinger, S., Rice, C.M., 1996. Alphavirus-based expression vectors: strategies and applications. Proc. Natl Acad. Sci. USA 93 (21), 11371–11377.

Fukuhara, H., Ino, Y., Todo, T., 2016. Oncolytic virus therapy: a new era of cancer treatment at dawn. Cancer Sci. 107, 1373–1379.

Furuyama, W., Reynolds, P., Haddock, E., Meade-White, K., Le, M.Q., Kawaoka, Y., Feldmann, H., Marzi, A., 2020. A single dose of a vesicular stomatitis virus-based influenza vaccine confers rapid protection against H5 viruses from different clades. NPJ Vaccines 5, 4.

Galanis, E., Hartmann, L.C., Cliby, W.A., Long, H.J., Peethambaram, P.P., Barrette, B.A., Kaur, J.S., Haluska Jr., P.J., Aderca, I., Zollman, P.J., Sloan, J.A., Keeney, G., Atherton, P.J., Podratz, K.C., Dowdy, S.C., Stanhope, C.R., Wilson, T.O., Federspiel, M.J., Peng, K.W., Russell, S.J., 2010. Phase I trial of intraperitoneal administration of an oncolytic measles virus strain engineered to express carcinoembryonic antigen for recurrent ovarian cancer. Cancer Res. 70, 875–882.

Garcia-Hernandez, M.L., Gray, A., Hubby, B., Kast, W.M., 2007. In vivo effects of vaccination with six-transmembrane epithelial antigen of the prostate: a candidate antigen for treating prostate cancer. Cancer Res. 67, 1344–1351.

Garcia-Hernandez, M.L., Gray, A., Hubby, B., Klinger, O.J., Kast, W.M., 2008. Prostate stem cell antigen vaccination induces a long-term protective immune response against prostate cancer in the absence of autoimmunity. Cancer Res. 68, 861–869.

Geall, A.J., Verma, A., Otten, G.R., Shaw, C.A., Hekele, A., Banerjee, K., Cu, Y., Beard, C.W., Brito, L.A., Krucker, T., O'Hagan, D.T., Singh, M., Mason, P.W., Valiante, N.M., Dormitzer, P.R., Barnett, S.W., Rappuoli, R., Ulmer, J.B., Mandl, C.W., 2012. Non-viral delivery of self-amplifying RNA vaccines. Proc. Natl. Acad. Sci. USA 109 (36), 14604–14609.

Geisbert, T.W., Feldmann, H., 2011. Recombinant vesicular stomatitis virus-based vaccines against Ebola and Marburg infections. J. Infect. Dis. 204 (Suppl 3), S1075–S1081.

Giraud, A., Ataman-Onal, Y., Battail, N., 1999. Generation of monoclonal antibodies to native human immunodeficiency virus type 1 envelope glycoprotein by immunization of mice with naked RNA. J. Virol. Methods 79 (1), 75–84.

Granot, T., Meruelo, D., 2012. The role of natural killer cells in combinatorial anti-cancer therapy using Sindbis viral vector and irinotecan. Cancer Gene Ther. 19, 588–591.

Granot, T., Yamanashi, Y., Meruelo, D., 2014. Sindbis viral vectors transiently deliver tumor-associated antigens to lymph nodes and elicit diversified antitumor CD8+ T-cell immunity. Mol. Ther. 22, 112–122.

Griffin, D.E., 2016. Neurotropic alphaviruses. In: Reiss, C., Springer, E. (Eds.), Neurotropic Viral Infections. Cham, Switzerland, pp. 175–204.

Grossardt, C., Engeland, C.E., Bossow, S., Halama, N., Zaoui, K., Leber, M.F., Springfeld, C., Jaeger, D., Von Kalle, C., Ungerechts, G., 2013. Granulocyte-macrophage colony-stimulating factor-armed oncolytic measles virus is an effective therapeutic cancer vaccine. Hum. Gene Ther. 24, 644–654.

Guo, C., Zhou, Z., Wen, Z., Liu, Y., Zeng, C., Xiao, D., Ou, M., Han, Y., Huang, S., Liu, D., Ye, X., Zou, X., Wu, J., Wang, H., Zeng, E.Y., Jing, C., Yang, G., 2017. Global Epidemiology of Dengue Outbreaks in 1990–2015: a systematic review and meta-analysis. Front. Cell. Infect. Microbiol. 7, 317.

Gupta, G., Giannino, V., Rishi, N., Glueck, R., 2016. Immunogenicity of next-generation HPV vaccines in non-human primates: measles-vectored HPV vaccine versus Pichia pastoris recombinant protein vaccine. Vaccine 34, 4724–4731.

Guy, B., Saville, M., Lang, J., 2010. Development of Sanofi Pasteur tetravalent dengue vaccine. Hum. Vaccines 6 (9), 696–705.

Hannaman, D., Dupuy, L.C., Ellefsen, B., Schmaljohn, C.S., 2016. A phase 1 clinical trial of a DNA vaccine for Venezuelan equine encephalitis delivered by intramuscular or intradermal electroporation. Vaccine 34 (31), 3607–3612.

Hasegawa, K., Nakamura, T., Harvey, M., Ikeda, Y., Oberg, A., Figini, M., Canevari, S., Hartmann, L.C., Peng, K.-W., 2006a. The use of a tropism-modified measles virus in folate receptor-targeted virotherapy of ovarian cancer. Clin. Cancer Res. 12, 6170–6178.

Hasegawa, K., Pham, L., O'Connor, M.K., Federspiel, M.J., Russel, S.J., Peng, K.-W., 2006b. Dual therapy of ovarian cancer using measles viruses expressing carcinoembryonic antigen and sodium iodide symporter. Clin. Cancer Res. 12, 1868–1875.

Hastle, E., Besmer, D.M., Shah, N.R., Murphy, A.M., Moredyk-Schauwecker, M., Molestina, C., Roy, L.R., Curry, J.M., Mukherjee, P., Grdzelishvili, V.Z., 2013. Oncolytic vesicular stomatitis virus in an immunocompetent model of MUC1-positive or MUC1-nulll pancreatic ductal adenocarcinoma. J. Virol. 87, 10283–10294.

Hayasaka, D., Yoshii, K., Ueki, T., Iwasaki, T., Takashima, I., 2004. Sub-genomic replicons of tick-borne encephalitis virus. Arch. Virol. 149 (6), 1245–1256.

Heikkilä, J.E., Vähä-Koskela, M.J., Ruotsalainen, J.J., Martikainen, M.W., Stanford, M.M., McCart, J.A., Bell, J.C., Hinkkanen, A.E., 2010. Intravenously administered alphavirus vector VA7 eradicates orthotopic human glioma xenografts in nude mice. PLoS One 5, e8603.

Henao-Restrepo, A.M., Longini, I.M., Egger, M., Dean, N.E., Edmunds, W.J., Camacho, A., Carroll, M.W., Doumbia, M., Draguez, B., Duraffour, S., 2015. Efficacy and effectiveness of an rVSV-vectored vaccine expressing Ebola surface glycoprotein: interim results from the Guinea ring vaccination cluster-randomised trial. Lancet 386 (9996), 857–866.

Henao-Restrepo, A.M., Camacho, A., Longini, I.M., Watson, C.H., Edmunds, W.J., Egger, M., Carroll, M.W., Dean, N.E., Diatta, I., Doumbia, M., Draguez, B., Duraffour, S., Enwere, G., Grais, R., Gunther, S., Gsell, P.-S., Hossmann, S., Viksmoen Watle, S., Kader Kondé, M., Keita, S., Kone, S., Kuisma, E., Levine, M.M., Mandal, S., Mauget, T., Norheim, G., Riveros, X., Soumah, A., Trelle, S., Vicari, A.S., Rottingen, J.-A., Kieny, M.-P., 2017. Efficacy and effectiveness of an rVSV-vectored vaccine in preventing Ebola virus disease: final results from the Guinea ring vaccination, open-label, cluster-randomised trial (Ebola Ca Suffit!). Lancet 389 (10068), 505–518.

Herbert, A.S., Kuehne, A.I., Barth, J.F., Ortiz, R.A., Nichols, D.K., Zak, S.E., Stonier, S.W., Muhammad, M.A., Bakken, R.R., Prugar, L.I., Olinger, G.G., Groebner, J.L., Lee, J.S., Pratt, W.D., Custer, M., Kamrud, K.I., Smith, J.F.,

Hart, M.K., J.M., 2013. Venezuelan equine encephalitis virus replicon particle vaccine protects nonhuman primates from intramuscular and aerosol challenge with ebolavirus. J. Virol. 2013 (87), 4852–4964.
Hoang-Le, D., Smeenk, L., Anraku, I., Pijlman, G.P., Wang, X.P., De Vrij, J., 2009. A Kunjin replicon vector encoding granulocyte macrophage colony-stimulating factor for intra-tumoral gene therapy. Gene Ther. 16, 190–199.
Hörner, C., Schürmann, C., Auste, A., Ebenig, A., Muraleedharan, S., Dinnon III, K.H., Scholz, T., Herrmann, M., Schnierle, B.S., Baric, R.S., Mühlebach, M.D., 2020. A highly immunogenic and effective measles virus-based Th1-biased COVID-19 vaccine. Proc. Natl. Acad. Sci. U. S. A. 117, 32657–32666.
Hu, H.M., Chen, H.W., Hsiao, Y., Wu, S.H., Chung, H.H., Hsieh, C.H., Chong, P., Leng, C.H., Pan, C.H., 2016. The successful induction of T-cell and antibody responses by a recombinant measles virus-vectored tetravalent dengue vaccine provides partial protection against dengue-2 infection. Hum. Vaccines Immunother. 12, 1678–1689.
Ickenstein, L.M., Garidel, P., 2019. Lipid-based nanoparticle formulations for small molecules and RNA drugs. Expert Opin. Drug Deliv. 16 (11), 1205–1226.
Ip, P.P., Boerma, A., Walczak, M., Oosterhuis, K., Haanen, J.B., Schumacher, T.N., Nijman, H.W., Daemen, T., 2015. Antigen design enhances the immunogenicity of Semliki Forest virus-based therapeutic human papillomavirus vaccines. Gene Ther. 22, 560–567. https://doi.org/10.1038/gt.2015.24.
Ito, T., Kumagai, T., Yamaji, Y., Sawada, A., Nakayama, T., 2020. Recombinant measles AIK-C vaccine strain expressing influenza HA protein. Vaccine 8, 149.
Jansen, K.A., 2013. The 2005–2007 Chikungunya epidemic in Reunion: ambiguous etiologies, memories, and meaning-making. Med. Anthropol. 32, 174–189.
Jiang, X., Dalebout, T.J., Bredenbeck, P.J., Carrion, R.J., Brasky, K., Patterson, J., Goicochea, M., Bryant, J., Salvato, M.S., Lukashevich, I.S., 2011. Yellow fever 17D-vectored vaccines expressing LASV GP1 and GP2 glycoproteins provide protection against fatal disease in guinea pigs. Vaccine 29 (6), 1248–1257.
Jiang, B., Huang, D., He, W., Guo, W., Yin, X., Forsyth, P., Lun, X., Wang, Z., 2021. Inhibition of glioma using a novel non-neurotoxic vesicular stomatitis virus. Neurosurg. Focus. 50, E9. https://doi.org/10.3171/2020.11.FOCUS20839.
Johnson, D.M., Jokinen, J.D., Wang, M., Pfeiffer, T., Tretyakova, I., Carrion Jr., R., Griffiths, A., Pushko, P., Lukashevich, I.S., 2020. Bivalent Junin and Machupo experimental vaccine based on alphavirus RNA replicon vector. Vaccine 38 (14), 2949–2959.
Jones, S.M., Feldmann, J., Strober, U., Geisbert, J.B., Fernando, L., Grolla, A., Klenk, H.D., Sullivan, N.J., Volchkov, V.E., Fritz, E.A., Daddario, K.M., Hensley, L.E., Jahrling, P.B., Geisbert, T.W., 2005a. Live attenuated recombinant vaccine protects nonhuman primates against Ebola and Marburg viruses. Nat. Med. 11 (7), 786–790.
Jones, C.T., Patkar, C.G., Kuhn, R.J., 2005b. Construction and applications of yellow fever virus replicons. Virology 331 (2), 247–259.
Jones, S.M., Stroher, U., Fernando, L., Qiu, X., Alimonti, J., Melito, P., Bray, M., Klenk, H.D., Feldmann, H., 2007. Assessment of a vesicular stomatitis virus-based vaccine by use of the mouse model of Ebola virus hemorrhagic fever. J. Infect. Dis. 196 (Suppl. 2), 5404–5412.
Kelvin, A.A., 2011. Outbreak of Chikungunya in the Republic of Congo and the global picture. J. Infect. Dev. Ctries. 5, 441–444.
Khalil, S.M., Tonkin, D.R., Mattocks, M.D., Snead, A.T., Johnston, R.E., White, L.J., 2014. A tetravalent alphavirus-vector based dengue vaccine provides effective immunity in an early life mouse model. Vaccine 32, 4068–4074.
Khromykh, A.A., Varnavski, A.N., Westaway, E.G., 1998. Encapsidation of the flavivirus Kunjin replicon RNA by using a complementation system providing Kunjin virus structural proteins in trans. J. Virol. 72 (7), 5967–5977.

Kimpel, J., Urbiola, C., Koske, I., Tober, R., Banki, Z., Wollmann, G., 2018. The oncolytic virus VSV-GP is effective against malignant melanoma. Viruses 10, 108.

Knudsen, M.L., Ljungberg, K., Tatoud, R., Weber, J., Esteban, M., Liljestrom, P., 2015. Alphavirus replicon DNA expressing HIV antigens is an excellent prime for boosting with recombinant modified vaccinia Ankara (MVA) or with HIV gp140 protein antigen. PLoS One 10, e0117042.

Komdeur, F.L., Singh, A., Van de Wall, S., Meulenberg, J.J.M., Boerma, A., Hoogeboom, B.N., Paijens, S.T., Oyarce, C., de Bruyn, M., Schuuring, E., Regts, J., Marra, R., Werner, N., Sluis, J., van der Zee, A.G.J., Wilschut, J.C., Allersma, D.P., van Zanten, C.J., Kosterink, J.G.W., Jorritsma-Smit, A., Yigit, R., Nijman, H.W., Daemen, T., 2021. First-in-human phase I clinical trial of an SFV based RNA replicon cancer vaccine against HPV-induced cancers. Mol. Ther. 29, 611–625.

Kong, L., Ju, B., Chen, Y., He, L., Ren, L., Liu, J., Kunxue Hong, K., Su, B., Wang, Z., Ozorowski, G., Ji, X., Hua, Y., Chen, Y., Deller, M.C., Hao, Y., Feng, Y., Garces, F., Wilson, R., Dai, K., O'Dell, S., McKee, K., Mascola, J.R., Ward, A.B., Wyatt, R.T., Li, Y., Wilson, I.A., Zhu, J., Shao, Y., 2016. Key gp120 glycans pose roadblocks to the rapid development of VRC01-class antibodies in an HIV-1-infected Chinese donor. Immunity 44 (4), 939–950.

Kramer, M.G., Masner, M., Casales, E., Moreno, M., Smerdou, C., Chabalgoity, J.A., 2015. Neoadjuvant administration of Semliki Forest virus expressing interleukin-12 combined with attenuated Salmonella eradicates breast cancer metastasis and achieves long-term survival in immunocompetent mice. BMC Cancer 15, 620.

Krishnavajhala, H.R., Willimas, J., Heidner, H., 2018. An influenza A virus vaccine based on an M2e-modified alphavirus. Arch. Virol. 163, 483–488.

Kurup, D., Wirblich, C., Lambert, R., Zabihi Diba, L., Leiby, B.E., Schnell, M.J., 2022. Measles-based Zika vaccine induces long-term immunity and requires NS1 antibodies to protect the female reproductive tract. NPJ Vaccines 7, 43.

Lachman, L.B., Rao, X.M., Kremer, R.H., Ozpolat, B., Kirjakova, G., Price, J.E., 2001. DNA vaccination against neu reduces breast cancer incidence and metastasis in mice. Cancer Gene Ther. 8, 259–268.

Lamb, Y.N., 2021. BNT162b2 mRNA COVID-19 vaccine: first approval. Drugs 81, 495–501.

Li, Q., Liu, Q., Huang, W., Li, X., Wang, Y., 2018. Current status on the development of pseudoviruses for enveloped viruses. Rev. Med. Virol. 28 (1), e1963.

Liao, J.B., Publicover, J., Rose, J.K., DiMaio, D., 2008. Single-dose, therapeutic vaccination of mice with vesicular stomatitis virus expressing human papillomavirus type 16 E7 protein. Clin. Vaccine Immunol. 15, 817–824.

Liljestrom, P., Garoff, H., 1991. A new generation of animal cell expression vectors based on the Semliki Forest virus replicon. Biotechnology 9, 1356–1361.

Liu, R., Wang, J., Shao, Y., Wang, X., Zhang, H., Shuai, L., Ge, J., Wen, Z., Bu, Z., 2018. A recombinant VSV-vectored MERS-CoV vaccine induces neutralizing antibody and T cell responses in rhesus monkeys after single dose immunization. Antivir. Res. 150, 30–38.

Ljungberg, K., Whitmore, A.C., Fluet, M.E., Moran, T.P., Shabman, R.S., Collier, M.L., Kraus, A.A., Thompson, J.M., Montefiori, D.C., Beard, C., Johnston, R.E.J., 2007. Increased immunogenicity of a DNA-launched Venezuelan equine encephalitis virus-based replicon DNA vaccine. Virology 81 (24), 13412–13423.

Lorin, C., Mollet, L., Delebecque, F., Combredet, C., Hurtrel, B., Charneau, P., Brahic, M., Tangy, F., 2004. A single injection of recombinant measles virus vaccine expressing human immunodeficiency virus (HIV) type 1 clade B envelope glycoproteins induces neutralizing antibodies and cellular immune responses to HIV. J. Virol. 78, 146–157.

Lukashevich, I.S., Pushko, P., 2016. Vaccine platforms to control Lassa fever. Exp. Rev. Vaccines 15 (9), 1135–1150.

Lundstrom, K., 2018. Viral vectors for gene therapy. Diseases 6, 42.
Lundstrom, K., 2021. Viral vectors for COVID-19 vaccine development. Viruses 13, 317.
Lyons, J.A., Sheahan, B.J., Galbraith, S.E., Mehra, R., Atkins, G.J., Fleeton, M.N., 2007. Inhibition of angiogenesis by a Semliki Forest virus vector expressing VEGFR-2 reduces tumour growth and metastasis in mice. Gene Ther. 14, 503–513.
Lythgoe, M.P., Middleton, P., 2021. Comparison of COVID-19 vaccine approvals at the US Food and Drug Administration, European Medicines Agency, and Health Canada. JAMA Netw. Open 4 (6), e2114531.
Määttä, A.M., Mäkinen, K., Ketola, A., Liimatainen, T., Yongabi, F.N., Vähä-Koskela, M., 2008. Replication competent Semliki Forest virus prolongs survival in experimental lung cancer. Int. J. Cancer 123, 1704–1711.
Martikainen, M., Niittykoski, M., von und zu Frauenberg, M., Immonen, A., Koponen, S., 2015. MicroRNA attenuated clone of virulent Semliki Forest virus overcomes antiviral type I interferon in resistant mouse CT-2A glioma. J. Virol. 89, 10637–10647.
Martin, A., Staeheli, P., Schneider, U., 2006. RNA polymerase II-controlled expression of antigenomic RNA enhances the rescue efficacies of two different members of the Mononegavirales independently of the site of viral genome replication. J. Virol. 80 (12), 5708–5715.
Marzi, A., Robertson, S.J., Haddock, E., Feldmann, F., Hanley, P.W., Scott, D.-P., Strong, J.E., Kobinger, G., Best, S.M., Feldmann, H., 2015. Ebola vaccine. VSV-EBOV rapidly protects macaques against infection with the 2014/2015 Ebola virus outbreak strain. Science 349 (6249), 739–742.
Mateo, M., Reynard, S., Carnec, X., Journeaux, A., Baillet, N., Schaeffer, J., Picard, C., Legras-Lachuer, C., Allan, R., Perthame, E., Hillion, K.-H., Pietrosemoli, N., Dillies, M.-A., Barrot, L., Valive, A., Barron, S., Fellmann, L., Gaillard, J.-C., Armengaud, J., Carbonelle, C., Raoul, H., Tangy, F., Baize, S., 2019. Vaccines inducing immunity to Lassa fever glycoprotein and nucleoprotein protect macaques after a single shot. Sci. Transl. Med. 11 (512), eaaw3163.
Mateo, M., Reynard, S., Carnec, X., Journeaux, A., Germain, C., Hortion, J., Carnec, X., Picard, C., Baillet, N., Borges-Cardoso, V., Merabet, O., Valive, A., Barron, S., Jourjon, O., Lacroix, O., Duthey, A., Dirheimer, M., Jouvion, G., Moreau, P.-H., Fellmann, L., Carbonelle, C., Raoul, H., Tangy, F., Baize, S., 2021. A single-shot Lassa vaccine induces long-term immunity and protects cynomolgus monkeys against heterologous strains. Sci. Transl. Med. 13 (597), eabf6348.
Maxmen, A., 2017. Ebola vaccine approved for use in ongoing outbreak. Nature. https://doi.org/10.1038/nature.2017.22024.
McAllister, A., Arbetman, A.E., Mandl, S., Pena-Rossi, C., Andino, R., 2000. Recombinant yellow fever viruses are effective therapeutic vaccines for treatment of murine solid tumors and pulmonary metastases. J. Virol. 74, 9197–9205.
McCoy, L.E., van Gils, M.J., Ozorowski, G., Messmer, T., Briney, B., Voss, J.E., Kulp, D.W., Macauley, M.S., Sok, D., Pauthner, M., Menis, S., Cottrell, C.A., Torres, J.L., Hsueh, J., Schief, W.R., Wilson, I.A., Ward, A.B., Sanders, R.W., Burton, D.R., 2016. Holes in the glycan shield of the native HIV envelope are a target of trimer-elicited neutralizing antibodies. Cell Rep. 16 (9), 2327–2338.
McKay, P.F., Hu, K., Blakney, A.K., Samnuan, K., Brown, J.C., Penn, R., Zhou, J., Bouton, C.R., Rogers, P., Polra, K., Lin, P.J.C., Barbosa, C., Tam, Y.K., Barclay, W.S., Shattock, R.J., 2020. Self-amplifying RNA SARS-CoV-2 lipid nanoparticle vaccine candidate induces high neutralizing antibodies in mice. Nat. Commun. 11, 3523.
Morse, M.A., Hobeika, A.C., Osada, T., Berglund, P., Hubby, B., Negri, S., Niedzwiecki, D., Devi, G.R., Burnett, B.K., Clay, T.M., Smith, J., Lyerly, H.K., 2010. An alphavirus vector overcomes the presence of neutralizing antibodies and elevated numbers of Tregs to induce immune responses in humans with advanced cancer. J. Clin. Investig. 120, 3234–3241.

Msaouel, P., Iankov, I.D., Allen, C., Morris, J.C., von Messling, V., Cattaneo, R., 2009. Engineered measles virus as a novel oncolytic therapy against prostate cancer. Prostate 69, 82–91.

Mühlebach, M.D., 2017. Vaccine platform recombinant measles virus. Virus Genes 53 (5), 733–740.

Muik, A., Dold, C., Geiß, Y., Volk, A., Werbizki, M., Dietrich, U., von Laer, D., 2012. Semireplication-competent vesicular stomatitis virus as a novel platform for oncolytic therapy. J. Mol. Med. (Berl.) 90, 959–970.

Murphy, A.M., Morris-Downes, M.M., Sheahan, B.J., Atkins, G.J., 2000. Inhibition of human lung carcinoma cell growth by apoptosis induction using Semliki Forest virus recombinant particles. Gene Ther. 7, 1477–1482.

Murphy, A.M., Besmer, D.M., Moerdyk-Schauwecker, M., Moestl, N., Ornelles, D.A., Mukherjee, P., 2012. Vesicular stomatitis virus as an oncolytic agent against pancreatic ductal adenocarcinoma. J. Virol. 86, 3073–3087.

Niavarani, S.R., Lawson, C., Boudaud, M., Simard, C., Tai, L.H., 2020. Oncolytic vesicular stomatitis virus-based cellular vaccine improves triple-negative breast cancer outcome by enhancing natural killer and CD8(+) T-cell functionality. J. Immunother. Cancer 8, e000465. https://doi.org/10.1136/jitc-2019-000465.

Olkkonen, V.M., Liljestrom, P., Garoff, H., Simons, K., Dotti, C.G., 1993. Expression of heterologous proteins in cultured rat hippocampal neurons using the Semliki Forest virus vector. J. Neurosci. Res. 35, 445–451.

Osada, T., Berglund, P., Morse, M.A., Hubby, B., Lewis, W., Niedzwiecki, D., Yang, X.Y., Hobeika, A., Burnett, B., Devi, G.R., Clay, T.M., Smith, J., Kim Lyerly, H., 2012. Co-delivery of antigen and IL-12 by Venezuelan equine encephalitis virus replicon particles enhances antigen-specific immune responses and anti-tumor effects. Cancer Immunol. Immunother. 61, 1941–1951. https://doi.org/10.1007/s00262-012-1248-y.

Osakada, F., Callaway, E.M., 2013. Design and generation of recombinant rabies virus vectors. Nat. Protoc. 8, 1583–1601.

Pang, X., Zhang, M., Dayton, A.I., 2001. Development of dengue virus type 2 replicons capable of prolonged expression in host cells. BMC Microbiol. 1, 18.

Parks, C.L., Picker, L.J., King, C.R., 2013. Development of replication-competent viral vectors for HIV vaccine delivery. Curr. Opin. HIV AIDS 8, 402–411.

Patel, M.R., Jacobson, B.A., Belgum, H., Raza, A., Sadiq, A., Drees, J., Wang, H., Jay-Dixon, J., Etchison, R., Federspiel, M.J., Russell, S.J., Kratzke, R.A., 2014. Measles vaccine strains for virotherapy of non-small cell lung carcinoma. J. Thorac. Oncol. 9, 1101–1110.

Patel, M.R., Jacobson, B.A., Ji, Y., Drees, J., Tang, S., Xiong, K., 2015. Vesicular stomatitis virus expressing interferon-β is oncolytic and promotes antitumor immune responses in a syngeneic murine model of non-small cell lung cancer. Oncotarget 6, 33165–33177.

Patil, S., Gao, Y.-G., Lin, X., Li, Y., Dang, K., Tian, Y., Zhang, W.-J., Jiang, S.-F., Qadir, A., Qian, A.-R., 2019. The development of functional non-viral vectors for gene delivery. Int. J. Mol. Sci. 20 (21), 5491.

Pielnaa, P., Al-Saadawe, M., Saro, A., Dama, M.F., Zhou, M., Huang, Y., Huang, J., Xia, Z., 2020. Zika virus spread, epidemiology, genome, transmission cycle, clinical manifestation, associated challenges, vaccine and antiviral drug development. Virology 543, 34–42.

Pijlman, G.P., Suhrbier, A., Khromykh, A.A., 2006. Kunjin virus replicons: an RNA-based, non-cytopathic viral vector system for protein production, vaccine and gene therapy applications. Expert. Opin. Biol. Ther. 6 (2), 134–145.

Pol, J.G., Zhang, L., Bridle, B.W., Stephenson, K.B., Resséquier, J., Hanson, S., Chen, L., Kazdhanm, N., Bramson, J.L., Strojdl, D.F., Wan, Y., Lichty, B.D., 2014. Maraba virus as a potent oncolytic vaccine vector. Mol. Ther. 22 (2), 420–429.

Pollock, K.M., Cheeseman, H.M., Szubert, A.J., Libri, V., Boffito, M., Owen, D., Bern, H., O'Hara, J., McFarlane, L.R., Lemm, N.-M., McKay, P.F., Rampling, T., Yim, Y.T.N., Milinkovic, A., Kingsley, C., Cole, T., Fagerbrink, S., Aban, M., Tanaka, M., Mehdipour, Sa., Robbins, A., Budd, W., Faust, S.N., Hassanin, H., Cosgrove, C.A., Winston, A., Fidler, S., Dunn, D.T., McCormack, S., Shattock, R.J.,, 2021. Safety and immunogenicity of a self-amplifying RNA vaccine against COVID-19: COVAC1, a phase I. Dose-ranging trial, Lancet. doi:https://doi.org/10.2139/ssrn.3859294 Available at SSRN.

Prince, G.A., Horswood, R.L., Chanock, R.M., 1985. Quantitative aspects of passive immunity to respiratory syncytial virus infection in infant cotton rats. J. Virol. 55, 517–520.

Pushko, P., Geisbert, J., Parker, M., Jahrling, P., Smith, J., 2001. Individual and bivalent vaccines based on alphavirus replicons protect guinea pigs against infection with Lassa and Ebola virus. J. Virol. 75 (23), 11677–11685.

Pyankov, O.V., Bodnev, S.A., Pyankova, O.G., Solodkyi, V.V., Pyankov, S.A., Setoh, Y.X., Volchokova, V.A., Suhrbier, A., Volchikov, V.V., Agafonov, A.A., Khromykh, A.A., 2015. A Kunjin replicon virus-like vaccine provides protection against Ebola virus infection in nonhuman primates. J. Infect. Dis. 212, S368–S371.

Rabinovich, S., Powell, R.I.R., Lindsay, R.W.B., Yuan, M., Carpov, A., Wilson, A., Lopez, M., Coleman, J.W., Wagner, D., Sharma, P., Kemelman, M., Wright, K.J., Seabrook, J.P., Arendt, H., Martinez, J., DeStefano, J., Chiuchiolo, M.J., Parks, C.L., 2014. A novel, live-attenuated vesicular stomatitis virus vector displaying conformationally intact, functional HIV-1 envelope trimers that elicits potent cellular and humoral responses in mice. PLoS One 9, e106597.

Racine, T., Kobinger, G.P., Arts, E.J., 2017. Development of an HIV vaccine using a vesicular stomatitis virus vector expressing designer HIV-1 envelope glycoproteins to enhance humoral responses. AIDS Res. Ther. 14 (1), 55.

Radecke, F., Spielhofer, P., Schneider, H., Kaelin, K., Huber, M., Dötsch, C., Christiansen, G., Billeter, M.A., 1995. Rescue of measles viruses from cloned DNA. EMBO J. 14 (23), 5773–5784.

Ramsauer, K., Schwameis, M., Firbas, C., Mullner, M., Putnak, R.J., Thomas, S.J., Despres, P., Tauber, E., Jilma, B., Tangy, F., 2015. Immunogenicity, safety, and tolerability of a recombinant measlesvirus-based chikungunya vaccine: a randomised, double-blind, placebo controlled, active-comparator, first-in-man trial. Lancet Infect. Dis. 15, 519–527.

Raouane, M., Desmaele, D., Urbinati, G., Massaad-Massade, L., Couvreur, P., 2012. Lipid conjugated oligonucleotides: a useful strategy for delivery. Bioconjug. Chem. 23, 1091–1104.

Ravel, G., Mantel, N., Silvano, J., Rogue, A., Guy, B., Jackson, N., Burdin, N., 2017. Biodistribution and safety of a live and attenuated tetravalent dengue vaccine in the cynomolgus monkey. Vaccine 35 (43), 5918–5923.

Reed, D.S., Glass, P.J., Bakken, R.R., Barth, J.F., Lind, C.M., da Silva, L., Hart, M.K., Rayner, J., Alterson, K., Custer, M., Dudek, J., Owens, G., Kamrud, K.I., Parker, M.D., Smith, J., 2014. Combined alphavirus replicon particle vaccine induces durable and cross-protective immune responses against equine encephalitis virus. J. Virol. 88, 12077–12086.

Reisinger, E.C., Tschismarov, R., Beubler, E., Wiedermann, U., Firbas, C., Loebermann, M., Pfeiffer, A., Muellner, M., Tauber, E., Ramsauer, K., 2018. Immunogenicity, safety, and tolerability of the measles-vectored chikungunya virus vaccine MV-CHIK: a double-blind, randomised, placebo-controlled and active-controlled phase 2 trial. Lancet 392 (10165), 2718–2727.

Ren, H., Boulikas, T., Lundstrom, K., Söling, P.C., Warnke, P.C., Rainov, N.G., 2003. Immunogene therapy of recurrent glioblastoma multiforme with a liposomally encapsulated replication-incompetent Semliki Forest virus vector carrying the human interleukin-12 gene—a phase I/II clinical protocol. J. Neuro-Oncol. 64, 147–154. https://doi.org/10.1007/BF02700029.

Reynard, O., Mokhonov, V., Mokhonova, E., Leung, J., Page, A., Mateo, M., Pyankova, O., Georges-Courbot, M.C., Raoul, H., Khromykh, A.A., Volchkov, V.E., 2011. Kunjin virus replicon-based vaccines expressing Ebola virus glycoprotein GP protect the guinea pig against lethal Ebola virus infection. J. Infect. Dis. 204 (Suppl. 3), S1060–S1065.

Reynolds, T.D., Buonocore, L., Rose, N.F., Rose, J.K., Robek, M.D., 2015. Virus-like vesicle-based therapeutic vaccine vectors for chronic hepatis B virus infection. J. Virol. 89 (20), 10407–10415.

Roche, F.P., Sheahan, B.J., O'Mara, S.M., Atkins, G.J., 2010. Semliki Forest virus-mediated gene therapy the RG2 rat glioma. Neuropathol. Appl. Neurobiol. 36, 648–660. https://doi.org/10.1111/j.1365-2990.2010.01110.x.

Rodriguez-Madoz, J.R., Prieto, J., Smerdou, C., 2005. Semliki forest virus vectors engineered to express higher IL-12 levels induce efficient elimination of murine colon adenocarcinomas. Mol. Ther. 12, 153–163.

Roldão, A., Mellado, M.C., Castilho, L.R., Carrondo, M.J., Alves, P.M., 2010. Virus-like particles in vaccine development. Expert Rev. Vaccines 9 (10), 1149–1176.

Rossi, S.L., Comer, J.E., Wang, E., Azar, S.R., Lawrence, W.S., Plante, J.A., Ramsauer, K., Schrauf, S., Weaver, S.C., 2019. Immunogenicity and efficacy of a measles virus-vectored chikungunya vaccine in nonhuman primates. J. Infect. Dis. 220, 735–742.

Ryder, A.B., Buonocore, L., Vogel, L., Nachbagauer, R., Krammer, F., Rose, J.K., 2015. A viable recombinant rhabdovirus lacking its glycoprotein gene and expressing influenza virus hemagglutinin and neuraminidase is a potent influenza vaccine. J. Virol. 89, 2820–2830.

Sabchareon, A., Wallace, D., Sirivichayakul, C., Limkittikul, K., Chanthavanich, P., Suvannadabba, S., Jiwariyavej, V., Dulyachai, W., Pengsaa, K., Wartel, T.A., Moureau, A., Saville, M., Bouckenooghe, A., Viviani, S., Tornieporth, N.G., Lang, J., 2012. Protective efficacy of the recombinant, live-attenuated, CYD tetravalent dengue vaccine in Thai schoolchildren: a randomised, controlled phase 2b trial. Lancet 380 (9853), 1559–1567.

Safronetz, D., Mire, C., Rosenke, K., Feldmann, F., Haddock, E., Geissbert, T., Feldmann, H., 2015. A recombinant vesicular stomatitis virus-based Lassa fever vaccine protects guinea pigs and macaques against challenge with geographically and genetically distinct Lassa viruses. PLoS Negl. Trop. Dis. 9 (4), e0003736.

Samsa, M.M., Dupuy, L.C., Beard, C.W., Six, C.M., Schmaljohn, C.S., Mason, P.W., Geall, A.J., Ulmer, J.B., Yu, D., 2019. Self-amplifying RNA vaccines for Venezuelan equine encephalitis virus induce robust protective immunogenicity in mice. Mol. Ther. 27 (4), 850–865.

Savar, N.S., Vallet, T., Azizi, M., Arashkia, A., Lundstrom, K., Vignuzzi, M., Niknam, H.M., 2021. Quantitative evaluation of PpSP15-LmSTI1 fusion gene expression following transfection with an alphavirus-derived self-amplifying mRNA and conventional DNA vaccine platforms. Mol. Cell. Probes 59, 101749.

Scaglione, A., Opp, S., Hurtado, A., Lin, Z., Pampeno, C., Noval, M.G., Thannickal, S.A., Stapleford, K.A., Meruelo, D., 2021. Combination of a Sindbis-SARS-CoV-2 spike vaccine and αOX40 antibody elicits protective immunity against SARS-CoV-2 induced disease and potentiates long-term SARS-CoV-2-specific humoral and t-cell immunity. Front. Immunol. 12, 719077.

Schneider, H., Spielhofer, P., Kaelin, K., Dötsch, C., Radecke, F., Sutter, G., Billetter, M.A., 1997. Rescue of measles virus using a replication-deficient vaccinia-T7 vector. J. Virol. Methods 64 (1), 57–64.

Schnell, M.J., Buonocore, L., Kretzschmar, E., Johnson, E., Rose, J.K., 1996. Foreign glycoproteins expressed from recombinant vesicular stomatitis viruses are incorporated efficiently into virus particles. Proc. Natl. Acad. Sci. U. S. A. 93, 11359–11365.

Scholle, I., Girard, Y.A., Zhao, Q., Higgs, S., Mason, P.W., 2004. Trans-packaged West Nile virus-like particles: infectious properties in vitro and in infected mosquito vectors. J. Virol. 78 (21), 11605–11614.

Schultz-Cherry, S., Dybing, J.K., Davis, N.L., Williamson, C., Suarez, D.L., Johnston, R., Perdue, M.L., 2000. Influenza virus (A/HK/156/97) hemagglutinin expressed by an alphavirus replicon system protects against lethal infection with Hong Kong-origin H5N1 viruses. Virology 278, 55–59.

Sebastian, S., Lambe, T., 2018. Clinical advances in viral-vectored influenza vaccines. Vaccine 6, 29.

Sheahan, T., Whitmore, A., Long, K., Ferris, M., Rockx, B., Funkhouser, B., Donaldson, E., Gralinski, L., Collier, M., Heise, M., Davis, N., Johnston, R., Baric, R.S., 2011. Successful vaccination strategies that protect aged mice from lethal challenge from influenza virus and heterologous severe acute respiratory syndrome coronavirus. J. Virol. 85, 217–230.

Slovin, S.F., Kehoe, M., Durso, R., Fernandez, C., Olson, W., Gao, J.P., Israel, R., Scher, H.I., Morris, S., 2013. A phase I dose escalation trial of vaccine replicon particles (VRP) expressing prostate-specific membrane antigen (PSMA) in subjects with prostate cancer. Vaccine 31, 943–949.

Son, H.A., Zhang, L., Cuong, B.K., Van Tong, H., Cuong, L.D., Hang, N.T., Nhung, H.T.M., Yamamoto, N., Toan, N.L., 2018. Combination of vaccine-strain measles and mumps viruses enhances oncolytic activity against human solid malignancies. Cancer Investig. 7, 106–117.

Spencer, A., McKay, P.F., Belij-Rammerstorfer, S., Ulaszewska, M., Bissett, C.D., Hu, K., Samnuan, K., Blakney, A.K., Wright, D., Sharpe, H.R., Gilbride, C., Truby, A., Allen, E.R., Gilbert, S.C., Shattock, R.J., Lambe, T., 2021. Heterologous vaccination regimens with self-amplifying RNA and adenoviral COVID vaccines induce robust immune responses in mice. Nat. Commun. 12 (1), 2893.

Su, K.Y., Balasubramaniam, V.R.M.T., 2019. Zika virus as oncolytic therapy for brain cancer: myth or reality? Front. Microbiol. 10, 2715.

Sugiyama, T., Yoneda, M., Kuraishi, T., Hattori, S., Inoue, Y., Sato, H., Kai, C., 2013. Measles virus selectively blind to signaling lymphocyte activation molecule as a novel oncolytic virus for breast cancer treatment. Gene Ther. 20, 338–347.

Sun, S., Liu, Y., He, C., Hu, W., Liu, W., Huang, X., Wu, J., Xie, F., Chen, C., Wang, J., Lin, Y., Zhu, W., Yan, G., Cai, J., Li, S., 2021. Combining Nanoknife with M1 oncolytic virus enhances anticancer activity in pancreatic cancer. Cancer Lett. 502, 9–24. https://doi.org/10.1016/j.canlet.2020.12.1018.

Tani, H., Morikawa, S., Matsuura, Y., 2012. Development and applications of VSV vectors based on cell tropism. Front. Microbiol. 2, 272.

Tao, J., Li, B., Shi, Y., Chen, J., Zhu, G., Shen, X., Liu, H., 2020. Attenuated porcine-derived type 2 bovine viral diarrhea virus as vector stably expressing viral gene. J. Virol. Methods 279, 113842.

Thomas, J.M., Moen, S.T., Gnade, B.T., Vargas-Inchaustegui, D.A., Foitz, S.M., Suarez, G., Heidner, H.W., König, R., Chopra, A.K., Peterson, J.W., 2009. Recombinant Sindbis virus vectors designed to express protective antigen of Bacillus anthracis protect animals from anthrax and display synergy with ciprofloxacin. Clin. Vaccine Immunol. 16 (11), 1696–1699.

Torresi, J., Ebert, G., Pellegrini, M., 2017. Vaccines licensed and in clinical trials for the prevention of dengue. Hum. Vaccines Immunother. 13 (5), 1059–1072.

Tretyakova, I., Tibbens, A., Jokinen, J.D., Johnson, D.M., Lukashevich, J.S., Pushko, P., 2019. Novel DNA-launched Venezuelan equine encephalitis virus vaccine with rearranged genome. Vaccine 37 (25), 3317–3325.

Tretyakova, I., Plante, K.S., Rossi, S.L., Lawrence, W.S., Peel, J.E., Gudjohnsen, S., Wang, E., Mirchandani, D., Tibbens, A., Lamichhane, T.N., 2020. Venezuelan equine encephalitis vaccine with rearranged genome resists reversion and protects non-human primates from viremia after aerosol challenge. Vaccine 38 (17), 3378–3386.
Tsuji, M., Bergmann, C.C., Takita-Sonoda, Y., Murata, K., Rodrigues, E.G., Nussenzweig, R.S., Zavala, F., 1998. Recombinant Sindbis viruses expressing a cytotoxic T-lymphocyte epitope of a malaria parasite or of influenza virus elicit protection against the corresponding pathogen in mice. J. Virol. 72, 6907–6910.
Unno, Y., Shino, Y., Kondo, F., Igarashi, N., Wang, G., Shimura, R., Yamaguchi, T., Asano, T., Saisho, H., Sekiya, S., Shirasawa, H., 2005. Oncolytic virotherapy for cervical and ovarian cancer cells by Sindbis virus strain AR339. Clin. Cancer Res. 11, 4553–4560.
Urbiola, C., Santer, F.R., Petersson, M., van der Pluijm, G., Horninger, W., Erlmann, P., 2018. Oncolytic activity of the rhabdovirus VSV-GP against prostate cancer. Int. J. Cancer 143, 1786–1796.
Van de Wall, S., Ljungberg, K., Ip, P.P., Boerma, A., Knudsen, M.L., Nijman, H.W., Liljeström, P., Daemen, T., 2018. Potent therapeutic efficacy of an alphavirus replicon DNA vaccine expressing human papilloma virus E6 and E7 antigens. Oncoimmunology 7, e1487913.
Vasilevska, J., Skrastina, D., Spunde, K., Garoff, H., Kozlovska, T., Zajakina, A., 2012. Semliki Forest virus biodistribution in tumor-free and 4T1 mammary tumor-bearing mice: a comparison of transgene delivery by recombinant virus particles and naked RNA replicon. Cancer Gene Ther. 19, 579–587.
Velders, M.P., McElhiney, S., Cassetti, M.C., Eiben, G.L., Higgins, T., Kovacs, G.R., 2001. Eradication of established tumors by vaccination with Venezuelan equine encephalitis virus replicon particles delivering human papillomavirus 16 E7 RNA. Cancer Res. 61, 7861–7867.
Villar, L., Dayan, G.H., Arredondo-Garcia, J.L., Rivera, D.M., Cunha, R., Deseda, C., Reynales, H., Costa, M.S., Morales-Ramırez, J.O., Carrasquilla, G., Rey, L.C., Dietze, R., Luz, K., Rivas, E., Miranda Montoya, M.C., Cortés Supelano, M., Zambrano, B., Langevin, E., Boaz, M., Tornieporth, N., Saville, M., Noriega, F., CYD15 Study Group, 2015. Efficacy of a tetravalent dengue vaccine in children in Latin America. N. Engl. J. Med. 372 (2), 113–123.
Vogel, A.B., Lambert, L., Kinnear, E., Busse, D., Erbar, S., Reufer, K.C., Wicke, L., Perkovic, M., Beissert, T., Haas, H., Reece, S.T., Sahin, U., Tregoning, J.S., 2018. Self- amplifying RNA vaccines give equivalent protection against influenza to mRNA vaccines but at much lower doses. Mol. Ther. 26 (2), 446–455.
Wahane, A., Waghmode, A., Kapphahn, A., Dhuri, K., Gupta, A., Bahal, R., 2020. Role of lipid-based and polymer-based non-viral vectors in nucleic acid delivery for next-generation gene therapy. Molecules 25 (12), 2866.
Wang, X., Wang, J.-P., Maughan, M.F., Lachman, L.B., 2005a. Alphavirus replicon particles containing the gene for HER2/neu inhibit breast cancer growth and tumorigenesis. Breast Cancer Res. 7, R145–R155.
Wang, X., Wang, J.P., Rao, X.M., Price, J.E., Zhou, H.S., Lachman, L.B., 2005b. Prime-boost vaccination with plasmid and adenovirus gene vaccines control HER2/neu+ metastatic breast cancer in mice. Breast Cancer Res. 7, R580–R588.
Wang, H.C., An, H.J., Yu, Y.Z., Xu, Q., 2015. Potentiation of anthrax vaccines using protective antigen-expressing viral replicon vectors. Immunol. Lett. 163 (2), 206–213.
Wang, M., Jokinen, J., Tretyakova, I., Pushko, P., Lukashevich, I.S., 2018. Alphavirus-based replicon particles expressing multivalent cross-protective Lassa virus glycoproteins. Vaccine 36 (5), 683–690.
Weaver, S.C., Salas, R., Rico-Hesse, R., Ludwig, G.V., Oberste, M.S., Boshell, J., Tesh, R.B., 1996. Re-emergence of epidemic Venezuelan equine encephalomyelitis in South America. VEE Study Group. Lancet 348, 436–440.

Wecker, M., Gilbert, P., Russell, N., Hural, J., Allen, M., Pensiero, M., Chulay, J., Chiu, Y.-L., Karim, S.S.A., Burke, D.S., 2020. Phase I safety and immunogenicity evaluations of an alphavirus replicon HIV-1 subtype C gag vaccine in healthy HIV-1-uninfected adults. Clin. Vaccine Immunol. 19, 1651–1660.

Xiong, C., Levis, R., Shen, P., Schlesinger, S., Rice, C.M., Huang, H.V., 1989. Sindbis virus: an efficient, broad host range vector for gene expression in animal cells. Science 243, 1188–1191.

Yahalom-Ronen, Y., Tamir, H., Melamed, S., Politi, B., Shifman, O., Achdout, H., Vitner, E.B., Israeli, O., Milrot, E., Stein, D., Cohen-Gihon, I., Lazar, S., Gutman, H., Glinert, I., Cherry, L., Vagima, Y., Lazar, S., Weiss, S., Ben-Schmuel, A., Avraham, R., Lupu, E., Bar-David, E., Sittner, A., Erez, N., Zichel, R., Mamroud, E., Mazor, O., Levy, H., Laskar, O., Yitzhaki, S., Shapira, S.C., Zvi, A., Beth-Din, A., Paran, N., Israely, T., 2020. A single dose of recombinant VSV-ΔG-spike provides protection against SARS-CoV-2 challenge. Nat. Commun. 2020 (11), 6402.

Yamaki, M., Shinozaki, K., Sakaguchi, T., Meseck, M., Ebert, O., Ohdan, H., Woo, S.L., 2013. The potential of recombinant vesicular stomatitis virus-mediated virotherapy against metastatic colon cancer. Int. J. Mol. Med. 31, 299–306.

Yamanaka, R., Xanthopoulos, K.G., 2005. Induction of antigen-specific immune responses against malignant brain tumors by intramuscular injection of sindbis DNA encoding gp100 and IL-18. DNA Cell Biol. 24, 317–324.

Yamanaka, R., Zullo, S.A., Ramsey, J., Onodera, M., Tanaka, R., Blaese, M., 2001. Induction of therapeutic antitumor anti-angiogenesis by intratumoral injection of genetically engineered endostatin-producing Semliki Forest virus. Cancer Gene Ther. 8, 796–802.

Yamanaka, R., Tsuchiya, N., Yajima, N., Honma, J., Hasegawa, H., Tanaka, R., Ramsey, J., Blaese, R.M., Xanthopoulos, K.G., 2003. Induction of an antitumor immunological response by an intratumoral injection of dendritic cells pulsed with genetically engineered Semliki Forest virus to produce interleukin-18 combined with the systemic administration of interleukin-12. J. Neurosurg. 99, 746–753.

Yang, B., Shi, J., Sun, Z., Zhu, D., Xu, X., 2020. Attenuated measles virus overcomes radio- and chemoresistance in human breast cancer cells by inhibiting the non-homologous end joining pathway. Oncol. Rep. 44, 2253–2264.

Yin, X., Wang, W., Zhu, X., Wang, Y., Wu, S., Wang, Z., 2015. Synergistic antitumor efficacy of combined DNA vaccines targeting tumor cells and angiogenesis. Biochem. Biophys. Res. Commun. 465, 239–244.

Ying, H., Zaks, T.Z., Wang, R.-F., Irvine, K.R., Kammula, U.S., Marincola, F.M., 1999. Cancer therapy using a self-replicating RNA vaccine. Nat. Med. 5, 823–827.

Zhang, Y.Q., Tsai, Y.C., Monie, A., Wu, T.C., Hung, C.F., 2010. Enhancing the therapeutic effect against ovarian cancer through a combination of viral oncolysis and antigen-specific immunotherapy. Mol. Ther. 18, 692–699.

Zhang, X., Mao, G., Van den Pol, A.N., 2018. Chikungunya-vesicular stomatitis chimeric virus targets and eliminates brain tumors. Virology 522, 244–259.

Zhang, J., Liu, Y., Tan, J., Zhang, Y., Wong, C.-W., Lin, Z., Liu, X., Sander, M., Yang, X., Liang, L., Song, D., Dan, J., Zhou, Y., Cai, J., Lin, Y., Liang, J., Hu, J., Yan, G., Zhu, W., 2021. Necroptotic virotherapy of oncolytic alphavirus M1 cooperated with doxorubicin displays promising therapeutic efficacy in TNBC. Oncogene 40, 4783–4795.

Zhao, D., Chen, P., Yang, H., Wu, Y., Zeng, X., Zhao, Y., Wen, Y., Zhao, X., Liu, X., Wei, Y., Li, Y., 2013. Live attenuated measles virus vaccine induces apoptosis and promotes tumor regression in lung cancer. Oncol. Rep. 29, 199–204.

Zhao, X., Huang, S., Luo, H., Wan, X., Gui, Y., Li, J., Wu, D., 2014. Evaluation of vesicular stomatitis virus mutant as an oncolytic agent against prostate cancer. Int. J. Clin. Exp. Med. 7, 1204–1213.

Zylberberg, C., Gaskill, K., Pasley, S., Matosevic, S., 2017. Engineering liposomal nanoparticles for targeted gene therapy. Gene Ther. 24 (8), 441–452.

CHAPTER FOUR

Present and future of lipid nanoparticle-mRNA technology in phenylketonuria disease treatment

Ramon Diaz-Trelles* and Carlos G. Perez-Garcia
Arcturus Therapeutics, Inc, San Diego, CA, United States
*Corresponding author: e-mail address: ramon@arcturusrx.com

Contents

Abstract

Phenylketonuria (PKU) is a metabolic rare disease characterized by a failure of the body to clear out the high levels of Phenylalanine (Phe), leading to devastating neurological defects and growth retardation. PKU was discovered in 1934 by AsbjrØrn FØlling, and even though there have been continuous efforts from the scientific community to find therapeutic approaches to modulate the high levels of phenylalanine found in the body of the PKU patients, an efficient therapy still needs to be developed. Current standard of care includes low phenylalanine diets, but the strict restrictions for patients and families makes it very difficult to adequately being implemented. FDA has approved two drugs to help reduce Phe levels in PKU patients: an enzyme substitution therapy, Palynziq® (PEGylated recombinant phenylalanine ammonia lyase), and Kuvan®, a supplemental tetrahydrobiopterin (BH4) cofactor that enhances residual enzyme activity. Both treatments are restricted to certain PKU patients' population, and, therefore, there are still high unmet needs for most of the patients. The present review will focus on current advancements in lipid nanoparticles (LNP)-mRNA technologies and their potential in treating the root cause of PKU, a therapeutic approach that will be analyzed in the context of other promising therapeutic approaches that are been developed for PKU.

International Review of Cell and Molecular Biology, Volume 372
ISSN 1937-6448
https://doi.org/10.1016/bs.ircmb.2022.04.008
Copyright © 2022 Elsevier Inc.
All rights reserved.

1. Introduction

Phenylketonuria (PKU) is defined by high levels of phenylalanine (Phe) in the blood due to the defective function of Phenylalanine Hydroxylase (PAH), a liver enzyme fundamental for phenylalanine metabolism (Woo et al., 1984) (Fig. 1A). PKU is an autosomal recessive inherited metabolic disorder caused by the deficiency of PAH enzymatic activity due to pathological genetic variants in the PAH gene (Blau, 2016) (PKU Disease Database, http://www.biopku.org/home/pah.asp). PAH is an enzyme that triggers the metabolic transformation of dietary Phe into Tyr. Its enzymatic activity requires the formation of a tetrameric structure and the action of the co-factor tetrahydrobiopterin (BH4) (Flydal et al., 2019), which deficiency can leads to non-classical PKU deficiency. Most recently, the co-chaperone DNAJC12, together with the 70kDa heatshock protein (HSP70) have also been identified as responsible for the proper folding of PAH, and could also induced hyperphenylalanemic crisis as has been recently reported for DNAJC12 (van Spronsen et al., 2021).

If untreated, high levels of Phe lead to severe brain damage resulting in severe cognitive and behavioral impairments, problems on skin with rashes and eczema, eyes and hair pigmentation, and osteopenia among others (OMIM Clinical Synopsis #261600). The lower levels of Tyr observed in PKU patients, as a result of the PAH deficiency, may explain some of the phenotypes observed in the PKU patients such as deficit in some neurotransmitters (serotonin, dopamine) and melanin, however, the mechanisms leading to the pathology are not entirely clear (van Spronsen et al., 2021).

Since the discovery of the correlation between hyperphenylalanemia and PKU deficiency, many approaches have been investigated to modulate the levels of Phe and the triggering of a hyperphenylalanemic crisis. The most successful one has been dietary Phe restrictions, which has been the base of the standard of care for PKU patients during the last 60 years (Daly et al., 2021). Dietary restrictions have been shown to be effective in lowering Phe levels and slowing down some of the symptoms observed in PKU patients. Although efficacious, it is a burden to patients and families not only for the time and effort required to comply but also economically for the cost of special protein preparations (Ford et al., 2018; MacDonald et al., 2020). Most recently, FDA has approved two additional therapeutic treatments for PKU: Kuvan® and Palynziq®. The synthetic BH4 cofactor (Kuvan®) is very

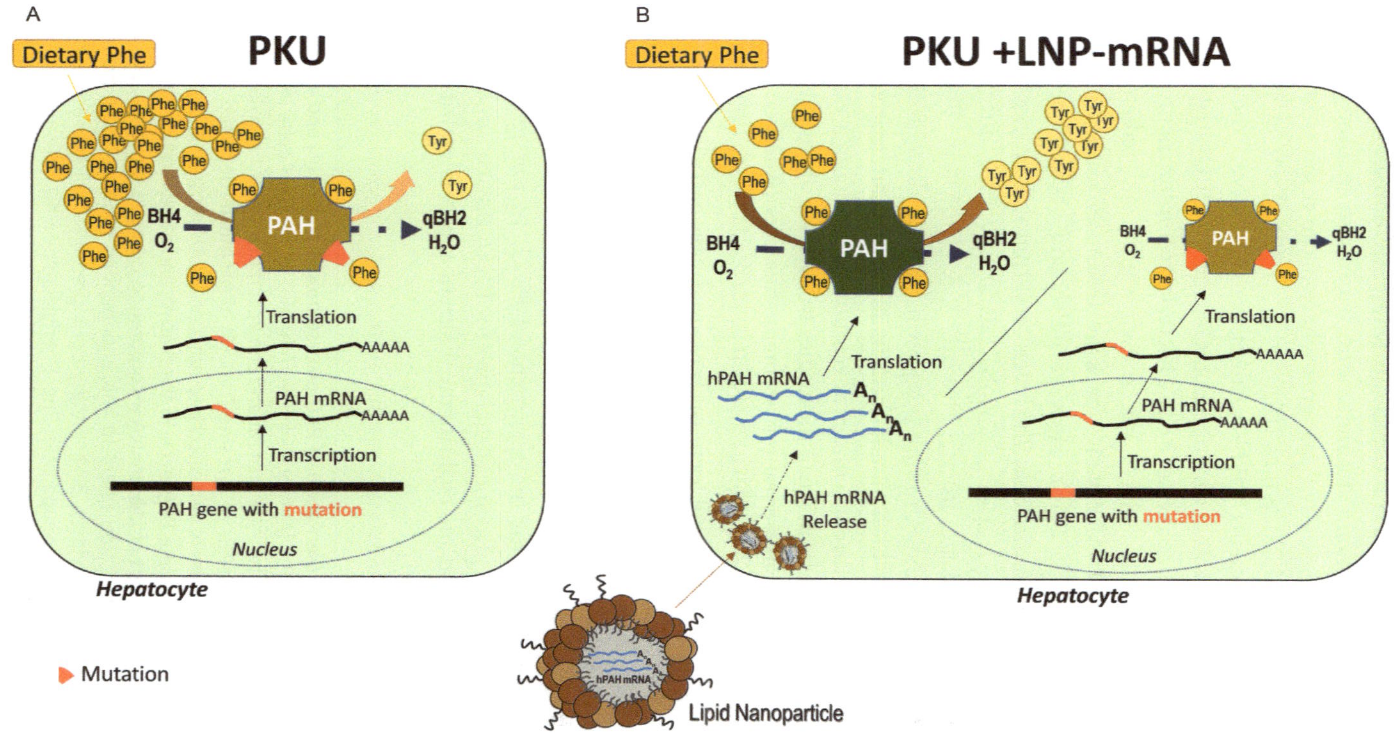

Fig. 1 See figure legend on next page.

effective in non-classical PKU patients where the human (h) PAH (hPAH) protein does not contain any mutations. It can also be used to treat PKU "BH4 treatment-responsive" patients where it can improve the activity of partially active PAH enzymes (Blau et al., 2009; Trefz et al., 2009). In the case of PKU patients "BH4-treatment unresponsive," the only alternative treatment available at this moment is Palynziq® (PEGyated recombinant phenylalanine ammonia lyase, PAL) a bacterial recombinant protein injected subcutaneously daily which has shown significant reduction in Phe levels up to 60% (PALYNZIQ package insert, Novato, CA: BioMarin Pharmaceutical Inc.; 2018). However, Palynziq® is only recommended for adults with uncontrollable Phe levels >600 μmol/L on existing management, with reported cases of anaphylaxis at any time during treatment (PALYNZIQ package insert, Novato, CA: BioMarin Pharmaceutical Inc.; 2018). Even though some progress have been done in the treatment of PKU, there is still high unmet needs in developing alternative treatments, some of those are currently under development including a variety of therapeutic approaches from small molecules to bacteria, from gene therapy to genome editing or non-viral replacement therapies (Adolfsen et al., 2021; Lichter-Konecki and Vockley, 2019; Perez-Garcia et al., 2022).

In this review, we will focus on how lipid nanoparticles (LNP)-mRNA technologies could be part of a new generation of drugs that could effectively treat PKU. The two main therapeutic approaches that could be addressed with LNP-mRNA technology are replacement and genome editing. Very few peer-reviewed studies have addressed the use of LNP-mRNA technology for the treatment of PKU. In this revision, we will explore non-viral replacement therapies for PKU, discuss some recent work published in the field and evaluate the different approaches that the scientific

Fig. 1 Schematic representation of PKU disease model before (A) and after treatment with LNP-mRNA PAH (B). (A) Mutation (red triangle) in the PAH gene leads to the low enzymatic activity of the PAH protein and the accumulation of Phenylalanine (Phe) in the plasma, and low levels of Tyrosine (Tyr). (B) hPAH mRNA delivered to hepatocytes using lipid nanoparticles is translated into functional PAH proteins that can metabolize Phe into Tyr, reducing the Phe levels in the plasma and increasing Tyr production. Left side refers to the effect of LNP-mRNA incorporating de novo functional protein in the system whereas right side refers to the endogenous production of defective PAH (phenylalanine hydroxylase). BH4: tetrahydrobiopterin; hPAH: human phenylalanine hydroxylase; LNP: lipid nanoparticle; Phe: Phenylalanine PKU: Phenylketonuria; qBH2: quinonoid dihydrobiopterin; Tyr: Tyrosine.

community and the pharmaceutical industry are developing to treat the root cause of the PKU deficiency, including the emerging field of gene editing.

2. Replacement therapies for PKU

Classic PKU patients are characterized by the severe reduction or total absence in PAH activity. An ideal replacement approach will target hepatocytes to either deliver a functional PAH protein or the genetic coding (e.g., DNA, mRNA) to produce it while using cell's translational machinery. Even though advances in protein replacement therapies have been made, its use still has challenges due to inefficient delivery and endosomal release, and undesired immune responses (Fu et al., 2014; Mitragotri et al., 2014). In the recent years, novel therapies using non-viral carriers such as liposomes or lipid nanoparticles (LNPs) have been proven to be successfully in delivering nucleic acids to the targeted system. Most recently, Perez-Garcia et al. (2022) have proven how an efficient delivery of LNP, carrying a healthy copy of the PAH mRNA, can control Phe levels in a mouse model of PKU (Fig. 1B). The advantage to use this approach is that the delivery of the mRNA to the cytosol of the hepatocytes is transient, and the protein is generated using cell's internal translational machinery, including the same post-translational modifications as the endogenous PAH protein. This therapeutic approach for PKU is currently being developed by several pharmaceutical companies, including Arcturus Therapeutics (Perez-Garcia et al., 2022), and Moderna (https://www.modernatx.com/pipeline).

How does an LNP-mRNA therapy for PKU come to fruition? What are the considerations when developing this strategy? How important is the LNP carrier when targeting the hepatocytes? How is the selection criteria for a lead PAH mRNA?

PAH mRNA. One of the main therapeutic advantages of an mRNA replacement therapy is the short-half-life of the mRNA that minimizes any potential immune response. However, this can also be a caveat since the mRNA should be translationally efficient, meaning that it needs to be enough time in the system to produce enough PAH protein to account for a therapeutic benefit. Therefore, different strategies to improve stability and translation of the mRNAs can be considered.

The 5′ and 3′ untranslated regions (UTRs) of the mRNA are key components regulating its stability and translation efficiency (Araujo et al., 2012; Mignone et al., 2002). Therefore, changes in both UTR components can potentially change the stability and translational profile of an mRNA. These have been demonstrated in multiple studies using in vitro screening approaches in human primary hepatocytes, where specific 5′ and 3′ UTR combinations have shown to be significantly advantagenous over natural UTRs (Araujo et al., 2012; Mignone et al., 2002; Perez-Garcia et al., 2022) (Fig. 2A).

Another modification on the mRNA sequence that can have an impact in ribosomal activity and translation of mRNA is the codon usage in the open reading frame, an approach that can improve translation but can also affect the folding and function of the protein (Mauro and Chappell, 2014). Established codon optimization algorithms will be beneficial to overcome these challenges by using optimized coding strategies in the mRNA that will result in improved expression profiles for the resulting PAH protein (Perez-Garcia et al., 2022) (Fig. 2A).

Chemically modified uridine nucleotides can be incorporated in the mRNA, during the in vitro transcription, to minimize the potential for any immune response to the mRNA, which will also increase its stability (Karikó et al., 2005). It needs to be considered that not all the chemically modified uridine nucleotides have the same impact on the mRNA. Thus, uridine nucleotides such as N1-Methoxy-Pseudouridine have been shown to improve the expression of the PAH protein, others such as 5-Methoxy Uridine did not, suggesting a potential sequence-dependent effect (Perez-Garcia et al., 2022). Different strategies to increase stability and half-life of the PAH mRNA have also been described by using inverted deoxythymidine (idT) dT residues at the 3′end of the polyA, with potential for improving mRNA efficacy in vivo (Moderna_Therapeutics_Annual_Science_Day, 2020) (Fig. 2A).

At the protein level, natural mutations and post-translational modifications of the PAH protein sequence have been shown to affect its stability and activity (Jaffe, 2017). Deletions at the N-terminal domain in the PAH protein have been described to generate constitutively active forms of the PAH protein (Roberts et al., 2014). However, incorporation of these deletions did not improve protein expression in vitro probably due to a negative effect on stability (Arturo et al., 2020, 2021; Perez-Garcia et al., 2022). PAH activity is also impacted by phosphorylation (Li and Fitzpatrick, 2013) and ubiquitination (Tyagi et al., 2020). Engineered phosphomimetics and modification of potential ubiquitination sites in the synthesized hPAH mRNA did not significantly improve the duration of protein expression in vitro (Perez-Garcia et al., 2022). However, protein engineering by introducing

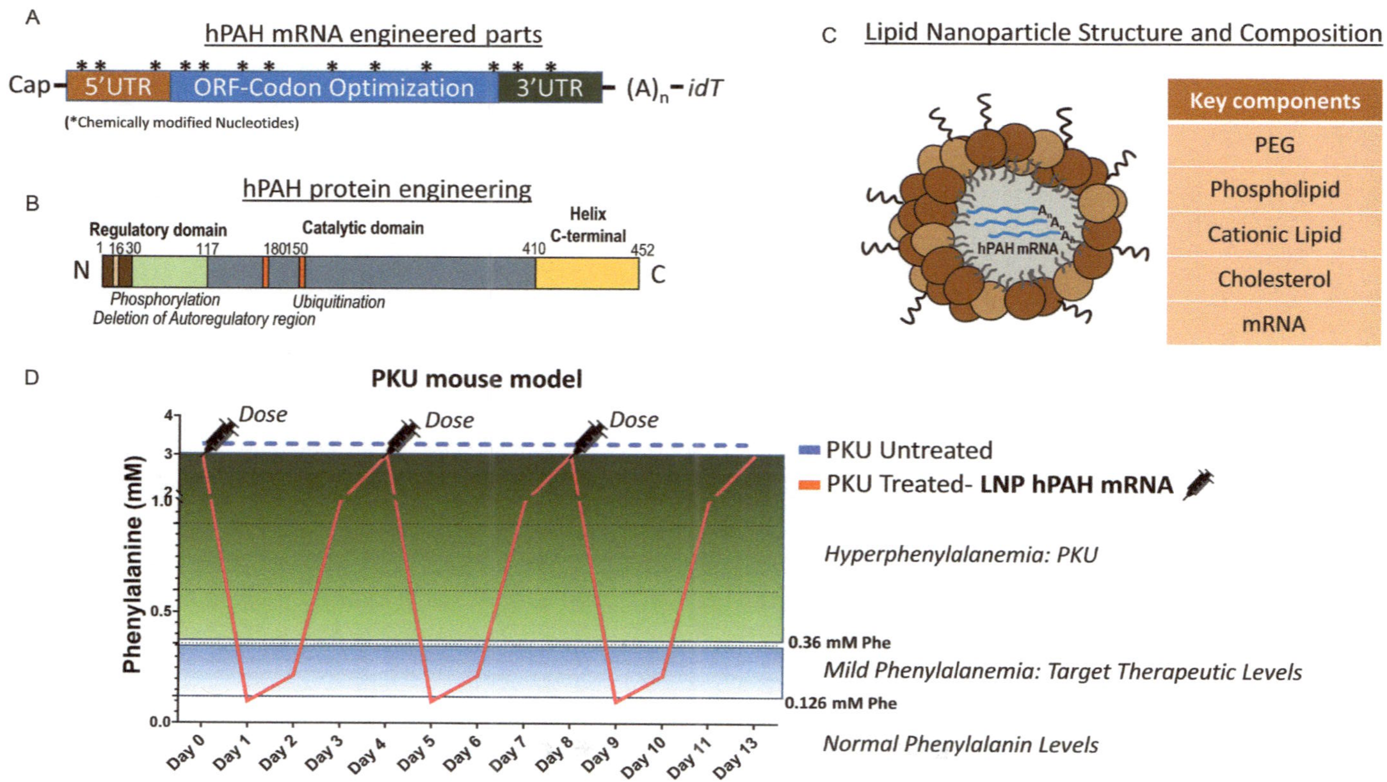

Fig. 2 See figure legend on next page.

specific mutations in key ubiquitination sites, such as K150T and M180T, have been shown to increase the duration of efficacy in vivo when comparing to the reference hPAH protein (Moderna_Therapeutics_Annual_Science_Day, 2020) (Fig. 2B). Altogether, engineering of both the hPAH mRNA and protein should be taken into consideration to prolong the duration and activity of the therapeutic hPAH protein.

Delivery Criteria: LNPs are highly efficient in delivering the cargo to the target cell, and their composition is one of the main drivers in establishing their tolerability profile and, therefore, it is preferred the use of highly biodegradable lipids in their composition. Thus, a balance between selectivity and tolerability will need to be considered when selecting an LNP for a therapeutic approach, as we will discuss below (Hou et al., 2021).

The properties of LNPs can be modulated to selectively target specific biological systems (Cheng et al., 2020), and the liver is a clear example, which makes it an ideal vehicle to deliver a hPAH mRNA (Perez-Garcia et al., 2022; Ramaswamy et al., 2017; Witzigmann et al., 2020). It has been recently shown how intravenous delivery of LNPs encapsulating a fluorescent mRNA are preferentially targeting hepatocytes in a rodent model (Perez-Garcia et al., 2022), supporting the notion that an intravenous delivery of this therapy could potentially target a large population of hepatocytes in human, which will be a therapeutic requirement for a disease such as PKU.

Any mRNA replacement therapy will require repeat dosing, and that is an important nonclinical criterion that needs to be factor in when selecting a lead LNP. As described above, the LNP components should preferable be biodegradable and with very low immunogenicity. A good biodegradability

Fig. 2 (A) Summary of mRNA engineering strategies: 5′ and 3′ UTR combinations, ORF codon optimization, idT polyA modification and inclusion of chemically modified nucleotides. (B) Summary of protein engineering approaches tested: phosphomimic (vertical pink line) and deletion of autoregulatory region (brown fragment in the N term), both in the regulatory region. Mutation of potential ubiquitination sites modifications (vertical red lines) in the catalytic region. No modifications tested in the helix-C terminal domain, critical for the interaction of PAH monomers to form PAH tetramers. (C) Typical composition of LNP used for mRNA delivery. (D) Graph representing the dynamics of Phe levels over time in PKU mice treated with a repeated dose of LNP-mRNA PAH therapy. Levels of normal (white), mild (blue) and PKU levels (green) of hyperphenylalanemia are shadow in different color gradients. Notice the standard target Phe levels for the therapy is below 0.36 mM. hPAH: human phenylalanine hydroxylase; LNP: lipid nanoparticle; ORF: Open reading frame; Phe: Phenylalanine; PKU: Phenylketonuria; PEG: polyethylene glycol; Tyr: Tyrosine; UTR: Untranslated region.

profile will ensure that lipids are cleared from the liver and non-hepatic organs before any subsequent repeat dosing to avoid any undesirable accumulation in the biological systems (Maier et al., 2013). A immunogenic response against any of the components of the LNPs could potentially reduce the efficacy of the treatment after every dose, but most importantly could induce undesired secondary effects to the patients (Igyártó et al., 2021). For the preclinical evaluation of hPAH mRNA in PKU mice, at Arcturus Therapeutics we have used LUNAR® formulations containing ionizable lipids previously screened as part of our platform pipeline and known to comply with both requirements of biodegradability and tolerability in mice (Fig. 2C) (Perez-Garcia et al., 2022; Ramaswamy et al., 2017).

In vivo *studies in PKU mice:* Once the linear hPAH mRNA has been engineered and optimized its production needs to be scaled up to be formulated with the chosen LNP and tested in vivo. There are different PKU mouse models with mutations in the PAH gene that lead to either no expression of PAH protein (Richards et al., 2020a) or expression of a PAH protein with very low enzymatic activity (Shedlovsky et al., 1993). In both cases, like in PKU patients, PAH mutant mice present high levels of Phe (>2 mM) and decreased levels of Tyr in plasma under a regular diet. The Pah^{enu2} mouse model is one of the preferred models to evaluate PKU therapies because has a type of mutation present in a broad part of PKU patients, it carries a missense mutation in exon 7 with a severe reduction in the activity of the enzyme PAH, and many of the same phenotypes found in the clinic (Dobrowolski et al., 2021).

It has been recently published the first LNP-based replacement therapy for PKU using the Pah^{enu2} mice as a model (Perez-Garcia et al., 2022). In this paper, an LNP-hPAH mRNA was intravenously administered to Pah^{enu2} mice, resulting in a significant fast drop in the levels of Phe in plasma. Even though the observed effect was dose dependent, it probed that a single dose of LNP-hPAH mRNA was able to lower Phe levels below 0.3 mM by 6 h, an effect that was maintained low up to 48 h. This quick effect correlates with the stability of the hPAH mRNA and its enhanced translation efficiency as shown by the threefold increase in PAH protein levels over endogenous PAH in wildtype mice (Perez-Garcia et al., 2022). At 6 h post-dose, Phe levels in the plasma were significantly lowered, which suggest that the PAH protein is almost immediately produced in hepatocytes. Interestingly, the presence of mutant forms of the mouse PAH protein could lead to the formation of chimeric tetramers with the corrected protein, and have a dominant negative effect, reducing the efficacy of the treatment

(Waters et al., 2001). However, this possibility did not impact the efficacy of the treatment in the mouse model, which may differ to what we will observe in patients (Fig. 2D). Similarly, the pharmaceutical Moderna have reported a similar profile on the pharmacodynamics of their LNP-hPAH mRNA using the same Pah^{enu2} mouse model where a fast drop of Phe plasma levels was observed almost immediately after treatment that slowly returns to the high pre-dose Phe levels after 2 days (Moderna_Therapeutics_Annual_Science_Day, 2020). The mRNA optimization and modifications done in this study differ to what other companies such as Moderna have done on their mRNAs, which may result in different outcomes, however the results were surprisingly similar. The synergistic effect of these optimization approaches with protein engineering may even improve further the efficacy profile of these therapies as has been recently suggested (Moderna_Therapeutics_Annual_Science_Day, 2020).

In terms of half-life, it has been reported that the endogenous PAH protein has a half-life of 8 h in rat hepatoma cells (Chang et al., 1979) and about 48 h in rat liver (Chang et al., 1979). The short half-life of the PAH protein is consistent with the efficacy data observed using an mRNA replacement (Perez-Garcia et al., 2022). Therefore, a replacement therapy would require repeated dosing. Preclinical studies in Pah^{enu2} mice have shown that it is possible to do repeat dosing of LNP-hPAH mRNA and obtain the same efficacy after each dose reflected in the same reduction of Phe without detecting any adverse effect from the treatment (Perez-Garcia et al., 2022). These data is critical since the lack of pathological findings in the liver after five doses (one dose every 3 days) reflects the potential of this treatment using well tolerated lipid formulations (Fig. 2D) (Perez-Garcia et al., 2022).

Another consideration to bring the LNP-mRNA technology to the patient is the route of administration. All the preclinical studies performed until now in PKU mouse model have been done by *intravenous* injections of LNP-hPAH mRNA in adult mice, which could potentially suggest the use of a similar route of administration in patients or a more patient-friendly delivery may be developed (e.g., subcutaneous).

As with any drug development therapy, we need to be cautious in terms of translating data generated in preclinical models that, although encouraging, may differ to what we will observe in humans. Clinical trials will be a good platform to better understand the potential therapeutic impact of these technologies not only in the adult population, but particularly for any pediatric population before any neurological deficit is observed (van Spronsen et al., 2021).

3. Emerging strategy: Genome editing on PKU

Until now, no peer review publication described the use of LNP-mRNA technologies to treat the root cause of PKU, but there are several publications using adeno-associated viruses (AAV) to deliver CRISPR-based technologies that have successfully shown a reduction of Phe levels in the Pah^{enu2} mouse model as will be discussed below. CRISPR technology can efficiently be delivered as LNP-driven mRNAs into patients as shown in the ongoing clinical trial for transthyretin amyloidosis sponsored by Intellia Therapeutics (ClinicalTrials.gov, NCT04601051) (Gillmore et al., 2021). An advantage in using LNP mRNA technology respect to a more traditional AAV for gene editing is the transient expression of the Cas nucleases encoded in the mRNA which reduces the chances of off-target, and the ability to carry large size mRNAs (i.e., ~11 Kb of replicon mRNA) (de Alwis et al., 2021).

Three different CRIPSR strategies have been tested in PKU mouse models delivered as AAV: (1) correction of PAH point mutations by base-editing (BE) (Villiger et al., 2018; Zhou et al., 2022), (2) rewriting the correct DNA sequence by prime editing (PE) (Böck et al., 2022), and (3) correcting the mutated sequence by homologous recombination by introducing a DNA donor with the correct sequence (Richards et al., 2020b). CRISPR-based and other genome editing tools have previously been delivered using LNP mRNA technology in different disease models (Rothgangl et al., 2021; Qiu et al., 2021) pointing to the possibility of incorporating this strategy in PKU disease. In the case of the integration of DNA donor strategy, the DNA donor was delivered with an AAV virus while the genome editing tool mRNA was delivered in an LNP (Conway et al., 2019).

In the case of base editing, a catalytically dead-Cas9 or Cas12 nuclease was linked to an APOBEC cytidine deaminase that converts the mutation present in the PAH^{enu2} mice, a cytosine (c835T->C, F263S) back to the correct thymidine. Dosing of AAV-Cas12-BE in adult PAH^{enu2} mice led to reduction of Phe levels below 0.360 mM by the fourth week post dosing and reaching 0.1 mM Phe by 6 weeks (Villiger et al., 2018). PAH^{enu2} newborn mice (post-natal day 2), were also treated with recombinant AAV carrying saCas9-BE and by 8 weeks, at the right viral titer levels, mutant mice showed normal levels of Phe like WT mice (Zhou et al., 2022). The lowering of Phe levels correlated with the frequency of base editing detected in the livers.

Similar strategy is followed for prime editing. In this case, a spCas9 nickase (making single strand breaks and not double strand DNA breaks) is fused to a reverse transcriptase; while the guide RNA is now a prime-editing RNA (also called pegRNA), adding to the guide RNA scaffold a primer binding site and a template sequence with the correct PAH sequence. Once the pegRNA binds to the target sequence, in this case next to the C835T of the PAH^{enu2} mouse, the template sequence will be reverse transcribed into DNA and copied into the PAH locus. Although the efficiency of the method is still low, Phe levels of 0.36–0.6 mM can be obtained when injecting adult mice and close to wildtype Phe levels if treating newborn (P1) mice (Böck et al., 2022). Because of the size of the cargo which includes the Cas protein (dead-Cas9 or spCas9-nickase) and the cytidine deaminase (base editing) or the reverse transcriptase (prime editing) and the guide RNA, in both studies two rAAV had to be used to deliver in two parts the completed genome editing tools. The size of the cargo would not be a limitation in the case of the LNP mRNA technology allowing the delivery of the complete genome editing cargo in a single LNP.

In the third genome editing approach, a regular CRISPR Cas9 machinery was used to introduce double strand DNA breaks near the PAH mutation in PAH^{enu2} mice where a ~2 Kb PAH DNA fragment delivered as a template would be introduced by homologous recombination. Both the CRISPR tools and the DNA donor template were delivered in two different viral vectors (Richards et al., 2020b).

The combination of LNP-mRNA delivery and new genome editing strategies becomes a very exciting therapeutic opportunity mainly for the transient expression of editing molecules that reduces the potential off-targets, their low immunogenicity, and the potential of a single dose treatment. Also, there is no risk of integration of the vectors delivered as may happen with AAV delivery. However, there are several limitations when thinking to apply LNP mRNA genome editing strategies to PKU patients. Due to the large number of different mutations in the PAH locus it is challenging to develop a single drug or strategy that would work for all the patients. Also, at this point the efficiency of prime editing or homologous recombination mechanisms that could include several mutations have a low efficiency in vivo. Although promising, it would be challenging to bring any genome editing strategies into PKU patients any time soon due to the complexity, low efficiency and risks associated with the technology.

4. Future perspective

PKU patients need new therapies that would improve their quality of life and can work independent or in synergy with other therapies. LNP-hPAH mRNA replacement therapy is mutation agnostic, and it has the potential in treating PKU patients from newborn to adults. These therapies have overcome many challenges and they are resurfacing as an opportunity to treat multiple genetic diseases, a technology that has been validated with the recent development of COVID vaccines (Chaudhary et al., 2021). An LNP-hPAH mRNA therapy for PKU could potentially be used to treat the chronic aspects of the disease, but it may also be considered in treating sudden hyperphenylalanemic crisis due to its fast-acting effect in lowering Phe-levels. Improved LNP delivery methods (i.e., subcutaneous over intravenous) could also facilitate self-administration by the patient in their homes without the tedious processing of going to clinical sites (e.g., intravenous). Inborn errors of metabolism can be devastating for patients and families, but we live in a period in which we understand better the deficiencies, the genetic causes, the patient needs, all this knowledge is allowing the development of multiple therapeutic approaches that soon could provide options to the many patients that are born with these genetic defects.

Acknowledgments and declaration of interests

R.D.-T and C.G.P.-G are employees of Arcturus Therapeutics. The opinions expressed on this review are the solely opinions of the authors and do not represent Arcturus Therapeutics position.

References

Adolfsen, K.J., Callihan, I., Monahan, C.E., Greisen Jr., P., Spoonamore, J., Momin, M., Fitch, L.E., Castillo, M.J., Weng, L., Renaud, L., Weile, C.J., Konieczka, J.H., Mirabella, T., Abin-Fuentes, A., Lawrence, A.G., Isabella, V.M., 2021. Improvement of a synthetic live bacterial therapeutic for phenylketonuria with biosensor-enabled enzyme engineering. Nat. Commun. 12, 6215.

Araujo, P.R., Yoon, K., Ko, D., Smith, A.D., Qiao, M., Suresh, U., Burns, S.C., Penalva, L.O.F., 2012. Before it gets started: regulating translation at the 5′ UTR. Comp. Funct. Genomics 2012, 475731.

Arturo, E.C., Merkel, G.W., Borne, E., Hansen, M.R., Lisowski, S., Gupta, K., Jaffe, E.K., 2020. Amino acid substitution at Phe80 of mammalian phenylalanine hydroxylase destabilizes both resting-state and activated conformations increasing the population of intermediates. FASEB J. 34, 1.

Arturo, E.C., Merkel, G.W., Hansen, M.R., Lisowski, S., Almeida, D., Gupta, K., Jaffe, E.K., 2021. Manipulation of a cation-π sandwich reveals conformational flexibility in phenylalanine hydroxylase. Biochimie 183, 63–77.

Blau, N., 2016. Genetics of phenylketonuria: then and now. Hum. Mutat. 37, 508–515.

Blau, N., Belanger-Quintana, A., Demirkol, M., Feillet, F., Giovannini, M., MacDonald, A., Trefz, F.K., van Spronsen, F.J., 2009. Optimizing the use of sapropterin (BH(4)) in the management of phenylketonuria. Mol. Genet. Metab. 96, 158–163.

Böck, D., Rothgangl, T., Villiger, L., Schmidheini, L., Matsushita, M, Mathis, N., Ioannidi, E., Rimann, N., Grisch-Chan, H.M., Kreutzer, S., Kontarakis, Z., Kopf, M, Thöny, B., Schwank, G., 2022. In vivo prime editing of a metabolic liver disease in mice. Sci. Transl. Med. 14 (636), eabl9238. https://doi.org/10.1126/scitranslmed.abl9238. In press.

Chang, N., Kaufman, S., Milstien, S., 1979. The mechanism of the irreversible inhibition of rat liver phenylalanine hydroxylase due to treatment with p-chlorophenylalanine. The lack of effect on turnover of phenylalanine hydroxylase. J. Biol. Chem. 254, 2665–2668.

Chaudhary, N., Weissman, D., Whitehead, K.A., 2021. mRNA vaccines for infectious diseases: principles, delivery and clinical translation. Nat. Rev. Drug Discov. 20, 817–838.

Cheng, Q., et al., 2020. Selective organ targeting (SORT) nanoparticles for tissue-specific mRNA delivery and CRISPR-Cas gene editing. Nat. Nanotechnol. 15 (4), 313–320. https://doi.org/10.1038/s41565-020-0669-6. In press.

Conway, A., Mendel, M., Kim, K., McGovern, K., Boyko, A., Zhang, L., Miller, J.C., DeKelver, R.C., Paschon, D.E., Mui, B.L., Lin, P.J.C., Tam, Y.K., Barbosa, C., Redelmeier, T., Holmes, M.C., Lee, G., 2019. Non-viral delivery of zinc finger nuclease mRNA enables highly efficient in vivo genome editing of multiple therapeutic gene targets. Mol. Ther. 27, 866–877.

Daly, A., Evans, S., Pinto, A., Ashmore, C., MacDonald, A., 2021. Protein substitutes in PKU; their historical evolution. Nutrients 13.

de Alwis, R., Gan, E.S., Chen, S., Leong, Y.S., Tan, H.C., Zhang, S.L., Yau, C., Low, J.G.H., Kalimuddin, S., Matsuda, D., Allen, E.C., Hartman, P., Park, K.-J.J., Alayyoubi, M., Bhaskaran, H., Dukanovic, A., Bao, Y., Clemente, B., Vega, J., Roberts, S., Gonzalez, J.A., Sablad, M., Yelin, R., Taylor, W., Tachikawa, K., Parker, S., Karmali, P., Davis, J., Sullivan, B.M., Sullivan, S.M., Hughes, S.G., Chivukula, P., Ooi, E.E., 2021. A single dose of self-transcribing and replicating RNA-based SARS-CoV-2 vaccine produces protective adaptive immunity in mice. Mol. Ther. 29, 1970–1983.

Dobrowolski, S.F., Sudano, C., Phua, Y.L., Tourkova, I.L., Spridik, K., Goetzman, E.S., Vockley, J., Blair, H.C., 2021. Mesenchymal stem cell energy deficit and oxidative stress contribute to osteopenia in the Pah(enu2) classical PKU mouse. Mol. Genet. Metab. 132, 173–179.

Flydal, M.I., Alcorlo-Pagés, M., Johannessen, F.G., Martínez-Caballero, S., Skjærven, L., Fernandez-Leiro, R., Martinez, A., Hermoso, J.A., 2019. Structure of full-length human phenylalanine hydroxylase in complex with tetrahydrobiopterin. Proc. Natl. Acad. Sci. U. S. A. 116, 11229–11234.

Ford, S., O'Driscoll, M., MacDonald, A., 2018. Living with phenylketonuria: lessons from the PKU community. Mol. Genet. Metab. Rep. 17, 57–63.

Fu, A., Tang, R., Hardie, J., Farkas, M.E., Rotello, V.M., 2014. Promises and pitfalls of intracellular delivery of proteins. Bioconjug. Chem. 25, 1602–1608.

Gillmore, J.D., Gane, E., Taubel, J., Kao, J., Fontana, M., Maitland, M.L., Seitzer, J., O'Connell, D., Walsh, K.R., Wood, K., Phillips, J., Xu, Y., Amaral, A., Boyd, A.P., Cehelsky, J.E., McKee, M.D., Schiermeier, A., Harari, O., Murphy, A., Kyratsous, C.A., Zambrowicz, B., Soltys, R., Gutstein, D.E., Leonard, J., Sepp-Lorenzino, L., Lebwohl, D., 2021. CRISPR-Cas9 in vivo gene editing for transthyretin amyloidosis. N. Engl. J. Med. 385, 493–502.

Hou, X., Zaks, T., Langer, R., Dong, Y., 2021. Lipid nanoparticles for mRNA delivery. Nat. Rev. Mater. 6, 1078–1094.

Igyártó, B.Z., Jacobsen, S., Ndeupen, S., 2021. Future considerations for the mRNA-lipid nanoparticle vaccine platform. Curr. Opin. Virol. 48, 65–72.
Jaffe, E.K., 2017. New protein structures provide an updated understanding of phenylketonuria. Mol. Genet. Metab. 121, 289–296.
Karikó, K., Buckstein, M., Ni, H., Weissman, D., 2005. Suppression of RNA recognition by toll-like receptors: the impact of nucleoside modification and the evolutionary origin of RNA. Immunity 23, 165–175.
Li, J., Fitzpatrick, P.F., 2013. Regulation of phenylalanine hydroxylase: conformational changes upon phosphorylation detected by H/D exchange and mass spectrometry. Arch. Biochem. Biophys. 535, 115–119.
Lichter-Konecki, U., Vockley, J., 2019. Phenylketonuria: current treatments and future developments. Drugs 79, 495–500.
MacDonald, A., van Wegberg, A.M.J., Ahring, K., Beblo, S., Belanger-Quintana, A., Burlina, A., Campistol, J., Coskun, T., Feillet, F., Gizewska, M., Huijbregts, S.C., Leuzzi, V., Maillot, F., Muntau, A.C., Rocha, J.C., Romani, C., Trefz, F., van Spronsen, F.J., 2020. PKU dietary handbook to accompany PKU guidelines. Orphanet J. Rare Dis. 15, 171.
Maier, M.A., Jayaraman, M., Matsuda, S., Liu, J., Barros, S., Querbes, W., Tam, Y.K., Ansell, S.M., Kumar, V., Qin, J., Zhang, X., Wang, Q., Panesar, S., Hutabarat, R., Carioto, M., Hettinger, J., Kandasamy, P., Butler, D., Rajeev, K.G., Pang, B., Charisse, K., Fitzgerald, K., Mui, B.L., Du, X., Cullis, P., Madden, T.D., Hope, M.J., Manoharan, M., Akinc, A., 2013. Biodegradable lipids enabling rapidly eliminated lipid nanoparticles for systemic delivery of RNAi therapeutics. Mol. Ther. 21, 1570–1578.
Mauro, V.P., Chappell, S.A., 2014. A critical analysis of codon optimization in human therapeutics. Trends Mol. Med. 20, 604–613.
Mignone, F., Gissi, C., Liuni, S., Pesole, G., 2002. Untranslated regions of mRNAs. Genome Biol. 3. reviews0004.1.
Mitragotri, S., Burke, P.A., Langer, R., 2014. Overcoming the challenges in administering biopharmaceuticals: formulation and delivery strategies. Nat. Rev. Drug Discov. 13, 655–672.
Moderna_Therapeutics_Annual_Science_Day, 2020. https://s29.q4cdn.com/745959723/files/doc_presentations/events-presentations/2020/06/Science-Day-Master-Final-(06.02.20)_updated-from-Ed.pdf.
Perez-Garcia, C.G., Diaz-Trelles, R., Vega, J.B., Bao, Y., Sablad, M., Limphong, P., Chikamatsu, S., Yu, H., Taylor, W., Karmali, P.P., Tachikawa, K., Chivukula, P., 2022. Development of an mRNA replacement therapy for phenylketonuria. Mol. Ther.–Nucleic Acids 28, 87–98. https://doi.org/10.1016/j.omtn.2022.02.020.
Qiu, M., Glass, Z., Chen, J., Haas, M., Jin, X., Zhao, X., Rui, X., Ye, Z., Li, Y., Zhang, F., Xu, Q., 2021. Lipid nanoparticle-mediated codelivery of Cas9 mRNA and single-guide RNA achieves liver-specific in vivo genome editing of *Angptl3*. Proc. Natl. Acad. Sci. U. S. A. 118, e2020401118.
Ramaswamy, S., Tonnu, N., Tachikawa, K., Limphong, P., Vega, J.B., Karmali, P.P., Chivukula, P., Verma, I.M., 2017. Systemic delivery of factor IX messenger RNA for protein replacement therapy. Proc. Natl. Acad. Sci. U. S. A. 114, E1941–E1950.
Richards, D.Y., Winn, S.R., Dudley, S., Fedorov, L., Rimann, N., Thöny, B., Harding, C.O., 2020a. A novel Pah-exon1 deleted murine model of phenylalanine hydroxylase (PAH) deficiency. Mol. Genet. Metab. 131, 306–315.
Richards, D.Y., Winn, S.R., Dudley, S., Nygaard, S., Mighell, T.L., Grompe, M., Harding, C.O., 2020b. AAV-mediated CRISPR/Cas9 gene editing in murine phenylketonuria. Mol. Ther. Methods Clin. Dev. 17, 234–245.

Roberts, K.M., Khan, C.A., Hinck, C.S., Fitzpatrick, P.F., 2014. Activation of phenylalanine hydroxylase by phenylalanine does not require binding in the active site. Biochemistry 53, 7846–7853.

Rothgangl, T., Dennis, M.K., Lin, P.J.C., Oka, R., Witzigmann, D., Villiger, L., Qi, W., Hruzova, M., Kissling, L., Lenggenhager, D., Borrelli, C., Egli, S., Frey, N., Bakker, N., Walker, J.A., Kadina, A.P., Victorov, D.V., Pacesa, M., Kreutzer, S., Kontarakis, Z., Moor, A., Jinek, M., Weissman, D., Stoffel, M., van Boxtel, R., Holden, K., Pardi, N., Thöny, B., Häberle, J., Tam, Y.K., Semple, S.C., Schwank, G., 2021. In vivo adenine base editing of PCSK9 in macaques reduces LDL cholesterol levels. Nat. Biotechnol. 39, 949–957.

Shedlovsky, A., McDonald, J.D., Symula, D., Dove, W.F., 1993. Mouse models of human phenylketonuria. Genetics 134, 1205–1210.

Trefz, F.K., Burton, B.K., Longo, N., Casanova, M.M., Gruskin, D.J., Dorenbaum, A., Kakkis, E.D., Crombez, E.A., Grange, D.K., Harmatz, P., Lipson, M.H., Milanowski, A., Randolph, L.M., Vockley, J., Whitley, C.B., Wolff, J.A., Bebchuk, J., Christ-Schmidt, H., Hennermann, J.B., Sapropterin Study, G, 2009. Efficacy of sapropterin dihydrochloride in increasing phenylalanine tolerance in children with phenylketonuria: a phase III, randomized, double-blind, placebo-controlled study. J. Pediatr. 154, 700–707.

Tyagi, A., Sarodaya, N., Kaushal, K., Chandrasekaran, A.P., Antao, A.M., Suresh, B., Rhie, B.H., Kim, K.S., Ramakrishna, S., 2020. E3 ubiquitin ligase APC/CCdh1 regulation of phenylalanine hydroxylase stability and function. Int. J. Mol. Sci. 21, 9076.

van Spronsen, F.J., Blau, N., Harding, C., Burlina, A., Longo, N., Bosch, A.M., 2021. Phenylketonuria. Nat. Rev. Dis. Primers 7, 36.

Villiger, L., Grisch-Chan, H.M., Lindsay, H., Ringnalda, F., Pogliano, C.B., Allegri, G., Fingerhut, R., Häberle, J., Matos, J., Robinson, M.D., Thöny, B., Schwank, G., 2018. Treatment of a metabolic liver disease by in vivo genome base editing in adult mice. Nat. Med. 24, 1519–1525.

Waters, P.J., Scriver, C.R., Parniak, M.A., 2001. Homomeric and heteromeric interactions between wild-type and mutant phenylalanine hydroxylase subunits: evaluation of two-hybrid approaches for functional analysis of mutations causing hyperphenylalaninemia. Mol. Genet. Metab. 73, 230–238.

Witzigmann, D., Kulkarni, J.A., Leung, J., Chen, S., Cullis, P.R., van der Meel, R., 2020. Lipid nanoparticle technology for therapeutic gene regulation in the liver. Adv. Drug Deliv. Rev. 159, 344–363.

Woo, S.L., Robson, J.H., Guttler, F., 1984. The possibility for prenatal diagnosis of PKU by linkage analyses based on phenylalanine hydroxylase locus specific DNA-polymorphisms. J. Inherit. Metab. Dis. 7 (Suppl. 2), 139–140.

Zhou, L., Su, J., Long, J., Tao, R., Tang, W., Qin, F., Liu, N., Wang, Y., Jiao, Y., Hu, Y., Jiang, L., Li, L., Yang, Y., Yao, S., 2022. A universal strategy for AAV delivery of base editors to correct genetic point mutations in neonatal PKU mice. Mol. Ther. Methods Clin. Dev. 24, 230–240.

Further reading

Ramon, Diaz-Trelles, et al., 2022. Lipid nanoparticle delivers phenylalanine ammonia lyase mRNA to the liver leading to catabolism and clearance of phenylalanine in a phenylketonuria mouse model. Molecular Genetics and Metabolism Reports 32. https://doi.org/10.1016/j.ymgmr.2022.100882.

CHAPTER FIVE

RNA gene editing in the eye and beyond: The neglected tool of the gene editing armatorium?

Ruofan Connie Han[a,*] and Robert E. MacLaren[a,b]
[a]Nuffield Laboratory of Ophthalmology, Department of Clinical Neurosciences, University of Oxford, John Radcliffe Hospital, Oxford, United Kingdom
[b]Oxford Eye Hospital, John Radcliffe, Oxford, United Kingdom
*Corresponding author: e-mail address: enquiries@ndcn.ox.ac.uk

Contents

Abstract

RNA editing allows correction of pathological point mutations without permanently altering genomic DNA. Theoretically targetable to any RNA type and site, its flexibility and reversibility makes it a potentially powerful gene editing tool. RNA editing offers a host of potential advantages in specific niches when compared to currently available alternative gene manipulation techniques. Unlike DNA editors, which are currently too large to be delivered in vivo using a viral vector, smaller RNA editors fit easily within the capabilities of an adeno-associated virus (AAV). Unlike gene augmentation, which is limited by gene size and viral packaging constraints, RNA editing may correct transcripts too long to fit within a viral vector. In this article we examine the development of RNA editing and discuss potential applications and pitfalls. We argue that, although in its infancy, an RNA editing approach can offer unique advantages for selected retinal diseases.

International Review of Cell and Molecular Biology, Volume 372
ISSN 1937-6448
https://doi.org/10.1016/bs.ircmb.2022.04.009
Copyright © 2022 Elsevier Inc.
All rights reserved.

1. Introduction

The genetic engineer's toolbox is expanding rapidly as never before. Adeno-associated viral vector (AAV) gene replacement is increasingly established as a therapeutic option, with 163 studies registered on ClinicalTrials.gov (Clinical Trials NIH, n.d.), and two FDA-licensed treatments: Luxturna® (voretigene neparvovec, AAV-RPE65) for the treatment of Leber's Congenital Amaurosis, and Zolgensma® (onasemnogene abeparvovec, AAV-SMN1) for the treatment of type 1 spinal muscular atrophy. Beyond gene replacement, the world of gene editing using CRISPR/Cas systems is also expanding. CRISPR (Clustered Regularly Interspaced Palindromic Repeats) associated proteins (Cas) are highly programmable constructs derived from bacterial innate immune systems, and guided to a target by a sequence-specific small guide RNA (sgRNA). A CRISPR/Cas9 editing strategy involving dual sgRNAs has been developed to excise an aberrant splice site in CEP290-associated Leber's Congenital Amaurosis (Maeder et al., 2019), and is undergoing phase 1/2 human trials. CRISPR base and prime editors have also been developed which allow targeted nucleotide edits: thus far, base editors are able to correct all four transition mutations (A > G, G > A, C > T and T > C), while prime editors are theoretically able to correct all mutations except large insertions, deletions, duplications and inversions (Kantor et al., 2020).

Nonetheless, key limitations remain. While gene replacement is able to correct recessive or haploinsufficiency disease by providing wild type protein, it is unlikely to be beneficial in dominant disease. It is also limited by viral packaging capabilities. For AAVs this is limited above 4.7 kB; after the addition of essential regulatory elements, only 3.5 kB of packaging space remain for gene delivery. Work is ongoing into the use of dual vectors to deliver long transcripts. However, delivery of transgene by this method remains unreliable and with the potential to generate unwanted transgene products due to undesirable recombination events between the split gene components (McClements and Maclaren, 2017).

Meanwhile, although CRISPR/Cas9 is able to efficiently cause double stranded breaks at a target DNA site, leading to the formation of insertions and deletions (indels) by non-homologous end-joining (NHEJ), the type of indel generated is random. This randomness means CRISPR/Cas9 editing via an NHEJ pathway is only suited for gene knockdown or, as in the case of CEP290, to excise an intronic segment acting as a cryptic splice site—a situation in which single nucleotide editing precision is not required (Maeder et al., 2019). In CRISPR/Cas9 homology-directed repair (HDR), a donor template is provided with homologous arms to the edited region, allowing integration of a desired sequence. Unfortunately, HDR is really only effective in actively dividing cells, rendering it inappropriate for photoreceptors. Furthermore, although both CRISPR base and prime editors show in vitro promise, both far exceed the packaging capabilities of AAVs and so lack a plausible delivery strategy (Kantor et al., 2020).

What then is available for recessive conditions with long transcripts that cannot be packaged into an AAV? In the eye, these account for an estimated 28% of all inherited retinal dystrophies (Stone et al., 2017). In cases where the disease causing mutation has created an aberrant splice site, antisense oligonucleotide therapy to redirect splicing is one effective option (Collin et al., 2012). For the remaining mutations, one potential approach is RNA editing.

RNA editing is an impermanent gene manipulation method, limited currently to only adenosine to inosine (A to I) or cytosine to uridine (C to U) edits. It may at first seem an unattractive proposition. Nonetheless, RNA editing possesses unique advantages. RNA editing does not involve permanent changes to the genome and is theoretically much easier to reverse. This is an attractive option given the novelty of gene-editing technologies and lack of longer term data, and is especially important in younger patients, as it avoids potential permanent off-targets edits in DNA. RNA editing is not limited by PAM recognition sites (as required with CRISPR/Cas9 systems), and thus can be targeted to any area of RNA. Further, many RNA editing constructs lie within AAV packaging capabilities. Might RNA editing be the missing tool for the currently untreatable 28%?

2. The advent of ADAR

2.1 ADAR: A natural RNA editor

Adenosine deaminase acting on RNA (ADAR) is a naturally occurring enzyme originally identified in *Xenopus* embryos (Woolf et al., 1995). The ADAR family is highly conserved across the animal kingdom. Three members of the ADAR family have been identified in vertebrates. The two main forms are ADAR1, which can be further subdivided into two isoforms (p110 and p150) (Bass, 2002), and ADAR2 (Melcher et al., 1996). ADAR1 is ubiquitously expressed, including in photoreceptors, while ADAR2 is most highly expressed in brain but can also be found elsewhere (Huntley et al., 2016). ADAR3 expression is limited to the brain, and its role and its RNA binding and deamination abilities are less clear (Chen et al., 2000), although it has been implicated as an inhibitor of ADAR2 editing by competitive binding (Oakes et al., 2017). All members of the family comprise a double stranded RNA (dsRNA) binding domain of around 65 amino acids in length, and a deaminase domain (Nishikura, 2016).

RNA is composed of four nucleotides, uracil, guanosine, adenosine, and cytosine. Deamination of adenosine by ADAR changes adenosine to inosine, which is then read as guanosine during translation (Basilio et al., 1962) (Fig. 1A). ADAR editing thus effects an A to G edit. During deamination, the targeted adenosine is "base-flipped" and placed into the ADAR

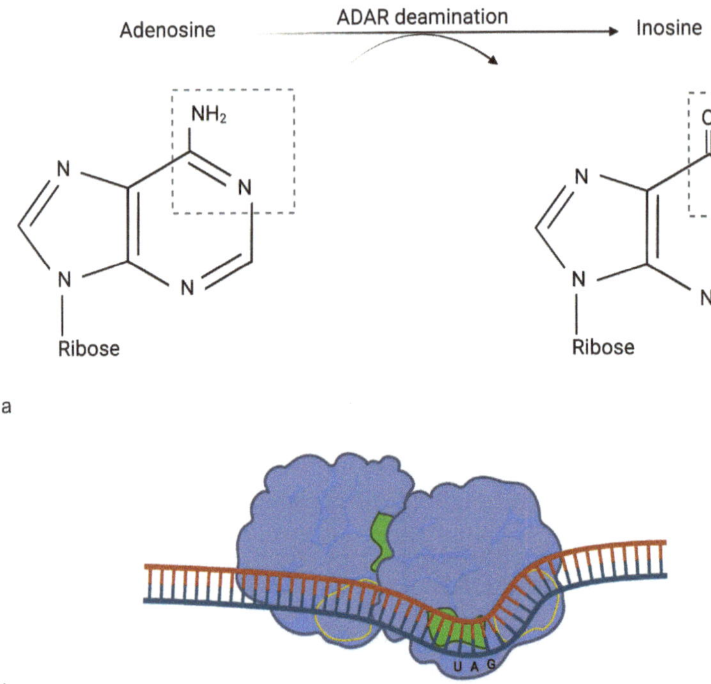

a

b

Fig. 1 (A) Deamination of adenosine to inosine. (B) Schematic showing ADAR binding to its substrate. In nature, ADAR1 and ADAR2 form homodimers via the interaction between the deaminase domain (shown in green) and part of the dsRNA-binding domain (dsRBD, shown in yellow) of one ADAR to an alpha-helix in the other. The dsRBD of the first ADAR molecule also binds a portion of the dsRNA. The second ADAR molecule is the main point of contact for the dsRNA via its dsRBD and achieves deamination via "base-flipping" of the target adenosine into the catalytic pocket (Thuy-Boun et al., 2020).

catalytic pocket (Matthews et al., 2016). Here, deamination of the adenosine C6 site occurs. Adenosines that are flanked by a 5′ uridine and a 3′ guanosine are preferentially targeted, and there is some suggestion that secondary structure plays a role in the selection of certain editing sites (Eggington et al., 2011; Kuttan and Bass, 2012) (Fig. 1B). The clearest example of this is in ADAR editing of the glutamate receptor GRIA2, which forms a secondary dsRNA structure comprised of its exonic target sequence and a downstream intronic complementary sequence. Due to this secondary structure, GRIA2 editing is specific to ADAR2 (Lomeli et al., 1994). In general, however, ADAR1 and ADAR2 show considerable overlap in activity, although they have individual preferences for certain editing sites (Costa Cruz et al., 2020; Melcher et al., 1996; Riedmann et al., 2008).

Our knowledge of natural ADAR function still remains limited. While an A to I edit can lead to an amino acid substitution and hence to modification of protein structure and function, only approximately 20% of ADAR A to I edits are made within coding sequences (Huntley et al., 2016; Nishikura, 2016). Thus the vast majority of ADAR editing does not lead to a "re-coding" event.

Where re-coding does occur, it can be significant. In the case of the GRIA2 receptor (also known as GluA2 or GluR2), A to I editing is essential for its proper function. GRIA2 is a subunit of a CNS glutamate receptor, AMPA. ADAR editing of GRIA2 is specific to ADAR2 due to its secondary structure (Wright and Vissel, 2012). A to I editing leads to substitution of a glutamine with an arginine, altering its calcium permeability (Lomeli et al., 1994). Mice lacking ADAR2 die in the first weeks of life after repeated seizures (Higuchi et al., 2000). Alterations in GRIA2 editing have been postulated as contributors to human CNS disease such as amyotrophic lateral sclerosis, astrocytoma, and transient forebrain ischaemia (Hideyama et al., 2012; Hogg et al., 2011; Oakes et al., 2017; van der Kwast et al., 2020). There is also a considerable body of work on the impact of ADAR editing on carcinogenesis and tumor progression, although the exact mechanism of ADAR edits on such events is not clear (Fritzell et al., 2018; Hogg et al., 2011). For other coding sequences, however, the impact of A to I editing is less clear, and the effects of many observed edits are unknown.

ADAR editing in intronic sequences, which account for 80% of all edits (Athanasiadis et al., 2004; Riedmann et al., 2008), results in even less well characterized and more diverse results. ADAR has a predilection for Alu repeats (Athanasiadis et al., 2004; Daniel et al., 2015; Kim et al., 2004), which play an important role in immune system modulation, gene expression, and microRNA (miRNA) regulation (Daniel et al., 2015). For some transcripts, editing of Alu repeats can create a new intronic splice site. In the case of ADAR2, self-editing of its own intron and creation of such a splice site results in a negative self-regulation of protein expression (Feng et al., 2006; Rueter et al., 1999). It is thought that extensively edited Alu ADAR also causes edits to miRNAs, resulting in a self-regulating loop, where the modified miRNAs regulate fully formed RNAs (Nishikura, 2016). Work done with engineered molecules mimicking edited Alu repeat regions suggests that that extensively edited Alu repeats lose the ability to activate innate inflammatory responses, leading to downregulation of the interferon pathway (Tossberg et al., 2020). This correlates with observations that ADAR1 mutations have been found in Aicardi-Goutières syndrome, an autosomal recessive inflammatory disorder associated with an increase in interferon alpha (Rice et al., 2012), and have been implicated in multiple sclerosis (Tossberg et al., 2020). ADAR1 meditated edits also play a role in regulation of apoptosis (Yang et al., 2017), and ADAR1 null mice die as embryos, possibly due to a dysregulated apoptotic response (Huntley et al., 2016; Riedmann et al., 2008; Wang et al., 2004).

2.2 Developing ADAR editing for therapeutics

The earliest attempt at harnessing endogeneous ADAR for RNA editing simply used a strand of antisense oligonucleotide complementary to the target region to create a section of double stranded RNA. This double stranded RNA could then recruit endogeneous ADAR (referred to in the paper as dsRAD or double stranded RNA adenosine deaminase) extracted from cell nuclei (Woolf et al., 1995). Using this technique, Woolf et al. were able to edit a UAG stop codon to a UGG tryptophan in the dystrophin gene both in vitro and in vivo (in xenopus embryos) with approximately 35% editing efficacy. The editing was highly non-specific, with several neighboring A nucleotides also edited to Gs (Woolf et al., 1995). However, this was a key proof of concept. It demonstrated ADAR could be harnessed to correct pathogenic RNA sequences.

Since then, a wide variety of constructs have been designed for programmable ADAR-based RNA editing systems. Although differing in specifics, they all hold two key features in common. The first is a guide RNA (gRNA) which allows targeting of a specific RNA transcript area. The gRNA is an oligonucleotide consisting of a section with a secondary structure which allows binding of the effector component (see below), and an antisense segment complementary to the target sequence. Due to the importance of binding to the complementary orphan base (that is, the base opposite the target adenosine, which is "base-flipped" into the ADAR catalytic pocket) in the process of ADAR deamination, the gRNA is commonly designed to incorporate a C mismatch at the A target domain, as pyrimidine orphan bases—i.e., C mismatches (and U pairs)—have been found to increase the efficacy of ADAR binding and thus of editing (Wong et al., 2001). The second is an effector component which allows RNA binding. The effector may be bound to an engineered ADAR or to the ADAR deaminase domain only. Alternatively, the effector may recruit endogeneously expressed ADAR. Single constructs have also been designed where the guide RNA and the effector are fused.

Targeted mutations to the ADAR protein and optimized gRNA design have improved ADAR activity and selectivity. Site-directed mutagenesis of the human ADAR2 catalytic domain found that a hyperactive mutant, E488Q, showed significantly higher A to I editing at G**A**C sites (target A in bold) typically unfavorable to wild-type ADAR editing, which prefers a U**A**G target, while still maintaining high editing efficacy of U**A**G targets (Kuttan and Bass, 2012). E488Q lies within a highly conserved loop in the ADAR RNA-binding site. Targeted mutagenesis around this site and subsequent examination of enzyme kinetics suggested that the substitution of E488 with a polar, uncharged amino acid with a side chain did not alter binding affinity but was more efficient at base-flipping the target A into the ADAR catalytic pocket for subsequent deamination and also possessed greater catalytic activity (Kuttan and Bass, 2012). The side chain has been identified as important in the base-flipping action, where it hydrogen-bonds to the complementary-strand "orphan base" (Matthews et al., 2016). Work on screening ADAR1 mutants for variants of E1008 (equivalent to the E488 residue in ADAR2) have shown that polar amino acids (i.e., E1008Q) perform an equivalent role as E488Q in increasing catalytic efficiency (Wang et al., 2015).

Hyperactive mutants and higher editing rates, however, come at the cost of increased off-targets (Katrekar et al., 2019), which is compounded by ADAR over-expression in systems using an exogeneous ADAR. Balancing the need for efficient RNA editing while minimizing off-target activity is a major challenge in all the ADAR constructs to date. Changes which increase on-target editing rates tend to also increase off-targets, and vice versa. For example, use of a nuclear localization signal (NLS; naturally occurring in human ADAR; Nie et al., 2004) can suppress off-target activity by targeting RNA editing toward the nucleus, and deletion of the NLS increases off-target activity; however, inclusion of an NLS tends to also decrease on-target editing rates (Vallecillo-Viejo et al., 2018). Conversely, use of a nuclear export signal (NES) simultaneously increases rate of editing and off-targets (Katrekar et al., 2019). Alterations have also been made to gRNA to increase editing specificity. Design features such as increasing the length of the gRNA (Katrekar et al., 2019), adding chemical modifications such as 2′O-methylation to gRNA antisense nucleotides flanking the desired target adenosine (Montiel-Gonź alez et al., 2016), and modifying levels of gRNA expression (Katrekar et al., 2019) have been tried to reduce off-targets.

The development of the REPAIR Cas13 system was a significant step in the battle against off-target activity. Developed by the Zhang lab, the REPAIR system fuses ADAR2-DD to a catalytically inactive Cas13 (dCas13). By rational mutagenesis of the dsRNA binding domains of ADAR2-DD, they were able to increase specificity significantly by reducing its ability to bind to duplex RNA, thus concentrating its activity at the target site where it was recruited at higher levels by the dCas13 (Cox et al., 2017). This managed to eliminate off-target effects at certain genes, although the problem of lowering on-target editing efficacy remained. The Zhang lab went on to develop the RESCUE system (RNA Editing for Specific C-to-U Exchange) by targeted mutation of the ADAR2 deaminase domain (ADAR2-DD) which caused it to become a cytidine deaminase, expanding the range of editable bases (Abudayyeh et al., 2019). Both RESCUE and REPAIR also come in a truncated version, allowing AAV-delivery.

The different approaches and their advantages and disadvantages are further explored below (summary: Fig. 2 and Table 1).

2.2.1 ADAR-SNAP

Vogel et al. created an effector construct consisting of ADAR with its naturally-occurring RNA substrate binding domain substituted by a SNAP-tag. The SNAP-Tag recognizes and covalently binds O6-benzylguanine (BG). The gRNA is then fused to BG at its 5′ terminus, thus allowing docking of the ADAR-DD-SNAP complex. Due to the essential BG modification of the gRNA, it cannot be encoded in a transgene. However, the gRNA can be delivered separately as a modified oligonucleotide. Vogel et al. were able to show that gRNA modified with 2′O-methylation and phosphorothioated bonds to enhance resistance to RNAse and improve intracellular uptake retained its ability to bind to the target site and recruit ADAR-SNAP (Vogel et al., 2014). Furthermore, by selectively modifying

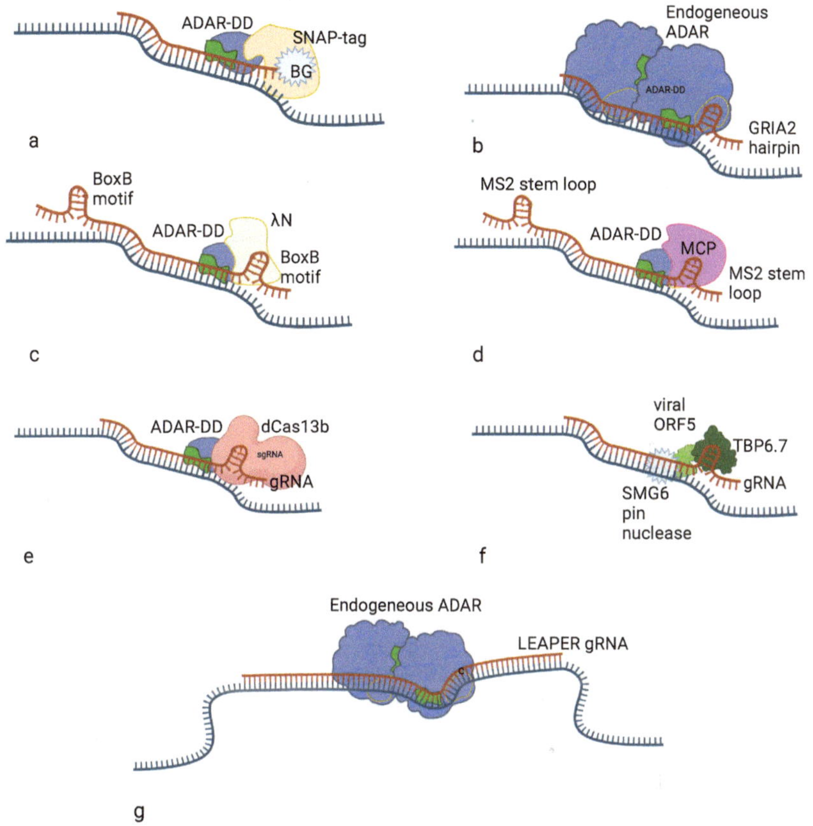

Fig. 2 Schematics of ADAR constructs. (A) ADAR-SNAP showing an ADAR deaminase domain (DD) fused to a SNAP-tag which recognizes a *O*6-benzyl-guanine (BG) fused to the gRNA. (B) ADAR-GRIA2 showing recruitment of endogenous ADAR via a GRIA2/GluR2 hairpin incorporated into the gRNA structure. (C) ADAR-BoxB showing an ADAR DD fused to a ⊠N protein which recognizes a BoxB hairpin motif in the gRNA. (D) ADAR-MS2 showing ADAR-DD fused to an MCP domain which recognizes an MS2 stem loop motif in the gRNA. (E) ADAR-dCAs13b. An ADAR-DD is fused to a deactivated Cas13b which recognizes its gRNA. (F) CIRTS1 showing the first iteration of CIRTS using a TBP6.7 domain (recognizing a TAR hairpin motif in the gRNA) bound to a viral ORF5 RNA binding segment and a human SMG6 Pin nuclease, a mRNA nonsense-mediated decay factor. (G) LEAPER, showing recruitment of endogenous ADAR via a long strand of LEAPER gRNA. gRNAs all shown in red, target RNAs shown in blue.

residues surrounding the target A, they were able to increase site specificity of editing even within an adenine-rich region. Although the need for repeated administration of gRNA for continued RNA editing is a potential drawback, the need for re-treatment depends both on the initial successful editing rate

Table 1 Summary of ADAR effector systems.

System	Effector	gRNA binding domain	Recognition domain	gRNA modifications/ design	Potential pitfalls	References
ADAR-SNAP	ADAR1-DDE1008Q	SNAP-tag	O6-benzylguanine (BG)	2′O-methylation and phosphorothioated bonds at each nucleotide apart from three surrounding target A	Need to deliver gRNA as an modified oligonucleotide (potential need for multiple treatments)	Vogel et al. (2014)
	ADAR2-DDE488Q					Vogel et al. (2018)
				C mismatch at target A site		
ADAR-GluR2	Endogeneous ADAR1 (p150 and p110)	–	GluR2	2′O-methylation of all nucleotides, phosphorothioated bonds at 5′ terminus C mismatch at target A site	Poor recruitment of endogeneous ADAR (thus poor efficacy)	Wettengel et al. (2017)
	Exogeneously delivered ADAR2-E488Q	–	GluR2	C mismatch at target A site	High level of off-targets Observed toxicity in vivo	Katrekar et al. (2019)
ADAR-BoxB	ADAR2-DD-E488Q	λN	BoxB	C mismatch at target A site	High level of off-targets	Montiel-Gonzalez et al. (2013, (2016)

ADAR MS2-MCP	ADAR2-DD (E488Q ADAR DD) and ADAR1-DD	MCP	MS2	C mismatch at target A site	High level of off-targets	Katrekar et al. (2019)
ADAR-dCas13b	ADAR2-EE-E488Q	Cas13b	Cas13b gRNA scaffold	C mismatch at target A site (for A to I editing)	Size of construct	Cox et al. (2017)
	ADAR2-DD cytidine deaminase			G mismatch at non-target A site (to diminish off-targets)		
ADAR-CIRTS	ADAR2E488Q	TBP	Tar hairpin	C mismatch at target A site	Low level of efficacy compared to Cas13b	Abudayyeh et al. (2019)
LEAPER	Endogeneous ADAR	–	–	“arRNA”: long strands (71 to 151 nt) of complementary RNA with a C mismatch at the target A site	Lower levels of efficacy	Qu et al. (2019)

and the half-life of the resultant corrected protein. For structural proteins with a long half-life, it may be that a single administration of a chemically modified gRNA is able to effect changes persisting for a year or even longer (Cideciyan et al., 2019).

Further work with the hyperactive ADAR mutants of ADAR1 (E1008Q) and ADAR2 (E488Q) combined with chemical modification of gRNA to improve target specificity increased editing rates up to 90% in vitro, with editing achieved in the 5′UTR, 3′UTR, and open reading frame (ORF). Suppression of translation with puromycin was able to boost ORF editing, an area relatively resistant to ADAR-editing. Although the authors indicated that off-targets, even in adenine-rich regions, were relatively low (usually below 10%), this still represents an unacceptably high off-target range, indicating further refinement is required (Vogel et al., 2018).

2.2.2 GluR2-ADAR (GRIA2, GluRB)

2.2.2.1 Endogeneous ADAR

As mentioned above (Section 2.1), endogeneous ADAR2 editing of the GRIA2 mRNA transcript occurs by recruitment of ADAR2 by a specific dsRNA secondary structure within GRIA2 mRNA. This GRIA2 hairpin secondary structure can be incorporated into a gRNA to recruit endogeneous or exogeneous ADAR2. While the idea of recruiting endogeneous ADAR is appealing, editing rates when only gRNA was delivered were disappointing when compared to co-delivery of gRNA with exogeneously expressed ADAR (Wettengel et al., 2017). By alteration of the ADAR-recruiting domains in the gRNA, in addition to chemical modification of the gRNA (delivered as an oligonucleotide as above with ADAR-SNAP), and using the common C mismatch at the desired A editing site, the Stafforst lab was able to increase recruitment of both endogeneous ADAR1 isoforms, in particular the p150 isoform. There was however no effect on endogeneous ADAR2. Concurrent editing with two modified gRNAs in HEK293FLpIn cells showed that two sites could be edited simultaneously without significant detriment to editing rates, which were up to 85–95% with ADAR1p150 (Merkle et al., 2019). Transfection of the modified gRNA (with 2′O-methylation and phosphorothioate at the 5′ end, and 2′O-methylation of all nucleosides) into fibroblasts, RPE and hepatocytes gave 12.7%, 9.3% and 27% editing levels of a UAG site in the ORF of GAPDH respectively. Promisingly, off-targets were low, most likely due to the lack of ADAR over-expression, with only 3 out of 10,156 sites differently edited compared to a gRNA-free control.

2.2.2.2 Exogeneous ADAR

Fortunately for those wishing to boost editing with exogeneous ADAR, the cDNA encoding full length ADAR2 protein is only 2.1 kB, easily within the packaging limits of an AAV vector system even with the inclusion of a gRNA with incorporated GRIA2 hairpin. In vivo work using AAV-8 delivered ADAR2-E488Q along with the GRIA2-gRNA (referred to by Katrekar et al. as GluR2-ADAR), up to 33% editing was seen in the spf^{ash} mice, a mouse of ornithine transcarbamylase (OTC) deficiency, with up to 5% OTC protein restoration in the liver (Katrekar et al., 2019). Disappointingly, toxicitywas observed in the RNA-edited mouse models. Although the authors did not examine specific off-target effects in the spf^{ash} mouse, their in-vitro work in HEK293 cells supported the well-known propensity of hyperactive ADAR variants to produce off-targets. Since off-targets are also strongly associated with ADAR over-expression, they have speculated the observed toxicity may be directly related to unwanted off-target activity.

2.2.3 ADAR bacteriophage effectors

2.2.3.1 ADAR BoxB

Rosenthal et al. describe an effector construct using the hyperactive E488Q deaminase domain of human ADAR2 fused to ʎN, a protein which naturally occurs in bacteriophages. ʎN binds BoxB, a short 5–6 bp stem with a 6–7 nt loop hairpin RNA structure. The guide RNA incorporates a BoxB motif in its sequence, allowing it to recruit the ADAR-DD-ʎN effector. Initial efforts in vitro resulted in poor editing efficacy in cells (Montiel-Gonzalez et al., 2013). Optimizations have included increasing the number of ʎN residues bound to a single deaminase domain, using a gRNA designs with two BoxB motifs, and the common optimization of including a C mismatch in the gRNA at the desired A editing site. Increasing the number of ʎN to four increased editing efficacies in HEK293 cells from 11% to 35%, while adding an extra BoxB domain increased editing up to 54% as measured by direct sequencing. Unfortunately, there was significant off-target activity, which the authors postulated was driven mainly by gRNA over-expression. Reducing gRNA over-expression reduced off-targets in the ADAR-(wt) DD-ʎN effector but had less effect on the hyperactive ADAR-(E488Q) DD-ʎN (Montiel-Gonźalez et al., 2016). Balancing editing efficiency with off-target control is the major challenge for this construct.

2.2.3.2 ADAR MS2-MCP

In the same paper as the GluR2-ADAR, Katrekar et al. described an effector using ADAR-DD fused to MS2 coat protein (MCP), a bacteriophage-derived protein which binds to an MS2 stem loop structure. The MS2 stem loop can then be incorporated into a RNA guide for recruitment of the ADAR-DD-MCP. In an optimized design, the antisense region was flanked by two MS2 stem loops. Using AAV8-mediated transduction of anterior tibialis muscle with MCP-ADAR-DD in the *mdx* mouse (a mouse model of Duchenne muscular dystrophy) lead to a 3.5% correction of the UAA stop codon to UGA (tryptophan) in exon 23, and a 1–2.5% *dystrophin* protein restoration (Katrekar et al., 2019). Although an important proof of principle, these editing rates are less impressive when compared with CRISPR-Cas9 dual gRNA excision of exon 23 in the mdx mouse, which restored 10% of normal protein expression in the authors' hands, and which in other studies have given up to 50% protein restoration, with concomitant functional improvement (el Refaey et al., 2017; Xu et al., 2016). On the other hand, it should be noted that the biological significance of this improved editing is uncertain, and is likely to vary widely for different proteins and different tissues. Restoration of even 1% of wild-type protein levels may be sufficient to significantly ameliorate phenotype where the disease is recessive and the mutation results in loss of function (Fry et al., 2020). Nonetheless, a higher rate of editing may be desirable for greater phenotypic correction, and to treat mutations resulting in a toxic product, where higher wild-type restoration may be required.

Off-targets remain an issue for this construct, despite modifications to the ADAR-DD-MCP system including addition of a nuclear-localizing signal, longer targeting gRNA sequences, use of WT as opposed to E488Q ADAR DD variants, and the use of ADAR1-DD-MCP as opposed to ADAR2-DD-MCP.

2.2.4 Cas based systems

2.2.4.1 Cas13-ADAR2-DD

CRISPR (Clustered regularly interspaced short palindromic repeats) Cas (CRISPR-associated protein) systems are naturally occurring bacterial defense mechanisms which target and create double stranded breaks in DNA and RNA. Two classes, 6 types, and 33 subtypes of Cas system have been described (Makarova et al., 2020). Earlier work used an inactivated Cas9 (a type II DNA targeting system which causes double stranded breaks in DNA) together with what was termed a "PAMmer"—an oligonucleotide carrying a DNA PAM recognition site essential for Cas9 binding and action (Batra et al., 2017; O'Connell et al., 2014) to create breaks in RNA and allow excision of pathogenic microsatellite repeats within RNA in patient fibroblasts. The discovery of Cas13a, a class 2, type IV RNA targeting system, meant the PAMmer could be dispensed with. Cas13 systems cleave their target RNA via two Higher Eukaryotes and Prokaryotes Nucleotide-binding endoRNAse (HEPN) domains (Liu et al., 2017). Mutation of HEPN domains can abrogate their endonuclease activity. The Cas protein is then simply turned into a gRNA recognizing effector molecule and can be fused to an ADAR2-DD domain, creating an effector which recognizes any gRNA which incorporates a Cas-scaffold (Cox et al., 2017).

Cas13 HEPN binding has a preference for certain protospacer flanking sites (PFS), much as Cas9 has a preference for protospacer adjacent motifs (PAM sites). Cas13a, however, does not have this requirement, making it a potentially highly flexible and programmable RNA editor (Cox et al., 2017). Although attempts to utilize a *L. wadei* Cas13a (LwaCas13a) gave knockdown rates of up to 85% for endogeneous genes in HEK297 cells with minimal off-target activity, the need for a monomeric superfolder GFP stabilization domain for efficient knockdown increased the size of the construct and led to a search for an alternative. Accordingly, the Zhang lab went on to develop a dCas13b derived from Prevotella sp. P5–125, which was functional without a stabilization domain, with an average knockdown of 62.9% in HEK293 cells. Testing revealed that PspdCas13b tolerated a wide variety of PFS combinations in mammalian cells but was highly intolerant of mismatches from base pairs 12–26, a key factor in minimizing off-target events (Cox et al., 2017).

Fusing the dCas13b to ADAR2-DD (E488Q) and use of a 30 nucleotide length guide RNA with a C mismatch at the target adenosine allowed A to I editing regardless of the identities of the flanking nucleotides (although editing efficacies differed from 12% to 35% according to the identity of the neighbors, with G**A**C the lowest and U**A**U the highest rate). Using dCas13b-ADAR2-DD (E488Q), the authors achieved 35% correction of an X-linked diabetes insipidus disease mutation (AVPR W293X) and 23% of a Fanconi anemia mutation (FANCC W506X) in HEK293FT cells. As mentioned previously, rational mutagenesis of the duplex RNA binding region was able to eliminate off-targets in the KRAS and PPIB target regions, although on-target efficiencies were correspondingly lower (27.1% for KRAS and 13% PPIB).

2.2.4.2 A programmable cytidine deaminase

An exciting development relevant beyond the Cas13-ADAR construct was the development of a cytidine deaminase capable of C to U editing. This was achieved by rational mutagenesis of the catalytic core of the human ADAR2-DD, based on *Escherichia coli* cytidine deaminase structure. Having validated its activity in yeast, the Zhang lab selected an ADAR2-DD cytidine deaminase fused to a dCas13b ortholog from *Riemerella anatipestifer* (here referred to as dRanCas13b), which they termed RESCUE (RNA Editing for Specific C-to-U Exchange). The resultant deaminase retained A to I editing ability, while gaining C to U editing activity. RESCUE was able to edit A to I and C to U simultaneously in the CTNNB1 transcript in HEK293FT cells at rates of 15% and 5% (Abudayyeh et al., 2019). However, there were some target limitations, with a 5′U or A preferred, and numerous off-targets were noted at adenosines surrounding the target A. The same rational mutagenesis of the duplex RNA binding region, along with gRNA design incorporating mismatched guanosines at non-target adenosines, was able to reduce but not completely eliminate off-targets.

A key limitation in the Cas-ADAR system is the size of the construct, which at 4473 bp (not including regulatory elements) is too large to be packaged within an AAV vector. However, a truncated version of REPAIR has been developed with a C-terminal truncation of 321 bp which retained editing ability and at 4152 bp lies within the packaging capability of an AAV. Similarly, a truncated C-terminal version of RESCUE has been developed. Their delivery and performance remains to be evaluated. Nonetheless, the development of a cytidine deaminase significantly increases the scope and applicability of RNA editing.

2.2.4.3 CIRTS

CIRTS (CRISPR-Cas-inspired RNA targeting system) is a designed, programmable, CRISPR-inspired molecule able to interact with gRNA and enact a range of effects in a site-specific manner, including ADAR A-to-I editing. CIRTS has two significant potential advantages over Cas13-ADAR. The first is that it is small and therefore able to fit into an AAV without further truncation. The second is more theoretical, and based on the observation that circulating antibodies to naturally-occurring Cas proteins are present in the population. Due to the need for potential re-treatment in an RNA-editing based therapy, the presence of endogeneous circulating antibodies raises the specter of future immune response. As a novel, engineered molecule, CIRTS circumvents the issue of pre-existing antibodies.

CIRTS is made up of a (1) a recognition domain that can bind specifically to a target on the guide RNA (e.g., a hair-pin binding protein) (2) an effector, i.e., an endonuclease (3) a charged protein to stabilize interaction with the gRNA and (4) the gRNA itself, containing a hairpin or other domain for binding, and an antisense region complementary to the target site. The first iteration of CIRTS involved the components (1) human hairpin-binding protein TBP6.7 fused to a (2) Pin nuclease domain of human SMG6, a nonsense-mediated mRNA decay factor (3) a single-stranded RNA binding protein comprised of ORF5 of a viral ssRNA binding protein, with a separate (4) gRNA including a TAR hairpin which is recognized by TBP6.7. This iteration (CIRTS1) was able to degrade target RNA transcripts in an vitro luciferase assay, although significantly less effectively than Cas13b. Further iterations involved the replacement of viral components with entirely human protein components, and the delivery of ADAR2-E488Q using CIRTS8 to correct a G to A mutation in firefly luciferase, resulting in quantifiable restored luciferase expression. CIRTS is interesting not only for the potential advantages in terms of immunogenicity and construct size, but the light it has shed on the interactions between Cas systems components with gRNA and target, which offer the temptation of custom-designed, programmable Cas-like molecules for any target (Rauch et al., 2019).

2.2.4.4 Cas7–11

Cas7–11 is a member of the class 1, type III CRISPR family. Class 1 CRISPR members comprise several functional units which combine to mediate DNA and RNA editing, while class 2 members consist of a single effector molecule (Makarova et al., 2020). For this reason, therapeutic research has focussed on class 2 Cas proteins such as Cas9 and Cas13b. The recent discovery of Cas7–11 has changed this landscape.

Cas7–11 is a type III-E Cas system. Discovered in *Desulfonema ishimotonii*, *Di*Cas7–11 comprises of 3 Cas7 units fused together with a Cas11 unit. In essence, it is a single effector molecule. Cas7 is known to have RNA cleavage activity. An in-vitro screen of *Di*Cas7–11 activity directed against a 60 nt RNA target using a crRNA showed effective cleavage at a number of sites, without the PFS constraints seen with HEPN domains (as with Cas13). Further analysis showed that Cas7–11 is active only against single stranded

RNA (ssRNA) and has no DNA cleavage activity. Targeting of a *Gaussia* luciferase (Gluc) in HEK293FT and U87 cells showed up to 85% knockdown with *Di*Cas7–11, comparable to Cas13-ADAR (Özcan et al., 2021).

*Di*Cas7–11 is 1602 amino acids long and thus does not easily package into an AAV, although using a dual-vector approach with hvsCas7–11 from a hydrothermal vent microbial metagenome sample, Özcan et al. were able to effect 25–40% Gluc knockdown in HEK293FT, HepG2, primary neuron and primary mouse embryonic stem cells (Özcan et al., 2021). It is also notable that although off-target activity remains an issue, *Di*Cas7–11 resulted in less off-target RNA editing and differential gene expression than *Lwa*Cas13a, *Psp*Cas13b and *Rfx*Cas13d (resulting in 30–50% cell death). More fundamentally, the existence and programmability of *Di*Cas7–11 provides a key insight into the existence and capabilities of class 1 Cas fusion molecules, which may further open up the RNA editing horizon.

2.2.5 Leaper

Similarly to the Stafforst lab's work on RESTORE, LEAPER (Leveraging Endogeneous ADAR for programmable editing of RNA) utilizes long strands of complementary RNA with a C mismatch at the desired editing site, and G mismatches at other A sites, in order to direct specific editing. As ADAR naturally binds to strands of double stranded RNA, it utilizes endogeneous ADAR binding activity without the need for a recruiting domain. Qu et al. termed these long strands of gRNA "arRNA": they vary in length from 71 nt up to 151 nt. Abrogation of endogeneous ADAR expression in cells transfected with arRNA also prevented editing activity, showing that native ADAR was responsible for the editing effect. Editing rates of around 13% were seen in HEK cells, and all combinations of nucleotide neighbors could be edited, although ADAR retained a preference for the UAG motif. Use of LEAPER to target endogeneous genes in human cell lines showed editing rates of 40%, 80% and 20% of the PPIB gene in human primary pulmonary fibroblasts, primary bronchial epithelial cells, and primary T cells respectively (Qu et al., 2019). Furthermore, the LEAPER arRNA was able to achieve editing whether it was delivered by lentiviral transduction (i.e., expressed from a transgene) or delivered as a modified oligonucleotide, which gives it a significant advantage over both RESTORE and ADAR-SNAP, both of which rely on chemical modification of oligonucleotides for editing activity and thus potentially require future re-treatment.

3. RNA editing in the eye

3.1 The eye as a site for RNA editing

The eye is an uniquely suitable body site for gene editing approaches. It is immunologically privileged and exists as a relatively self-contained entity, allowing localized delivery of therapy. Furthermore, each potential component of an ADAR-construct delivery system has already been tested in the eye. The surgical techniques of intravitreal or subretinal delivery of oligonucleotides or an AAV vector (required to deliver an RNA editing construct) (Xue et al., 2017) and its safety and lack of immunogenicity have been well established (Barker et al., 2009; Li et al., 2009). The efficacy of AAVs for transducing photoreceptors and RPE have been shown in a number of ongoing clinical trials, including those for X-linked RPGR, AMD and choroideremia (Clinical trials.gov NCT03116113, 2022a; Clinical trials.gov NCT04704921, 2022b; Dimopoulos et al., 2018), and AAV-RPE65 is an FDA approved therapy for Leber Congenital Amaurosis (FDA approval for Luxturna, n.d.). Meanwhile, chemically-modified RNA similar to those seen in the ADAR-SNAP and ADAR-BoxB systems has been delivered intravitreally with successful modification of splicing within photoreceptors, and long-lasting effects (Cideciyan et al., 2019). The way forward for an RNA-editing construct targeting the eye has been well prepared.

3.2 Potential targets

As discussed in the introduction, the most rational targets for RNA editing are monogenic insufficiency diseases with transcripts too long to be delivered via AAV. For other conditions, RNA editing is less suitable. For example, in recessive conditions involving a small gene, AAV gene replacement is a straightforward and effective one-off treatment with a strong body of evidence on its safety and efficacy. For autosomal dominant disease, correcting the transcript through RNA editing would require continual robust expression and high efficacy of any RNA editing construct, lest the abnormal gene product reaccumulate.

For recessive or X-linked conditions (or perhaps autosomal dominant disease caused by haploinsufficiency) with long transcripts, RNA editing would aim to restore native protein function. One might think that this is subject to the same problem discussed above in treatment of autosomal dominant disease, in that ongoing robust RNA editing is likewise required. However, it is likely that much lower levels of efficacy would be required in diseases of insufficiency compared to diseases of toxicity. One study relating mRNA transcript levels to disease phenotype in rare slowly-progressing mild cases of choroideremia showed that even levels of wildtype CHM mRNA at 1% those of normal can preserve the majority of retinal structure and function (Fry et al., 2020). Furthermore, depending on the half-life of the protein product, a single editing event may have a prolonged duration of effect (Cideciyan et al., 2019).

The three most common inherited recessive retinopathies with transcripts exceeding the packaging limitations of AAV are ABCA4 (17.3%), USH2A (7.6%), and CEP290 (1.8%) (Stone et al., 2017). Mutations in ABCA4 (the ATP-binding cassette transporter gene found in photoreceptors) are estimated to be the most common gene implicated of recessive cone-rod dystrophy (Maugeri et al., 2000), the most common manifestation of which is Stargardt disease, a disease with onset in childhood and leading to severe central vision impairment in later life. Mutations in USH2a (coding for usherin, a matrix protein with important structural functions, associated with the cilium in photoreceptors and in hair bundles of the cochlea (Liu et al., 2007)) are estimated to be the most common gene implicated in recessive rod-cone dystrophy (Dryja et al., 2002). All three have onset in childhood, making the potential reversibility of RNA editing particularly attractive, as the long-term effects of genome editing are currently unknown.

A search of targetable mutations (G to A or T to C) in the LOVD database reveals these make up 26% of all unique pathogenic or likely pathogenic variants in ABCA4, and 18% in USH2a, and 5% in CEP290 (Fokkema et al., 2011). That targetable mutations make up a significant percentage of single nucleotide variants is in keeping with the finding that transversion mutations are far more common than transition mutations in nature, although the reason for this is unclear. Indeed, the general principle is applicable to genetic conditions beyond the eye: in the OTC gene, for example, out of 30 unique pathogenic variants, 11 (37%) are amenable to RNA editing. Thus there is no paucity of potential targets.

4. Challenges and pitfalls

4.1 Off-targets

Currently the biggest obstacle to the implementation of RNA editing are the high levels of off-targets, and the safety concerns associated with these. The level of off-targets is closely linked with editing efficacy: as one increases, so does the other, and balancing the two is a major challenge. Off-target effects from RNA editing within the eye would be unlikely to create systemic toxicity, as seen with ADAR2E488Q editing in the spf^{ash} mouse (Katrekar et al., 2019). Nonetheless, given the role of ADAR editing in tumorigenesis and immune regulation, a generalized increase in ADAR editing can have potentially serious local effects. Since the use of hyperactive ADAR mutants is a prerequisite for achieving meaningful levels of editing, the potential link between increased editing, toxicity, tumorigenesis and immune regulation is concerning.

4.1.1 Tumorigenesis

Increased ADAR editing has been linked to changes in apoptosis and tumorigenesis (Fritzell et al., 2018; Yang et al., 2017). ADAR1 over-expression has been found in tumors, and there is widespread increased RNA editing in tumor cells (Paz-Yaacov et al., 2015). While the majority of ADAR edits are within non-coding regions, re-coding events have been implicated in the development and progression of diverse cancers from hepatocellular carcinoma to breast cancer to melanoma (Chen et al., 2013; Sagredo et al., 2018; Shoshan et al., 2015). Increased ADAR editing have been associated with cell proliferation, generation of metastases, decreased tumor suppression, and altered drug sensitivity (Chen et al., 2013; Han et al., 2015; Paz-Yaacov et al., 2015; Shoshan et al., 2015).

4.1.2 Immunogenicity

Dysregulation of RNA editing has been associated with autoimmune disease. Mutations in ADAR1 leading to decreased activity have been linked to Aicardi-Goutieres syndrome (Rice et al., 2012, 2013) and implicated in multiple sclerosis (Tossberg et al., 2020), and ADAR$^{-/-}$ mice experience embryonic lethality as a consequence of widespread apoptosis linked to over-expression of interferon (IFN) (Wang et al., 2004). ADAR1 editing of Alu repeats in endogenous dsRNA sequences is thought to downregulate our immune response to self dsRNA.

There is, however, speculation that *increased* RNA editing may also be associated with autoimmune disease activity. Studies have found that overall RNA editing rates (both A to I and C to U) were significantly increased in patients with systemic lupus erythematosus (SLE) compared to controls (Roth et al., 2018). Increased ADAR activity has been linked to altered gene expression in T cells (Laxminarayana et al., 2002, 2007), and in genetic loci which represent autoantibody targets in SLE (Roth et al., 2018). The authors have suggested that increased editing may contribute to disease by generation of self-antigens, alteration of autoantibody targets, and dysregulation of T cell responses.

4.1.3 Tackling off-targets

To combat the problem of off-target activity while maintaining on-target editing efficacy, modifications in guide RNA design, the use of nuclear localizing sequences, and the alteration of the native ADAR binding domain have all been utilized (Table 2, summary). Nonetheless, significant improvements remain necessary.

Table 2 Summaries of optimizations to reduce off-target effects.

Optimization	Effect	Reference
NLS (nuclear localization signal)	Directs editing toward transcripts in the nucleus and thus limits editing of transcripts in the cytoplasm	Vallecillo-Viejo et al. (2018) Katrekar et al. (2019)
Modified gRNA	gRNA chemically modified with 2′*O*-methylation, among other modifications, has an inhibitory effect on A to I editing. Thus, modifying nucleotides surrounding the target region helps confine editing to the target region only	Vogel et al. (2014)
C mismatch	A C mismatch at the target A helps the base flipping mechanism of the ADAR into the catalytic pocket	Wong et al. (2001)
G mismatch	A G mismatch inhibits the base flipping mechanism	Vogel et al. (2014)
Length of gRNA sequence	Mismatches inhibit gRNA binding to off-target areas. Mismatches are less well tolerated within the central region of the gRNA. An optimal length of binding sequence varies depending on the construct (LEAPER uses exceptionally long RNA sequences, for example)	Cox et al. (2017) Qu et al. (2019) Katrekar et al. (2019)
Modification of ADAR double-stranded RNA binding domain	Modification of this natural binding area weakens ADAR binding. The aim is to reduce the likelihood of off-target binding to naturally-occurring double-stranded RNA, and concentrate ADAR activity in areas where the construct is recruited by gRNA bearing a specific binding domain	Cox et al. (2017,2018)
Modification of ADAR recruiting domain	Improves the recruitment of endogeneous ADAR, which reduces the need for ADAR over-expression to maintain efficacy. ADAR over-expression is linked to more off-targets	Merkle et al. (2019)

4.2 Editing limitations

A more stringent limitation regards the A to I and C to U editing capabilities. Both are deamination reactions. Deamination occurs spontaneously in nature and accounts for the high number of C to T mutations: however, no equivalent "amination" step exists in which an amine group is added. Conversion of a purine to a pyrimidine or vice versa to address transversion mutations is similarly fraught with difficulty, as it requires the addition or removal of an imidazole ring. For the foreseeable future, it seems likely that RNA editing will be limited to A to I and C to U edits, limiting its potential applications.

4.3 Delivery challenges

4.3.1 Packaging limitations

Most ADAR constructs fit within an AAV. Delivery limitations are a consideration, however, with Cas13b-ADAR constructs. Although a truncated version exists which is active in vitro, it has not yet been successfully delivered with editing success in vivo. More work is required to show if the truncated Cas13b-ADAR retains sufficient stability and editing efficacy to make it a viable therapeutic option.

4.3.2 The need for retreatment

4.3.2.1 Safety concerns

Finally, a theoretical pitfall often raised with any discussion of RNA editing is the need for repeated treatment. However, this may be less problematic than it first appears. Where the RNA editing construct can be delivered by AAV (relevant for most ADAR constructs), transgene expression can remain robust in a range of tissues for months and perhaps years (Berraondo et al., 2005; Feng et al., 2004; Guy et al., 1999; Li et al., 2003; Petersen-Jones et al., 2009) and thus RNA editing can be ongoing. The effects of wild-type protein may also have a persistent effect on structure and function independently of ongoing AAV transgene expression, depending on the half-life of the protein in question (Mookherjee et al., 2015), again limiting the need for re-treatment.

If further administration is required, this is unlikely to be a problem, as repeated subretinal delivery of AAV has been shown to be minimally immunogenic (Barker et al., 2009). *Re*-delivery issues are more pertinent in constructs such as ADAR-SNAP where the gRNA is delivered as a modified oligonucleotide and thus cannot be expressed as a transgene. Nonetheless, repeated administration of modified antisense oligonucleotides intravitreally has been shown to be both safe and effective in correction of a cryptic splice site in CEP290 (Cideciyan et al., 2019).

4.3.2.2 Costs

Another potential pitfall associated with the need for retreatment is the cost-effectiveness of therapy. As a benchmark, one dose of Voretigene Neparvovec (Luxturna, Novartis Pharmaceuticals) in the United Kingdom costs £613,410 per patient (National Institute for Health and Care Excellence, 2019). Such a cost would obviously be prohibitive were frequent retreatments necessary. Indeed, the rationale for funding Voretigene Neparvovec within the National Health Service (UK) is based on the likelihood of a one-off treatment (National Institute for Health and Care Excellence, 2019). As discussed above, however, if an ADAR construct can be delivered by AAV, then depending on the duration of AAV transgene expression and the half-life of the protein in question, retreatment may not be required for many years and perhaps not at all. In constructs such as ADAR-SNAP, where a separate chemically-modified gRNA cannot be delivered via AAV, repeated deliveries of the gRNA are likely to be needed. Despite this, cost need not be prohibitive. Synthesis and chemical modification of separate gRNAs is relatively cheap and easy, and again, depending on gRNA survival and protein half-life, retreatment may not be required as frequently as imagined.

5. Conclusion

RNA editing offers key advantages for the treatment of recessive disease caused by mutations in genes with long transcripts. Another attractive quality is its relative ease of reversibility compared to editing genomic DNA. Currently, it is hampered by the challenges of controlling off-target effects. While these may be optimized, it is likely that the major limitation of its editing capabilities (addressing G to A and U to C mutations only) will remain. Clearly, many hurdles must be cleared before its implementation in clinical practice. Nevertheless, RNA editing remains an exciting area of innovation. The development of CIRTS and our increasing ability to rationally design elements of the RNA editing construct to suit our purposes is a major step forward in our ability to tackle an increasing number of human genetic diseases. Although in its infancy, we argue it is a potentially powerful and worthy addition to the gene manipulator's toolbox.

Acknowledgments

Figures created using BioRender.

References

Abudayyeh, O.O., Gootenberg, J.S., Franklin, B., Koob, J., Kellner, M.J., Ladha, A., Joung, J., Kirchgatterer, P., Cox, D.B.T., Zhang, F., 2019. A cytosine deaminase for programmable single-base RNA editing. Science 538, eaax 7063. https://doi.org/10.1126/science.aax7063.

Athanasiadis, A., Rich, A., Maas, S., 2004. Widespread A-to-I RNA editing of Alu-containing mRNAs in the human transcriptome. PLoS Biol. 2, e391. https://doi.org/10.1371/journal.pbio.0020391.

Barker, S.E., Broderick, C.A., Robbie, S.J., Duran, Y., Natkunarajah, M., Buch, P., Balaggan, K.S., Mac Laren, R.E., Bainbridge, J.W.B., Smith, A.J., Ali, R.R., 2009. Subretinal delivery of adeno-associated virus serotype 2 results in minimal immune responses that allow repeat vector administration in immunocompetent mice. J. Gene Med. 11 (6), 486–497. https://doi.org/10.1002/jgm.1327.

Basilio, C., Wahba, A.J., Lengyel, P., Speyer, J.F., Ochoa, S., 1962. Synthetic polynucleotides and the amino acid code. V. Proc. Natl. Acad. Sci. U. S. A. 48, 613–616. https://doi.org/10.1073/pnas.48.4.613.

Bass, B.L., 2002. RNA editing by adenosine deaminases that act on RNA. Annu. Rev. Biochem. 71, 817–846. https://doi.org/10.1146/annurev.biochem.71.110601.135501.

Batra, R., Nelles, D.A., Pirie, E., Blue, S.M., Marina, R.J., Wang, H., Chaim, I.A., Thomas, J.D., Zhang, N., Nguyen, V., Aigner, S., Markmiller, S., Xia, G., Corbett, K.D., Swanson, M.S., Yeo, G.W., 2017. Elimination of toxic microsatellite repeat expansion RNA by RNA-targeting Cas9. Cell 170, 899–912.e10. https://doi.org/10.1016/j.cell.2017.07.010.

Berraondo, P., Ochoa, L., Crettaz, J., Rotellar, F., Vales, A., Martínez-Ansó, E., Zaratiegui, M., Ruiz, J., González-Aseguinolaza, G., Prieto, J., 2005. IFN-alpha gene therapy for woodchuck hepatitis with adeno-associated virus: differences in duration of gene expression and antiviral activity using intraportal or intramuscular routes. Mol. Ther. 12, 68–76. https://doi.org/10.1016/j.ymthe.2005.02.017.

Chen, C.X., Cho, D.S., Wang, Q., Lai, F., Carter, K.C., Nishikura, K., 2000. A third member of the RNA-specific adenosine deaminase gene family, ADAR3, contains both single- and double-stranded RNA binding domains. RNA (New York, N.Y.) 6, 755–767. https://doi.org/10.1017/s1355838200000170.

Chen, L., Li, Y., Lin, C.H., Chan, T.H.M., Chow, R.K.K., Song, Y., Liu, M., Yuan, Y.-F., Fu, L., Kong, K.L., Qi, L., Li, Y., Zhang, N., Tong, A.H.Y., Kwong, D.L.-W., Man, K., Lo, C.M., Lok, S., Tenen, D.G., Guan, X.-Y., 2013. Recoding RNA editing of AZIN1 predisposes to hepatocellular carcinoma. Nat. Med. 19, 209–216. https://doi.org/10.1038/nm.3043.

Cideciyan, A.V., Jacobson, S.G., Drack, A.V., Ho, A.C., Charng, J., Garafalo, A.V., Roman, A.J., Sumaroka, A., Han, I.C., Hochstedler, M.D., Pfeifer, W.L., Sohn, E.H., Taiel, M., Schwartz, M.R., Biasutto, P., de Wit, W., Cheetham, M.E., Adamson, P., Rodman, D.M., Platenburg, G., Tome, M.D., Balikova, I., Nerinckx, F., de Zaeytijd, J., van Cauwenbergh, C., Leroy, B.P., Russell, S.R., 2019. Effect of an intravitreal antisense oligonucleotide on vision in Leber congenital amaurosis due to a photoreceptor cilium defect. Nat. Med. 25, 225–228. https://doi.org/10.1038/s41591-018-0295-0.

Clinical Trials NIH. WWW Document. https://www.clinicaltrials.gov/, n.d. (accessed 1.6.21).

Clinical trials.gov, 2022a. NCT03116113 [WWW Document]. clinicaltrials.gov/ct2/show/NCT03116113.

Clinical trials.gov, 2022b. NCT04704921.

Collin, R.W., den Hollander, A.I., der Velde-Visser, S.D.V., Bennicelli, J., Bennett, J., Cremers, F.P., 2012. Antisense oligonucleotide (AON)-based therapy for leber congenital amaurosis caused by a frequent mutation in CEP290. Mol. Ther.–Nucleic Acids 1, e14. https://doi.org/10.1038/mtna.2012.3.

Costa Cruz, P.H., Kato, Y., Nakahama, T., Shibuya, T., Kawahara, Y., 2020. A comparative analysis of ADAR mutant mice reveals site-specific regulation of RNA editing. RNA (New York, N.Y.) 26, 454–469. https://doi.org/10.1261/rna.072728.119.

Cox, D.B.T., Gootenberg, J.S., Abudayyeh, O.O., Franklin, B., Kellner, M.J., Joung, J., Zhang, F., 2017. RNA editing with CRISPR-Cas13. Science (New York, N.Y.) 358, 1019–1027. https://doi.org/10.1126/science.aaq0180.

Cox, D.B.T., Gootenberg, J.S., Abudayyeh, O.O., Franklin, B., Kellner, M.J., Joung, J., Zhang, F., 2018. Cas13 adar fusion. Science 358, 1019–1027. https://doi.org/10.1126/science.aaq0180.RNA.

Daniel, C., Behm, M., Öhman, M., 2015. The role of Alu elements in the cis-regulation of RNA processing. Cell. Mol. Life Sci. 72, 4063–4076. https://doi.org/10.1007/s00018-015-1990-3.

Dimopoulos, I.S., Hoang, S.C., Radziwon, A., Binczyk, N.M., Seabra, M.C., MacLaren, R.E., Somani, R., Tennant, M.T.S., MacDonald, I.M., 2018. Two-year results after AAV2-mediated gene therapy for choroideremia: the Alberta experience. Am. J. Ophthalmol. 193, 130–142. https://doi.org/10.1016/j.ajo.2018.06.011.

Dryja, T.P., Jian-Seyed-Ahmadi, B., Rivolta, C., Keene, J.A., Berson, E.L., 2002. Mutations in the USH2A gene are a frequent cause of recessive nonsyndromic retinitis pigmentosa as well as usher syndrome type II. Invest. Ophthalmol. Vis. Sci. 43, 795.

Eggington, J.M., Greene, T., Bass, B.L., 2011. Predicting sites of ADAR editing in double-stranded RNA. Nat. Commun. 2, 319. https://doi.org/10.1038/ncomms1324.

el Refaey, M., Xu, L., Gao, Y., Canan, B.D., Adesanya, T.M.A., Warner, S.C., Akagi, K., Symer, D.E., Mohler, P.J., Ma, J., Janssen, P.M.L., Han, R., 2017. In vivo genome editing restores dystrophin expression and cardiac function in dystrophic mice. Circ. Res. 121, 923–929. https://doi.org/10.1161/CIRCRESAHA.117.310996.

FDA approval for Luxturna. WWW Document. https://www.fda.gov/vaccines-blood-biologics/cellular-genetherapy-products/luxturna, n.d. (accessed 9.3.21).

Feng, X., Eide, F.F., Jiang, H., Reder, A.T., 2004. Adeno-associated viral vector-mediated ApoE expression in Alzheimer's disease mice: low CNS immune response, long-term expression, and astrocyte specificity. Front. Biosci. 9, 1540–1546. https://doi.org/10.2741/1323.

Feng, Y., Sansam, C.L., Singh, M., Emeson, R.B., 2006. Altered RNA editing in mice lacking ADAR2 autoregulation. Mol. Cell. Biol. 26, 480–488. https://doi.org/10.1128/MCB.26.2.480-488.2006.

Fokkema, I.F.A.C., Taschner, P.E.M., Schaafsma, G.C.P., Celli, J., Laros, J.F.J., den Dunnen, J.T., Celli, J., Taschner, P.E.M., 2011. LOVD v.2.0: the next generation in gene variant databases. Hum. Mutat. 32, 557–563. https://doi.org/10.1002/humu.21438.

Fritzell, K., Xu, L.-D., Lagergren, J., Öhman, M., 2018. ADARs and editing: the role of A-to-I RNA modification in cancer progression. Semin. Cell Dev. Biol. 79, 123–130. https://doi.org/10.1016/j.semcdb.2017.11.018.

Fry, L.E., Patrício, M.I., Williams, J., Aylward, J.W., Hewitt, H., Clouston, P., Xue, K., Barnard, A.R., MacLaren, R.E., 2020. Association of messenger RNA level with phenotype in patients with choroideremia: potential implications for gene therapy dose. JAMA Ophthalmol. 138, 128–135. https://doi.org/10.1001/jamaophthalmol.2019.5071.

Guy, J., Qi, X., Muzyczka, N., Hauswirth, W.W., 1999. Reporter expression persists 1 year after adeno-associated virus-mediated gene transfer to the optic nerve. Arch. Ophthalmol. 117, 929–937. https://doi.org/10.1001/archopht.117.7.929.

Han, L., Diao, L., Yu, S., Xu, X., Li, J., Zhang, R., Yang, Y., Werner, H.M.J., Eterovic, A.K., Yuan, Y., Li, J., Nair, N., Minelli, R., Tsang, Y.H., Cheung, L.W.T., Jeong, K.J., Roszik, J., Ju, Z., Woodman, S.E., Lu, Y., Scott, K.L., Li, J.B., Mills, G.B., Liang, H., 2015. The genomic landscape and clinical relevance of A-to-I RNA editing in human cancers. Cancer Cell 28, 515–528. https://doi.org/10.1016/j.ccell.2015.08.013.

Hideyama, T., Yamashita, T., Aizawa, H., Tsuji, S., Kakita, A., Takahashi, H., Kwak, S., 2012. Profound downregulation of the RNA editing enzyme ADAR2 in ALS spinal motor neurons. Neurobiol. Dis. 45, 1121–1128. https://doi.org/10.1016/j.nbd.2011.12.033.

Higuchi, M., Maas, S., Single, F.N., Hartner, J., Rozov, A., Burnashev, N., Feldmeyer, D., Sprengel, R., Seeburg, P.H., 2000. Point mutation in an AMPA receptor gene rescues lethality in mice deficient in the RNA-editing enzyme ADAR2. Nature 406, 78–81. https://doi.org/10.1038/35017558.

Hogg, M., Paro, S., Keegan, L.P., O'Connell, M.A., 2011. RNA editing by mammalian ADARs. Adv. Genet. 73, 87–120. https://doi.org/10.1016/B978-0-12-380,860-8.00003-3.

Huntley, M.A., Lou, M., Goldstein, L.D., Lawrence, M., Dijkgraaf, G.J.P., Kaminker, J.S., Gentleman, R., 2016. Complex regulation of ADAR-mediated RNA-editing across tissues. BMC Genomics 17, 61. https://doi.org/10.1186/s12864-015-2291-9.

Kantor, A., McClements, M.E., MacLaren, R.E., 2020. CRISPR-Cas9 DNA base-editing and prime-editing. Int. J. Mol. Sci. 21, 6240. https://doi.org/10.3390/ijms21176240.

Katrekar, D., Chen, G., Meluzzi, D., Ganesh, A., Worlikar, A., Shih, Y.-R.R., Varghese, S., Mali, P., 2019. In vivo RNA editing of point mutations via RNA-guided adenosine deaminases. Nat. Methods 16, 239–242. https://doi.org/10.1038/s41592-019-0323-0.

Kim, D.D.Y., Kim, T.T.Y., Walsh, T., Kobayashi, Y., Matise, T.C., Buyske, S., Gabriel, A., 2004. Widespread RNA editing of embedded alu elements in the human transcriptome. Genome Res. 14, 1719–1725. https://doi.org/10.1101/g.2855504.

Kuttan, A., Bass, B.L., 2012. Mechanistic insights into editing-site specificity of ADARs. Proc. Natl. Acad. Sci. U. S. A. 109. https://doi.org/10.1073/pnas.1212548109.

Laxminarayana, D., Khan, I.U., Kammer, G.M., 2002. Transcript mutations of the α regulatory subunit of protein kinase A and up-regulation of the RNA-editing gene transcript in lupus T lymphocytes. Lancet 360, 842–849. https://doi.org/10.1016/S0140-6736(02)09966-X.

Laxminarayana, D., O'Rourke, K.S., Maas, S., Olorenshaw, I., 2007. Altered editing in RNA editing adenosine deaminase ADAR2 gene transcripts of systemic lupus erythematosus T lymphocytes. Immunology 121, 359–369. https://doi.org/10.1111/j.1365-2567.2007.02582.x.

Li, J., Wang, D., Qian, S., Chen, Z., Zhu, T., Xiao, X., 2003. Efficient and long-term intracardiac gene transfer in delta-sarcoglycan-deficiency hamster by adeno-associated virus-2 vectors. Gene Ther. 10, 1807–1813. https://doi.org/10.1038/sj.gt.3302078.

Li, W., Kong, F., Li, X., Dai, X., Liu, X., Zheng, Q., Wu, R., Zhou, X., Lü, F., Chang, B., Li, Q., Hauswirth, W.W., Qu, J., Pang, J.J., 2009. Gene therapy following subretinal AAV5 vector delivery is not affected by a previous intravitreal AAV5 vector administration in the partner eye. Mol. Vis. 15, 267–275.

Liu, X., Bulgakov, O.V., Darrow, K.N., Pawlyk, B., Adamian, M., Liberman, M.C., Li, T., 2007. Usherin is required for maintenance of retinal photoreceptors and normal development of cochlear hair cells. Proc. Natl. Acad. Sci. U. S. A. 104, 4413–4418. https://doi.org/10.1073/pnas.0610950104.

Liu, L., Li, X., Ma, J., Li, Z., You, L., Wang, J., Wang, M., Zhang, X., Wang, Y., 2017. The molecular architecture for RNA-guided RNA cleavage by Cas13a. Cell 170, 714–726. e10. https://doi.org/10.1016/j.cell.2017.06.050.

Lomeli, H., Mosbacher, J., Melcher, T., Höger, T., Geiger, J.R., Kuner, T., Monyer, H., Higuchi, M., Bach, A., Seeburg, P.H., 1994. Control of kinetic properties of AMPA receptor channels by nuclear RNA editing. Science (New York, N.Y.) 266, 1709–1713. https://doi.org/10.1126/science.7992055.

Maeder, M.L., Stefanidakis, M., Wilson, C.J., Baral, R., Barrera, L.A., Bounoutas, G.S., Bumcrot, D., Chao, H., Ciulla, D.M., DaSilva, J.A., Dass, A., Dhanapal, V., Fennell, T.J., Friedland, A.E., Giannoukos, G., Gloskowski, S.W., Glucksmann, A., Gotta, G.M., Jayaram, H., Haskett, S.J., Hopkins, B., Horng, J.E., Joshi, S., Marco, E., Mepani, R., Reyon, D., Ta, T., Tabbaa, D.G., Samuelsson, S.J., Shen, S., Skor, M.N., Stetkiewicz, P., Wang, T., Yudkoff, C., Myer, V.E., Albright, C.F., Jiang, H., 2019. Development of a gene-editing approach to restore vision loss in Leber congenital amaurosis type 10. Nat. Med. 25, 229–233. https://doi.org/10.1038/s41591-018-0327-9.

Makarova, K.S., Wolf, Y.I., Iranzo, J., Shmakov, S.A., Alkhnbashi, O.S., Brouns, S.J.J., Charpentier, E., Cheng, D., Haft, D.H., Horvath, P., Moineau, S., Mojica, F.J.M., Scott, D., Shah, S.A., Siksnys, V., Terns, M.P., Venclovas, Č., White, M.F., Yakunin, A.F., Yan, W., Zhang, F., Garrett, R.A., Backofen, R., van der Oost, J., Barrangou, R., Koonin, E.V., 2020. Evolutionary classification of CRISPR–Cas systems: a burst of class 2 and derived variants. Nat. Rev. Microbiol. 18, 67–83. https://doi.org/10.1038/s41579-019-0299-x.

Matthews, M.M., Thomas, J.M., Zheng, Y., Tran, K., Phelps, K.J., Scott, A.I., Havel, J., Fisher, A.J., Beal, P.A., 2016. Structures of human ADAR2 bound to dsRNA reveal base-flipping mechanism and basis for site selectivity. Nat. Struct. Mol. Biol. 23, 426–433. https://doi.org/10.1038/nsmb.3203.

Maugeri, A., Klevering, B.J., Rohrschneider, K., Blankenagel, A., Brunner, H.G., Deutman, A.F., Hoyng, C.B., Cremers, F.P., 2000. Mutations in the ABCA4 (ABCR) gene are the major cause of autosomal recessive cone-rod dystrophy. Am. J. Hum. Genet. 67, 960–966. https://doi.org/10.1086/303079.

McClements, M.E., Maclaren, R.E., 2017. Adeno-associated virus (AAV) dual vector strategies for gene therapy encoding large transgenes. Yale J. Biol. Med. 90, 611–623.

Melcher, T., Maas, S., Herb, A., Sprengel, R., Seeburg, P.H., Higuchi, M., 1996. A mammalian RNA editing enzyme. Nature 379, 460–464. https://doi.org/10.1038/379460a0.

Merkle, T., Merz, S., Reautschnig, P., Blaha, A., Li, Q., Vogel, P., Wettengel, J., Li, J.B., Stafforst, T., 2019. Precise RNA editing by recruiting endogenous ADARs with antisense oligonucleotides. Nat. Biotechnol. 37, 133–138. https://doi.org/10.1038/s41587-019-0013-6.

Montiel-Gonzalez, M.F., Vallecillo-Viejo, I., Yudowski, G.A., Rosenthal, J.J.C., 2013. Correction of mutations within the cystic fibrosis transmembrane conductance regulator by site-directed RNA editing. Proc. Natl. Acad. Sci. U. S. A. 110, 18,285–18,290. https://doi.org/10.1073/pnas.1306243110.

Montiel-González, M.F., Vallecillo-Viejo, I.C., Rosenthal, J.J.C., 2016. An efficient system for selectively altering genetic information within mRNAs. Nucleic Acids Res. 44, 1–12. https://doi.org/10.1093/nar/gkw738.

Mookherjee, S., Hiriyanna, S., Kaneshiro, K., Li, L., Li, Y., Li, W., Qian, H., Li, T., Khanna, H., Colosi, P., Swaroop, A., Wu, Z., 2015. Long-term rescue of cone photoreceptor degeneration in retinitis pigmentosa 2 (RP2)-knockout mice by gene replacement therapy. Hum. Mol. Genet. 24, 6446–6458. https://doi.org/10.1093/hmg/ddv354.

National Institute for Health and Care Excellence, 2019. Voretigene neparvovec for treating inherited retinal dystrophies caused by RPE65 gene mutations highly specialized technologies guidance. NICE. https://www.nice.org.uk/guidance/hst11. (Accessed 18 January 2021).
Nie, Y., Zhao, Q., Su, Y., Yang, J.H., 2004. Subcellular distribution of ADAR1 isoforms is synergistically determined by three nuclear discrimination signals and a regulatory motif. J. Biol. Chem. 279, 13,249–13,255. https://doi.org/10.1074/jbc.M312753200.
Nishikura, K., 2016. A-to-I editing of coding and non-coding RNAs by ADARs. Nat. Rev. Mol. Cell Biol. 17, 83–96. https://doi.org/10.1038/nrm.2015.4.
Oakes, E., Anderson, A., Cohen-Gadol, A., Hundley, H.A., 2017. Adenosine deaminase that acts on RNA 3 (ADAR3) binding to glutamate receptor subunit B pre-mRNA inhibits RNA editing in glioblastoma. J. Biol. Chem. 292, 4326–4335. https://doi.org/10.1074/jbc.M117.779868.
O'Connell, M.R., Oakes, B.L., Sternberg, S.H., East-Seletsky, A., Kaplan, M., Doudna, J.A., 2014. Programmable RNA recognition and cleavage by CRISPR/Cas9. Nature 516, 263–266. https://doi.org/10.1038/nature13769.
Özcan, A., Krajeski, R., Ioannidi, E., Lee, B., Gardner, A., Makarova, K.S., Koonin, E.V., Abudayyeh, O.O., Gootenberg, J.S., 2021. Programmable RNA targeting with the single-protein CRISPR effector Cas7–11. Nature. https://doi.org/10.1038/s41586-021-03886-5.
Paz-Yaacov, N., Bazak, L., Buchumenski, I., Porath, H.T., Danan-Gotthold, M., Knisbacher, B.A., Eisenberg, E., Levanon, E.Y., 2015. Elevated RNA editing activity is a major contributor to transcriptomic diversity in tumors. Cell Rep. 13, 267–276. https://doi.org/10.1016/j.celrep.2015.08.080.
Petersen-Jones, S.M., Bartoe, J.T., Fischer, A.J., Scott, M., Boye, S.L., Chiodo, V., Hauswirth, W.W., 2009. AAV retinal transduction in a large animal model species: comparison of a self-complementary AAV2/5 with a single-stranded AAV2/5 vector. Mol. Vis. 15, 1835–1842.
Qu, L., Yi, Z., Zhu, S., Wang, C., Cao, Z., Zhou, Z., Yuan, P., Yu, Y., Tian, F., Liu, Z., Bao, Y., Zhao, Y., Wei, W., 2019. Programmable RNA editing by recruiting endogenous ADAR using engineered RNAs. Nat. Biotechnol. 37, 1059–1069. https://doi.org/10.1038/s41587-019-0178-z.
Rauch, S., He, E., Srienc, M., Zhou, H., Zhang, Z., Dickinson, B.C., 2019. Programmable RNA-guided RNA effector proteins built from human parts. Cell 178, 122–134.e12. https://doi.org/10.1016/j.cell.2019.05.049.
Rice, G.I., Kasher, P.R., Forte, G.M.A., Mannion, N.M., Greenwood, S.M., Szynkiewicz, M., Dickerson, J.E., Bhaskar, S.S., Zampini, M., Briggs, T.A., Jenkinson, E.M., Bacino, C.A., Battini, R., Bertini, E., Brogan, P.A., Brueton, L.A., Carpanelli, M., de Laet, C., de Lonlay, P., del Toro, M., Desguerre, I., Fazzi, E., Garcia-Cazorla, A., Heiberg, A., Kawaguchi, M., Kumar, R., Lin, J.-P.S.-M., Lourenco, C.M., Male, A.M., Marques, W.J., Mignot, C., Olivieri, I., Orcesi, S., Prabhakar, P., Rasmussen, M., Robinson, R.A., Rozenberg, F., Schmidt, J.L., Steindl, K., Tan, T.Y., van der Merwe, W.G., Vanderver, A., Vassallo, G., Wakeling, E.L., Wassmer, E., Whittaker, E., Livingston, J.H., Lebon, P., Suzuki, T., McLaughlin, P.J., Keegan, L.P., O'Connell, M.A., Lovell, S.C., Crow, Y.J., 2012. Mutations in ADAR1 cause Aicardi-Goutières syndrome associated with a type I interferon signature. Nat. Genet. 44, 1243–1248. https://doi.org/10.1038/ng.2414.
Rice, G.I., Forte, G.M.A., Szynkiewicz, M., Chase, D.S., Aeby, A., Abdel-Hamid, M.S., Ackroyd, S., Allcock, R., Bailey, K.M., Balottin, U., Barnerias, C., Bernard, G., Bodemer, C., Botella, M.P., Cereda, C., Chandler, K.E., Dabydeen, L., Dale, R.C., de Laet, C., de Goede, C.G.E.L., del Toro, M., Effat, L., Enamorado, N.N., Fazzi, E., Gener, B., Haldre, M., Lin, J.-P.S.-M., Livingston, J.H., Lourenco, C.M.,

Marques, W.J., Oades, P., Peterson, P., Rasmussen, M., Roubertie, A., Schmidt, J.L., Shalev, S.A., Simon, R., Spiegel, R., Swoboda, K.J., Temtamy, S.A., Vassallo, G., Vilain, C.N., Vogt, J., Wermenbol, V., Whitehouse, W.P., Soler, D., Olivieri, I., Orcesi, S., Aglan, M.S., Zaki, M.S., Abdel-Salam, G.M.H., Vanderver, A., Kisand, K., Rozenberg, F., Lebon, P., Crow, Y.J., 2013. Assessment of interferon-related biomarkers in Aicardi-Goutières syndrome associated with mutations in TREX1, RNASEH2A, RNASEH2B, RNASEH2C, SAMHD1, and ADAR: a case-control study. Lancet Neurol. 12, 1159–1169. https://doi.org/10.1016/S1474-4422(13)70258-8.

Riedmann, E.M., Schopoff, S., Hartner, J.C., Jantsch, M.F., 2008. Specificity of ADAR-mediated RNA editing in newly identified targets. RNA (New York, N.Y.) 14, 1110–1118. https://doi.org/10.1261/rna.923308.

Roth, S.H., Danan-Gotthold, M., Ben-Izhak, M., Rechavi, G., Cohen, C.J., Louzoun, Y., Levanon, E.Y., 2018. Increased RNA editing may provide a source for autoantigens in systemic Lupus erythematosus. Cell Rep. 23, 50–57. https://doi.org/10.1016/j.celrep.2018.03.036.

Rueter, S.M., Dawson, T.R., Emeson, R.B., 1999. Regulation of alternative splicing by RNA editing. Nature 399, 75–80. https://doi.org/10.1038/19992.

Sagredo, E.A., Blanco, A., Sagredo, A.I., Pérez, P., Sepúlveda-Hermosilla, G., Morales, F., Müller, B., Verdugo, R., Marcelain, K., Harismendy, O., Armisén, R., 2018. ADAR1-mediated RNA-editing of 3′UTRs in breast cancer. Biol. Res. 51, 36. https://doi.org/10.1186/s40659-018-0185-4.

Shoshan, E., Mobley, A.K., Braeuer, R.R., Kamiya, T., Huang, L., Vasquez, M.E., Salameh, A., Lee, H.J., Kim, S.J., Ivan, C., Velazquez-Torres, G., Nip, K.M., Zhu, K., Brooks, D., Jones, S.J.M., Birol, I., Mosqueda, M., Wen, Y., Eterovic, A.K., Sood, A.K., Hwu, P., Gershenwald, J.E., Robertson, A.G., Calin, G.A., Markel, G., Fidler, I.J., Bar-Eli, M., 2015. Reduced adenosine-to-inosine miR-455-5p editing promotes melanoma growth and metastasis. Nat. Cell Biol. 17, 311–321. https://doi.org/10.1038/ncb3110.

Stone, E.M., Andorf, J.L., Whitmore, S.S., DeLuca, A.P., Giacalone, J.C., Streb, L.M., Braun, T.A., Mullins, R.F., Scheetz, T.E., Sheffield, V.C., Tucker, B.A., 2017. Clinically focused molecular investigation of 1000 consecutive families with inherited retinal disease. Ophthalmology 124, 1314–1331. https://doi.org/10.1016/j.ophtha.2017.04.008.

Thuy-Boun, A.S., Thomas, J.M., Grajo, H.L., Palumbo, C.M., Park, S., Nguyen, L.T., Fisher, A.J., Beal, P.A., 2020. Asymmetric dimerization of adenosine deaminase acting on RNA facilitates substrate recognition. Nucleic Acids Res. 48, 7958–7972. https://doi.org/10.1093/nar/gkaa532.

Tossberg, J.T., Heinrich, R.M., Farley, V.M., Crooke 3rd, P.S., Aune, T.M., 2020. Adenosine-to-inosine RNA editing of Alu double-stranded (ds)RNAs is markedly decreased in multiple sclerosis and unedited Alu dsRNAs are potent activators of proinflammatory transcriptional responses. J. Immunol. 205, 2606–2617. https://doi.org/10.4049/jimmunol.2000384.

Vallecillo-Viejo, I.C., Liscovitch-Brauer, N., Montiel-Gonzalez, M.F., Eisenberg, E., Rosenthal, J.J.C., 2018. Abundant off-target edits from site-directed RNA editing can be reduced by nuclear localization of the editing enzyme. RNA Biol. 15, 104–114. https://doi.org/10.1080/15476286.2017.1387711.

van der Kwast, R.V.C.T., Parma, L., van der Bent, M.L., van Ingen, E., Baganha, F., Peters, H.A.B., Goossens, E.A.C., Simons, K.H., Palmen, M., de Vries, M.R., Quax, P.H.A., Nossent, A.Y., 2020. Adenosine-to-inosine editing of vasoactive microRNAs alters their targetome and function in ischemia. Mol. Ther.–Nucleic Acids 21, 932–953. https://doi.org/10.1016/j.omtn.2020.07.020.

Vogel, P., Schneider, M.F., Wettengel, J., Stafforst, T., 2014. Improving site-directed RNA editing in vitro and in cell culture by chemical modification of the guideRNA. Angew. Chem. Int. Ed. 53, 6267–6271. https://doi.org/10.1002/anie.201402634.

Vogel, P., Moschref, M., Li, Q., Merkle, T., Selvasaravanan, K.D., Li, J.B., Stafforst, T., 2018. Efficient and precise editing of endogenous transcripts with SNAP-tagged ADARs. Nat. Methods 15, 535–538. https://doi.org/10.1038/s41592-018-0017-z.

Wang, Q., Miyakoda, M., Yang, W., Khillan, J., Stachura, D.L., Weiss, M.J., Nishikura, K., 2004. Stress-induced apoptosis associated with null mutation of ADAR1 RNA editing deaminase gene. J. Biol. Chem. 279, 4952–4961. https://doi.org/10.1074/jbc.M310162200.

Wang, Y., Havel, J., Beal, P.A., 2015. A phenotypic screen for functional mutants of human adenosine deaminase acting on RNA 1. ACS Chem. Biol. 10, 2512–2519. https://doi.org/10.1021/acschembio.5b00711.

Wettengel, J., Reautschnig, P., Geisler, S., Kahle, P.J., Stafforst, T., 2017. Harnessing human ADAR2 for RNA repair—recoding a PINK1 mutation rescues mitophagy. Nucleic Acids Res. 45, 2797–2808. https://doi.org/10.1093/nar/gkw911.

Wong, S.K., Sato, S., Lazinski, D.W., 2001. Substrate recognition by ADAR1 and ADAR2. RNA (New York, N.Y.) 7, 846–858. https://doi.org/10.1017/s135583820101007x.

Woolf, T.M., Chase, J.M., Stinchcomb, D.T., 1995. Toward the therapeutic editing of mutated RNA sequences. Proc. Natl. Acad. Sci. U. S. A. 92, 8298–8302. https://doi.org/10.1073/pnas.92.18.8298.

Wright, A., Vissel, B., 2012. The essential role of AMPA receptor GluR2 subunit RNA editing in the normal and diseased brain. Front. Mol. Neurosci. 5 (34), 1–13.

Xu, L., Park, K.H., Zhao, L., Xu, J., el Refaey, M., Gao, Y., Zhu, H., Ma, J., Han, R., 2016. CRISPR-mediated genome editing restores dystrophin expression and function in mdx mice. Mol. Ther. 24, 564–569. https://doi.org/10.1038/mt.2015.192.

Xue, K., Groppe, M., Salvetti, A.P., MacLaren, R.E., 2017. Technique of retinal gene therapy: delivery of viral vector into the subretinal space. Eye (Basingstoke) 31, 1308–1316. https://doi.org/10.1038/eye.2017.158.

Yang, C.-C., Chen, Y.-T., Chang, Y.-F., Liu, H., Kuo, Y.-P., Shih, C.-T., Liao, W.-C., Chen, H.-W., Tsai, W.-S., Tan, B.C.-M., 2017. ADAR1-mediated 3' UTR editing and expression control of antiapoptosis genes fine-tunes cellular apoptosis response. Cell Death Dis. 8, e2833. https://doi.org/10.1038/cddis.2017.12.

CHAPTER SIX

mRNA delivery technologies: Toward clinical translation

Itziar Gómez-Aguado[a,b], Julen Rodríguez-Castejón[a,b], Marina Beraza-Millor[a,b], Alicia Rodríguez-Gascón[a,b], Ana del Pozo-Rodríguez[a,b], and María Ángeles Solinís[a,b,*]

[a]Pharmacokinetic, Nanotechnology and Gene Therapy Group (PharmaNanoGene), Faculty of Pharmacy, Centro de investigación Lascaray ikergunea, University of the Basque Country UPV/EHU, Vitoria-Gasteiz, Spain

[b]Bioaraba, Microbiology, Infectious Disease, Antimicrobial Agents, and Gene Therapy, Vitoria-Gasteiz, Spain

*Corresponding author: e-mail address: marian.solinis@ehu.eus

Contents

Abstract

Messenger RNA (mRNA)-therapies have recently taken a huge step toward clinic thanks to the first mRNA-based medicinal products marketed. mRNA features for clinical purposes are improved by chemical modifications, but the inclusion in a delivery system is a regular requirement. mRNA nanomedicines must be designed for the specific therapeutic purpose, protecting the nucleic acid and facilitating the overcoming of biological barriers. Polymers, polypeptides, and cationic lipids are the main used materials to design mRNA delivery systems. Among them, lipid nanoparticles (LNPs) are the most advanced ones, and currently they are at the forefront of preclinical and clinical evaluation in several fields, including immunotherapy (against infectious diseases and cancer), protein replacement, gene editing and regenerative medicine. This chapter includes an overview on mRNA delivery technologies, with special interest in LNPs,

International Review of Cell and Molecular Biology, Volume 372
ISSN 1937-6448
https://doi.org/10.1016/bs.ircmb.2022.04.010

Copyright © 2022 Elsevier Inc. All rights reserved.

and the most recent advances in their clinical application. Liposomes are the mRNA delivery technology with the highest clinical translation among LNPs, whereas the first clinical trial of a therapeutic mRNA formulated in exosomes has been recently approved for protein replacement therapy. The first mRNA products approved by the regulatory agencies worldwide are LNP-based mRNA vaccines against viral infections, specifically against the 2019 coronavirus disease (COVID-19). The clinical translation of mRNA-therapies for cancer is mainly focused on three strategies: anti-cancer vaccination by means of delivering cancer antigens or acting as an adjuvant, mRNA-engineered chimeric antigen receptors (CARs) and T-cell receptors (TCRs), and expression of antibodies and immunomodulators. Cancer immunotherapy and, more recently, COVID-19 vaccines spearhead the advance of mRNA clinical use.

1. Introduction

Nucleic acid medicinal products are making great progress, leading the way in the development of advanced therapies. These medicines can be applied for gene augmentation, gene silencing, or gene editing (Anguela and High, 2019). DNA has been classically studied to induce protein expression; however, due to the recent advance in understanding the structure of the messenger RNA (mRNA) and its role in genetic processes, and due to the progress on delivery carriers, mRNA is being studied for several applications, such as vaccination against infectious diseases, cancer immunotherapy, protein replacement, gene editing or regenerative medicine (Shuai et al., 2021; Xiong et al., 2018). In fact, after many years of research, in 2020 mRNA medicinal products reached the clinic worldwide in the form of mRNA-based vaccines for the prevention of Coronavirus Disease Pandemic (COVID-19) (European Medicines Agency, 2021a; U.S. Food and Drug Administration, 2021).

Synthetic mRNA presents several characteristics that support its use as an alternative to DNA. In contrast to DNA, mRNA molecules do not need to reach the nucleus to start the protein translation; therefore, mRNA is more effective to express the target protein, and it can be functional in mitotic and non-mitotic cells (Hajj and Whitehead, 2017; Meng et al., 2017; Patel et al., 2019a; Sahin et al., 2014; Xu et al., 2020; Zarghampoor et al., 2019; Zhong et al., 2018) Moreover, mRNA does not integrate into the host genome, which makes this molecule safer than DNA (Hajj and Whitehead, 2017; Meng et al., 2017; Patel et al., 2019a; Sahin et al., 2014; Xu et al., 2020; Zarghampoor et al., 2019). From the point of view of large-scale production, mRNA synthesis is fast, easy to standardize with high reproducibility and inexpensive (Xu et al., 2020).

Finally, protein expression is initiated earlier with mRNA; the protein can be detected since the first hour after transfection, and the maximum level of expression is usually observed after 5–7 h (Björklund, 2016).

Clinical application of mRNA can be implemented in vivo or ex vivo. In the latter case, cells are extracted from the patient and modified with the nucleic acid in the laboratory before being reinfused into the patient. In the former, the nucleic acid is administered into the patient via the routes commonly used for conventional drugs (Alnasser, 2021; Thorne et al., 2018). However, the use of synthetic mRNA for clinical purposes, both ex vivo and in vivo, needs to overcome important limitations to result effective: physical instability, immunogenicity and entry into the cell, which is difficult due to the high molecular weight, the anionic nature and the hydrosolubility of the nucleic acid (Ramanathan et al., 2016). Chemical modifications of mRNA structure improve stability and translation efficacy, and reduce immunogenicity, but the inclusion of synthetic mRNA in properly designed delivery systems is frequently essential for overcoming all these barriers. The most studied systems for mRNA delivery are non-viral vectors, and more specifically lipidic systems.

The aim of this work is to review the main mRNA delivery techniques, with special focus on lipid-based systems, and the recent and near future clinical applications of mRNA medicines.

2. Synthetic mRNA

2.1 Synthetic mRNA structure

Synthetic mRNA mimics the structure of natural mRNA and is formed by a single strand consisting of five basic structures (Fig. 1): 5′ Cap, 5′ untranslated region (UTR), an open reading frame (ORF), 3′ UTR, and a polyadenylated tail (poly(A) tail).

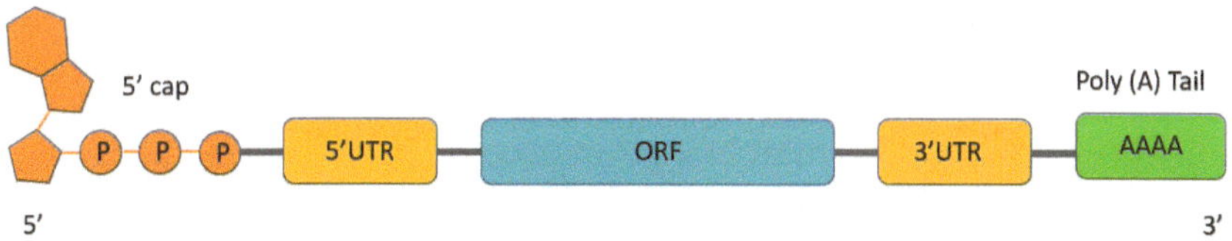

Fig. 1 Scheme of synthetic mRNA structure. 5′ UTR: 5′ untranslated region; 3′ UTR: 3′ untranslated region; ORF: open reading frame; Poly(A) tail: polyadenylated tail; A: adenine; P: phosphate.

5′ Cap structure is formed by an inverted 7-methyl guanosine (m7G) connected to the first transcribed RNA nucleotide by a 5′ to 5′ triphosphate bridge. This structure is known as cap0, and its functions include: mRNA stabilization in cellular processes (translation, splicing, polyadenylation or nuclear exportation), protection from exonucleases, interaction with cap binding proteins (CBPs), participation in nuclear exportation and in binding to the translation initiation factor 4E (eIF4E) (Jang and Paek, 2016; Xiong et al., 2018).

The ORF is the most crucial component of mRNA molecule because it consists of the nucleotide sequence encoding the target protein, and the codon composition influences the translation efficiency and the stability of the nucleic acid.

UTRs are non-coding regions located at 5′ and 3′ sides of ORF. They do not directly contribute to protein expression but they contain specific regulatory elements that modulate mRNA replication, stability and translation efficiency. In cooperation with RNA-binding proteins, they can slow down mRNA decay and improve transfection efficacy (Schlake et al., 2012). In general, UTRs regulate protein expression depending on three structural parameters: sequence, length and secondary structure (Matoulkova et al., 2012).

The 5′ UTR is an important element for ribosome recruitment and for choosing the start codon (Hinnebusch et al., 2016; Matoulkova et al., 2012). Moreover, it is a key regulatory factor in the initiation of protein translation as it presents two major domains: Kozak sequence and the internal ribosomal entry sites (IRES). In eukaryotes, Kozak consensus sequence is represented by the nucleotides RCCAUGG, where R is a purine nucleotide, adenine (A) or guanine (G). In this structure, the initiation codon AUG is included. In order to facilitate the recognition of the starting codon, an A or G nucleotide must be in the position −3 respect to the A of AUG, and a G in the position +4 (Kozak, 1984, 1987, 1991a). IRES is responsible for the recruitment of the ribosome, and it allows a cap-independent translation initiation process (Gómez-Aguado et al., 2020c; Hellen and Sarnow, 2001; Johnson et al., 2017; Yang and Wang, 2019).

The 3′ UTR sequence plays an important role in protein translation, particularly in stabilizing mRNA (Matoulkova et al., 2012). The stability and translation efficiency of mRNA is highly conditioned by the presence of AU rich elements in 3′ UTR. AU rich sequences are present in mRNA molecules with a high rate of degradation (Mayr, 2017; Xiong et al., 2018).

In eukaryotes, the poly(A) tail decorates the 3′ end of mature mRNA. It consists of a 100–250 adenosine residues chain. It is well known that its sequence plays a key role in regulating stability and translation aspects of mRNA (Kwon et al., 2018). In this sense, it presents several functions, such as interaction with initiation translation factors, protection from nucleases and mRNA exportation from the nucleus. It slows down the degradation process of RNA exonucleases, which increases the stability, in vivo half-life and the translation efficiency of mRNA (Linares-Fernández et al., 2020).

Synthetic mRNA is usually produced by a cell-free transcription system, which requires a linearized DNA template. Since its first successful report in 1990 (Wolff et al., 1990) in vitro transcription technology of mRNA has matured (Pardi et al., 2018). mRNA can be transcribed from a linearized DNA plasmid or a PCR product (Grier et al., 2016; Oh and Kessler, 2018). The DNA contains a promoter which is recognized by a bacteriophage RNA polymerase such as T7, T3 or SP6 (Kowalski et al., 2019). In addition to the ORF sequence, it may also contain the sequences that encode UTRs. The poly(A) tail and the 5′ Cap can be incorporated during the transcription process or post-transcriptionally. At the end of the process, the DNA template is degraded by DNases (Weissman, 2014). The use of cell-free systems can be carried out under Good Manufacturing Practices (GMPs), and it allows for easy standardization of clinical grade manufacturing. As a consequence, fabrication of synthetic mRNA under GMPs is more economical than production of recombinant proteins in eukaryotic cells (Steinle et al., 2017).

Once the mRNA is transcribed, a set of purification steps should be done to remove aberrant molecules, which are associated to immunogenicity (Baiersdörfer et al., 2019). In particular, double-stranded RNA (dsRNA) contaminants are similar to viral RNA replication intermediaries, and are considered pathogen-associated molecular patterns (PAMPs) (Linares-Fernández et al., 2020), so they can promote the production of type I interferon (IFN-1) (Xu et al., 2020). Purification processes include concentration, precipitation, and extraction and chromatography techniques (high performance liquid chromatography (HPLC), affinity and anion exchange) and size exclusion columns (Baiersdörfer et al., 2019; Karikó et al., 2011) may be applied. In addition to purification methods and in order to down-regulate the production of dsRNA, in vitro transcription process can be adjusted. Most recommended strategies include modified nucleosides, reduction of the Mg^{+2} concentration, which is needed for the elongation process, or the use of thermostable T7 RNA polymerase (Baiersdörfer et al., 2019; Mu et al., 2018).

2.2 Synthetic mRNA modifications

The design of the mRNA may include specific modifications aimed at improving stability, reducing immunogenicity and maximizing protein-producing potential. Table 1 summarizes the main chemical modifications of mRNA and the purpose of these modifications.

The incorporation of modified nucleosides into mRNA is a common strategy to optimize protein expression efficiency and to reduce its immunostimulatory activity. mRNA interacts with cytoplasmic sensors, such as retinoic acid-inducible protein I (RIG-I) (Hornung et al., 2010; Karikó et al., 2008), which is responsible of blocking the translational machinery. Specifically, uridine (U)-rich regions interact with RIG-I. It has been documented that the U depletion in mRNA molecules leads to an improvement of protein expression (Vaidyanathan et al., 2018). Alternatively, RIG-I activation can be also abolished with the incorporation into the transcript of the modified nucleosides pseudouridine (Ψ) or 2-thiouridine (s2U) (Diebold et al., 2006). Modified nucleosides (Ψ, s2U, N1-methylpseudouridine (N1mΨ), 5-methylcytidine (m5C), N6-methyladenosine (m6A), or 5-methyluridine (m5U)) can also avoid the activation of Toll-like receptors (TLRs), which is responsible of the innate immune responses associated to the administration of exogenous mRNA (Andries et al., 2015; Karikó et al., 2005; Lorenz et al., 2011; Martini and Guey, 2019; Svitkin et al., 2017).

Apart from the use of modified nucleosides, in 5′ Cap region cap0 can be replaced by cap1 or cap2 by means of the methylation of the second or third ribonucleotide at the 2′O or 3′O position of the riboses, respectively. The objective of this modification is reducing the immunogenicity and enhancing the translation efficiency (Schlake et al., 2019). This methylation can be performed enzymatically by the enzyme m7G-specific 2′O methyltransferase (2′O MTase). Alternatively, addition of cap1 and cap2 during transcription is simpler than enzymatic capping; however, this process usually results in not capped molecules, which are immunostimulant, and immediately digested by nucleases. This limitation can be avoided by applying two different strategies. On the one hand, the use of a phosphatase to eliminate the triphosphates of the 5′ end of uncapped mRNA

Table 1 Main modifications of mRNA structure.

Modified structure	Modification	Purpose
5′ Cap	Cap1 and cap2	Efficacy, translation stability and immunogenicity
	Elimination of triphosphate groups	Stability, recognition by ribosomes and initiation of translation
	ARCA (anti-reverse cap analogue)	Translation efficacy
	β-S-ARCA	Translation efficacy
	Chemically modified analogues (imidiphosphate, phosphorothiate, boranophosphate)	Efficacy and translation stability. Decapping protection
5′ UTR	Viral IRES	Translation efficacy and cap-independent translation
	m6A IRES substitution	Cap-independent translation
	No secondary or tertiary structures	Stability and translation efficacy
ORF	Codon optimization and frequent codons use	Stability and translation efficacy
	Reduction of UA and UU dinucleotides	Stability
3′ UTR	No AU rich regions	Stability
	α- and β-globin conjugates	Translation efficacy and stability
	Length	Location of the expressed protein
Poly A tail	Length	Stability, translation efficacy and protection
	Adenosine modifications	Protection
Modified nucleotides	Pseudouridine, N'methylpseudouridine, *N*-methyladenosine, 2-thiouridine	Immunogenicity, stability
circRNAs	Incorporation of IRES or m6A modification to 5′ UTR	Stability and avoid immune system

5′ UTR, 5′ untranslated region; 3′ UTR, 3′ untranslated region; ORF, open reading frame; A, adenine; P, phosphate; UA, uridine-adenine dinucleotides; UU, uridine-uridine dinucleotides; ARCA, anti-reverse cap analogue; β-S-ARCA, ARCA-β sulfured; IRES, internal ribosomal entry sites; circRNAs, circular RNAs.

(McCracken et al., 1997; Zhong et al., 2018) reduces the immunogenicity. On the other hand, co-transcriptional capping technology CleanCap™, developed by TriLink BioTechnologies, is able to incorporate cap1 or cap2 in the 94% of the mRNA molecules (Henderson et al., 2021; TriLink BioTechnologies, 2020). Other drawback during the capping process is the possibility that the cap analogs dispose in reverse orientation, resulting in the inability to bind to the CBP (Islam et al., 2015; Meng et al., 2017; Patel et al., 2019a). This effect can be avoided by using anti-reverse cap analogue (ARCA) (Islam et al., 2015;), which is a methyl group in the 3′-OH of the m7G nucleotide. ARCA enables the appropriate orientation of the cap, and prevents the elongation of the mRNA along the wrong site (Islam et al., 2015; Muttach et al., 2017). The substitution of an S group at β-position of the triphosphate bridge results in a new modification known as β-S-ARCA, which leads to in high affinity of the cap to EIF4E and low susceptibility to the decapping complex DCP1/DCP2 (Kowalska et al., 2008; Kuhn et al., 2010; Warminski et al., 2021). Removal of cap by this complex inhibits the translation initiation and makes the mRNA to be committed to full degradation by exonucleases (Te Chang et al., 2014; Pelletier et al., 2021). Chemically modified cap analogs, such as phosphorothioate, phosphorothiolate, imidiphosphate and boranophosphate, among others (Kuhn et al., 2010; Pelletier et al., 2021; Rydzik et al., 2012; Strenkowska et al., 2010; Warminski et al., 2013; Ziemniak et al., 2015; Zytek et al., 2014), are usually incorporated to synthetic mRNA to get protected from decapping.

mRNA can be also optimized by modifications in the UTRs. Namely, IRES viral sequences can be included in 5′ UTR segments as a strategy to improve translation efficacy. These viral IRES can begin the translation process without cap0, and are able to start translation in cells with low levels of translation initiation factors. Thus, the incorporation of IRES viral sequences represents a feasible strategy to improve translation rate on cells in which it is normally low (Vivinus et al., 2001). Other possibility to favor cap-independent translation is the insertion of the modified nucleoside m6A in the 5′ UTR as an alternative to IRES (Kozak, 1991a,b; Meyer et al., 2015; Vivinus et al., 2001). Regarding 3′ UTR modifications, the presence in this structure of the specific sequence of α-globin mRNA improves the stability of synthetic mRNA, whereas the addition of the sequence corresponding to the β-globin mRNA extends the protein expression (Jiang et al., 2006; Ross and Sullivan, 1985; Volloch and Housman, 1981). Moreover, the localization of the protein expressed may be modulated depending on the length of the 3′ UTR sequence. For example, the CD47 protein membrane is expressed on the cell surface when the 3′ UTR of the mRNA is long, but the same protein appears in the reticulum endoplasmic when the 3′ UTR is short (Berkovits and Mayr, 2015; Mayr, 2017). In addition, AU rich sequences in 3′ UTR can induce repression of translation, decapping and cleavage, which finally lead to mRNA decay. Thus, it is recommended to avoid this structure (Gergen and Petsch, 2020; Koh et al., 2019).

ORF sequence or codons are frequently optimized to increase the rate and efficiency of the protein translation (Kowalski et al., 2019; Presnyak et al., 2015; Weissman, 2014). Codon optimization can be conducted by using more frequent codons for each amino acid instead of multiple rare codons. However, this strategy can affect protein functions, and alterations in translation kinetics can lead to alterations in protein conformation (Bali and Bebok, 2015; Hunt et al., 2014; McCarthy et al., 2017; Presnyak et al., 2015). Dicodon optimization is other alternative to regulate the ORF because some pairs of codons work together better than others (Alexaki et al., 2019). A third approach is to include in the ORF sequence the same proportion of each codon that is naturally found in the highly expressed proteins of the target species and cells (Linares-Fernández et al., 2020). Finally, it is important to reduce the frequency of UU and UA dinucleotides, as it has been reported that a high proportion of both enhances mRNA decay (Al-Saif and Khabar, 2012).

A relatively long poly(A) tail is appropriate to optimize stability and translation efficiency of the mRNA (Karikó et al., 2005; Lorenz et al., 2011), although the optimal length may be different depending on the cell type. In human epithelial (HeLa) cells, a poly(A) tail composed by 98 residues has shown to increase protein expression (Peng and Schoenberg, 2005). However, in dendritic cells (DCs), the translation efficiency was improved with 120 adenosine residues (Holtkamp et al., 2006; Mockey et al., 2006), and in human primary T cells, efficient translation was achieved with a poly-A tail longer than 300 nucleotides (Grier et al., 2016). In terms of stability, long poly(A) tails are preferable: 120 adenosine-long molecules provide more stable mRNAs and more efficient translation processes than shorter ones (Grier et al., 2016). Apart from the tail length, it is also important to mention that mRNA stability can be increased by the incorporation of adenosine analogues inside the poly(A) tail sequence, which protect mRNA from 3′-exonuclease degradation (Xiong et al., 2018) (e.g., 8-azaadenosine and 3′-deoxyadenosine).

mRNA secondary structures play a major role in the translation process, stability and immunogenicity. Highly stable secondary structures in 5′ UTR and hairpin loops hamper ribosome entry and scanning, interfere in the elongation step and are recognized as PAMPs (Kozak, 1991a,b; Linares-Fernández et al., 2020; Vivinus et al., 2001). Recently, Wesselhoeft et al. (Wesselhoeft et al., 2019) have documented that circular RNAs (circRNAs) are an alternative to linearized mRNA. CircRNAs are a class of single-stranded non-coding RNA transcripts that are closed by covalent interactions through a splicing process called back-splicing, forming a loop-like structure. Thousands of circRNAs have been found in eukaryotic cells and they are preserved across species. Unlike linear mRNA, circRNAs are very stable against degradation by exonucleases thanks to the circular structure (Barrett and Salzman, 2016; Kristensen et al., 2019) In addition, it has been demonstrated that unmodified exogenous circRNA is able to bypass cellular RNA sensors, such as RIG-I and TLRs, and, therefore, immune response is avoided (Wesselhoeft et al., 2019). Not many endogenous circRNAs have shown to induce protein expression, but exogenous circRNAs can be engineered to enable protein translation through the incorporation of IRES or the m6A modification to the 5′ UTR region (Wesselhoeft et al., 2019, 2018; Yang et al., 2017).

To sum up, the final objective of the treatment should be considered in order to select the modifications that would improve the desired final effect. For most applications, an optimize protein expression efficiency, reduction of immunostimulatory activity and better mRNA stability is desired. Moreover, in the case of immunotherapy application, certain immunostimulation is relatively adequate up to a point, since the effect that is targeting is the induction of the immune response. The main recommendable modifications are the incorporation of cap1 or cap2; the inclusion of viral IRES or m6A IRES substitution in 5′, the incorporation of frequent codon; the reduction of UA and UU dinucleotides; the incorporation of α- and β-globin conjugate in 3′ UTR and the presence of long length of poly(A) tail. Among them, modified nucleosides are one of the most common applied modifications. In particular, the recent preclinical and/or clinical trials of immunotherapy incorporate N1mΨ, which has demonstrated to increase mRNA translation in vivo (Karikó et al., 2008; Laczkó et al., 2020; Parr et al., 2020) (Gao et al., 2021). In addition, a reduction in immunogenicity was observed when the chemically modified Ψ and m5C proportionally replaced the canonical U and C in synthetic mRNAs in cell reprogramming (Warren et al., 2010). Moreover, the strategy based on the inclusion of modified nucleosides s2U and m5C is useful to reduce immunogenicity and improve translation efficiency in protein replacement therapy (Kormann et al., 2011).

3. mRNA delivery systems

Chemical modifications improve the features of mRNA for clinical purposes, but even so, the therapeutic effect of the naked nucleic acid is limited. For an effective delivery of mRNA in the target tissues or organs, it must be protected from extracellular agents, mainly ribonucleases, and prevent the activation of the immune system (except in vaccination). In addition to this, mRNA has to overcome the cellular mechanisms that tackle with external nucleic acid molecules including cytosolic and endosomic receptors; for example, pattern recognition receptors, which are specifically involved in the recognition of non-self RNA and the subsequent activation of pro-inflammatory pathways (Hajj and Whitehead, 2017; Ibba et al., 2021; Kowalski et al., 2019).

Naked mRNA has been administered as a direct injection principally for the prevention or treatment of infectious diseases; the presence of naked mRNA stimulates the immune response and activates the antigen presentation process (Pardi et al., 2018). Depending on the clinical application, the preferred routes of administration include subcutaneous, intradermal, intranodular, intramuscular, intratumoral and intravenous injection (Xu et al., 2020). Nevertheless, the efficacy of naked mRNA is limited, because the nucleic acid is exposed to further degradation after injection (Xu et al., 2020). Indeed, the half-life of naked mRNA after intravenous administration has been shown to be less than 5 min, with a 10% reduction in serum after 5 min and 1% 1 h later.

3.1 Physical administration methods

Transfection efficacy of naked mRNA can be improved by using physical administration methods. They are based on the direct disruption of cell membranes, which allows the internalization of naked mRNA. Nucleic acids present high molecular weight and they are negatively charged and highly soluble in water due to the presence of the phosphate group in their backbone. All these features confer the impossibility to diffuse passively through the cell membranes, contrary to small, neutral and hydrophobic molecules (Schlich et al., 2021). The most common physical methods are electroporation, sonoporation, gene gun and microinjection (Mirón-Barroso et al., 2021; Mohammad, 2021; Xu et al., 2020). Electroporation and sonoporation form pores in the cell membranes using electric pulses or ultrasounds, respectively. As a consequence, permeability increase transiently and the entrance of the nucleic acid is allowed (Sharma et al., 2021). Electroporation has been applied as a therapeutic strategy in malignant, infectious and autoimmune diseases. Specifically, the most common clinical use of mRNA electroporation is related with antigen-presenting DCs carrying tumor-associated antigens (TAAs) alone or in combination with immunomodulatory molecules (Campillo-Davo et al., 2021). Gene gun, also known as particle bombardment method, pushes mRNA molecules coated with gold particles into target cells by using pressurized helium gas. However, the gene gun is not widely used in large animals and humans (Xu et al., 2020). Lastly, microinjection consists of the direct injection of the nucleic acid by microneedles. It has been mainly used to study development in the oocyte and early embryo (Jones et al., 2018; Lane et al., 2022).

3.2 Nucleic acid delivery systems

The inclusion of synthetic mRNA in a delivery system is a regular requirement for a successful therapy. In addition to the protection from extracellular agents, once the mRNA reaches the target cell, the delivery system also promotes overcoming of the cell membrane and several intracellular barriers, as it is observed in Fig. 2. First, the delivery system interacts with

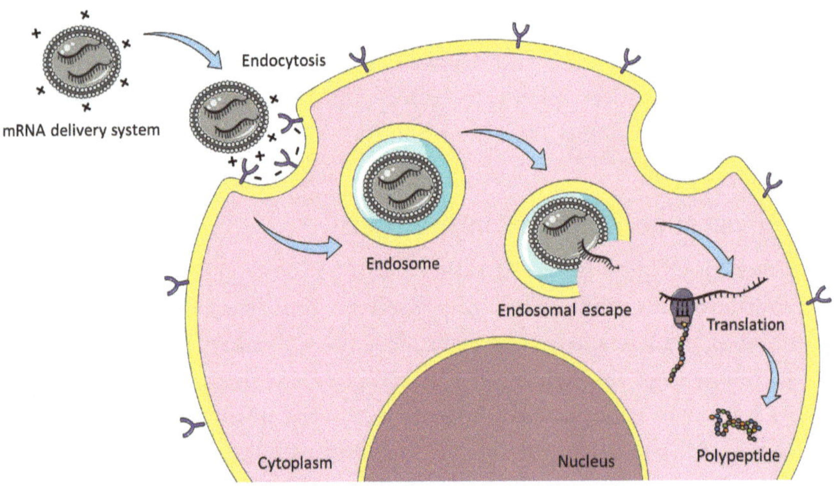

Fig. 2 Intracellular barriers for mRNA delivery system: interaction between mRNA delivery system and cell membrane, endocytosis, endosome incorporation, endosomal escape and release to cytoplasm to start translation process.

the negative charges of the target cell membrane through electrostatic interactions. Thus, cationic systems are preferred (Gómez-Aguado et al., 2020b). Moreover, the interaction between the delivery system and cell surface receptors can be favored by the incorporation of different ligands. Then, the delivery systems start to be surrounded by invaginations that cell membrane produces during the endocytosis process, which is the principal mechanism of cellular entry (Schlich et al., 2021). Once inside the cell, nucleic acids enter in endosome-lysosome pathway. The nucleic acid moves gradually from early to late endosomes where pH decreases, until reaching the lysosomes, in which the acidic pH and the presence of digestive enzymes can degrade mRNA (De Haes et al., 2012).

Taking into account, the importance of the endosomal escape, numerous strategies are being studied with the aim of boosting the endosomal escape before nucleic acid degradation, including "proton sponge" effect, pore-forming effect, membrane fusion, and photochemical disruption (Meng et al., 2021; Selby et al., 2017; Varkouhi et al., 2011). In "proton sponge" effect, ionisable moieties of the delivery systems are protonated during endosomal acidification. Consequently, and in order to equilibrate the charge and the increase of the osmotic pressure, chloride ions are transported into the endosome until endosomal membrane is ruptured (Behr, 1997). Pore formation in the endosomal membrane consists of the dynamic interplay between a membrane tension that opens the pore and a line tension that closes it. The use of peptides with high affinity for the pore edge, reduces the line tension of the membrane and maintain the pore radius stable (Huang et al., 2004). Membrane fusion is based on the destabilization of the endosomal membrane by the presence of fusogenic peptides (Du et al., 2014). Photochemical disruption is produced by small light-sensitive molecules that generate reactive oxidative substances after light exposition that destroy the endosomal membrane (Berg et al., 1999).

Hence, for a successful mRNA therapy, the design of suitable mRNA delivery systems specifically adapted to the characteristics of the nucleic acid and the therapeutic objective, is essential. A formulation designed to administer mRNA must address a number of requirements to obtain the maximum therapeutic effect: protection of the genetic material, specificity to produce targeted delivery, and appropriate intracellular disposition to favor endosomal escape and to allow the translation process; all this without activating the immune response (Meng et al., 2021; Patel et al., 2021).

Nucleic acid delivery systems are classically divided in viral and non-viral vectors. Unlike the progress experienced with viral vectors in DNA therapies, these delivery systems do not have a significant impact in the case of mRNA-based therapies. Non-viral vectors offer a safer, cheaper and more versatile alternative to viral vectors. In addition, they can transport nucleic acids of large size and are easier to produce. Although non-viral vectors do not have the natural capacity of viruses to pass through cellular barriers, the advances in nanotechnology and material sciences have led to safe and effective mRNA-medicinal products. Vaccines against severe acute respiratory syndrome coronavirus-2 (SARS-CoV-2) (Moderna mRNA-1273 and Pfizer/BioNTech BNT162b2) are clear examples of this (Baden et al., 2021; Jackson et al., 2020; Mulligan et al., 2020; Polack et al., 2020).

The incorporation of mRNA in nanocarriers is at the forefront in mRNA therapeutics (Meng et al., 2021). Chemical methods, and in particular organic nanocarriers, made up by synthetic or natural biocompatible materials, depict the most widely used non-viral gene carriers (del Pozo-Rodríguez et al., 2019a; Wadhwa et al., 2020). A large number of synthetic or natural-derived materials have been designed for mRNA therapeutics, such as lipids, lipid-like materials, polymers, inorganic particles, protein derivatives, and dendrimers; among them, lipids and lipid-derived materials are the most commonly used materials for mRNA delivery (Hou et al., 2021; Weng et al., 2020).

3.2.1 Polymeric delivery systems

The first successful transfection of pDNA using the cationic polymer polyl-L-ysine (PLL) was reported in 1987 (Wu and Wu, 1987), but the use of cationic polymers for nucleic acid delivery has increased dramatically in the few recent years. The combination of negative charges of mRNA and cationic polymers leads to the formation of polyplexes. Polymers present several advantages as nucleic acid delivery systems: high stability, and ease of preparation, purification and chemical modification. Moreover, they protect mRNA from ribonucleases, favor cellular uptake and endosomal escape. These advantages provide polyplexes high efficacy both in vitro and in vivo (Meng et al., 2021).

A large number of polymers have been studied for mRNA complexation, including polyethylenimine (PEI) (Chiper et al., 2017; Jere et al., 2009; Weng et al., 2020), polyacrylates (Shuai et al., 2021; Ulkoski et al., 2019), poly(β-amino) esters (PBAEs) (Kaczmarek et al., 2016; Patel et al., 2019; Wahane et al., 2020), poly(aspartamides) (PAsp) (Miyata et al., 2008; Uchida et al., 2011) and chitosan (McCullough et al., 2014; McKinlay et al., 2018; McKinlay et al., 2017). However, the clinical use of polymers is less extended than that of lipid-based systems for mRNA delivery (Gómez-Aguado et al., 2020b).

3.2.2 Polypeptidic delivery systems

Polypeptides are made up by one or more different aminoacids organized in block or random sequences. The possibility of adapting their cationic and endosomolytic properties makes them interesting to use as nucleic acid delivery systems (Ulkoski et al., 2019). Polypeptidic systems include protamine, cell penetrating peptides (CPPs), and virus-like particles (VLPs).

The positive charges of arginine groups of protamine interact with the negative charges of nucleic acids by electrostatic interactions leading to tight association in nanoparticles. Protamine protects the naked mRNA with the additional advantage of acting as an adjuvant to induce immune responses, due to the ability of this peptide to activate the TLR7/8 signal (Kallen et al., 2013; Scheel et al., 2005; Wang et al., 2021). Taking into account these properties, protamine-based systems have been applied when the activation of innate immune responses is needed, including infectious diseases and cancer (Meng et al., 2021). For example, RNActive® vaccine is an mRNA-based vaccine technology developed by CureVac, which uses sequence-optimized, unmodified mRNA complexed with the protamine. RNActive® vaccine composed by 50% free mRNA and 50% the complexed mRNA-protamine in a weight ratio of 2:1, has been under clinical evaluation against some cancers, such as melanoma (Weide et al., 2009), prostate cancer (Kübler et al., 2015) and non-small cell lung cancer (Papachristofilou et al., 2019), with promising results.

The use of CPPs is based on their capacity to enter inside the cells in an efficiently and safety way, without interrupting the integrity of cell membranes. Among them, nanosystems formulated with the amphipathic RALA motif, rich in arginine peptides, are the most used CPPs (Udhayakumar et al., 2017).

VLPs are structures similar to viruses, which simulate the organization and conformation of authentic native viruses, but without viral pathogenicity, as they do not contain virus genome. Recently, in a preclinical study in mice with a mRNA vaccine encoding SARS-CoV-2 formulated in VLPs, a potent antiviral-like immune response was detected after intramuscular administration (Lu et al., 2020).

3.2.3 Dendrimers

Dendrimers are nanoscale hyperbranched macromolecules with homogeneous and well-defined tree-like structure formed by a central core, inner branches and functional groups in the surface (Dammes and Peer, 2020; Sharma et al., 2021). Polyamidoamine (PAMAM) and polypropylenimine-based dendrimers have been widely used as nucleic acids delivery systems, due to their hydrophilic, biocompatible and non-immunogenic properties (Kowalski et al., 2019). Chahal et al. developed an adjuvant-free vaccine based on a PAMAM dendrimer containing antigens encoded by mRNA (Chahal et al., 2016). This study showed that, after the intramuscular injection of a single dose into mice, the vaccine generated protective immunity against lethal Ebola, H1N1 influenza, and *Toxoplasma gondii* challenges. Later, the same dendrimer-system was used as vaccine candidate to produce Zika virus E protein-specific IgG responses (Chahal et al., 2017).

3.2.4 Gold nanoparticles

Gold nanoparticles (AuNPs) show interesting biomedical application due to their stability, inertness and reduced cytotoxicity. In addition, they are highly biocompatibility, their surface modification is easy and they have the ability to conjugate with biological ligands, including polymers or nucleic acids. Yeom et al. (Yeom et al., 2013) evaluated the effect on mouse xenograft tumor growth of intratumoral production of the proapoptotic factor Bcl-2-associated X protein (BAX). For that purpose, they injected in the xenograft tumor an mRNA encoding BAX, loaded in AuNP-DNA oligonucleotide conjugates. BAX protein was produced and, consequently, tumor growth was inhibited.

3.2.5 Lipidic systems

As mentioned above, lipid-based systems are the most used for mRNA delivery, both in preclinical and clinical evaluation. Lipids and lipid-like materials are usually formulated as colloidal systems, which in combination with mRNA form the lipoplexes. Colloidal systems involve nanoemulsions and LNPs (del Pozo-Rodríguez et al., 2019b), and these latter include liposomes, solid lipid nanoparticles (SLNs), nanostructured lipid carriers (NLCs) and exosomes. Additionally, hybrid systems have been included in this group, as they consist of lipids combined with other components. Fig. 3 shows a representative scheme of the structures of the different lipid-based nanocarriers used for mRNA delivery.

3.2.5.1 Components of lipidic systems

The components most frequently used in lipid-based systems are cationic lipids. Positive charges of cationic lipids interact with the negative charges of mRNA by electrostatic interactions leading to lipoplexes (Sharma et al., 2021). Typically, the structure of cationic lipids consists of a positive charged headgroup and a hydrophobic chain joined by a linker (Fig. 4).

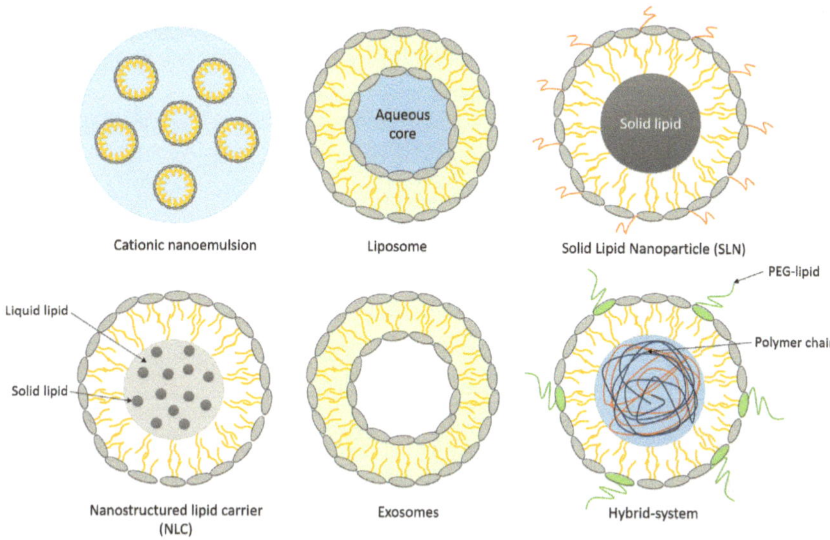

Fig. 3 Scheme of the structures of the different lipid-based nanocarriers used for mRNA delivery. PEG: Polyethylene glycol.

Concretely, the alkylated quaternary ammonium group confers the positive charge to the cationic lipids, which remains constant and pH-independent (Kowalski et al., 2019; Zhao et al., 2019). Cationic lipids have been widely used in lipoplexes for mRNA delivery. The first cationic lipid used for mRNA delivery was N-[1-(2,3-dioleyloxy)propyl]-*N*,*N*,*N*-trimethylammonium chloride (DOTMA). The system was able to deliver mRNA encoding luciferase in human, rat, mouse, xenopus (frog) and drosophila cells in vitro. Later, DOTMA was commercialized in combination with 1,2-dioleoyl-sn-glycero-3-phosphoethanolamine (DOPE) as Lipofectamine, one of the most commercialized transfection reagents (Malone et al., 1989). Additionally, 1,2-dioleoyl-3-trimethylammonium-propane (DOTAP), a biodegradable derivate of DOTMA, has been also used for mRNA delivery studies alone or in combination with DOTMA and other materials.

Cationic lipids have demonstrated high delivery efficacy, but in several in vitro and in vivo studies they have shown toxicity and immunogenicity. Moreover, they present short circulation half-life, and unspecific association to negatively charged cellular and extracellular components, (Ibba et al., 2021). In fact, these systems have produced liver damage and increase of the IFN-γ response in mice (Landesman-Milo and Peer, 2014; Ma et al., 2005; Platanias, 2005). Moreover, cationic lipids have low efficiency in

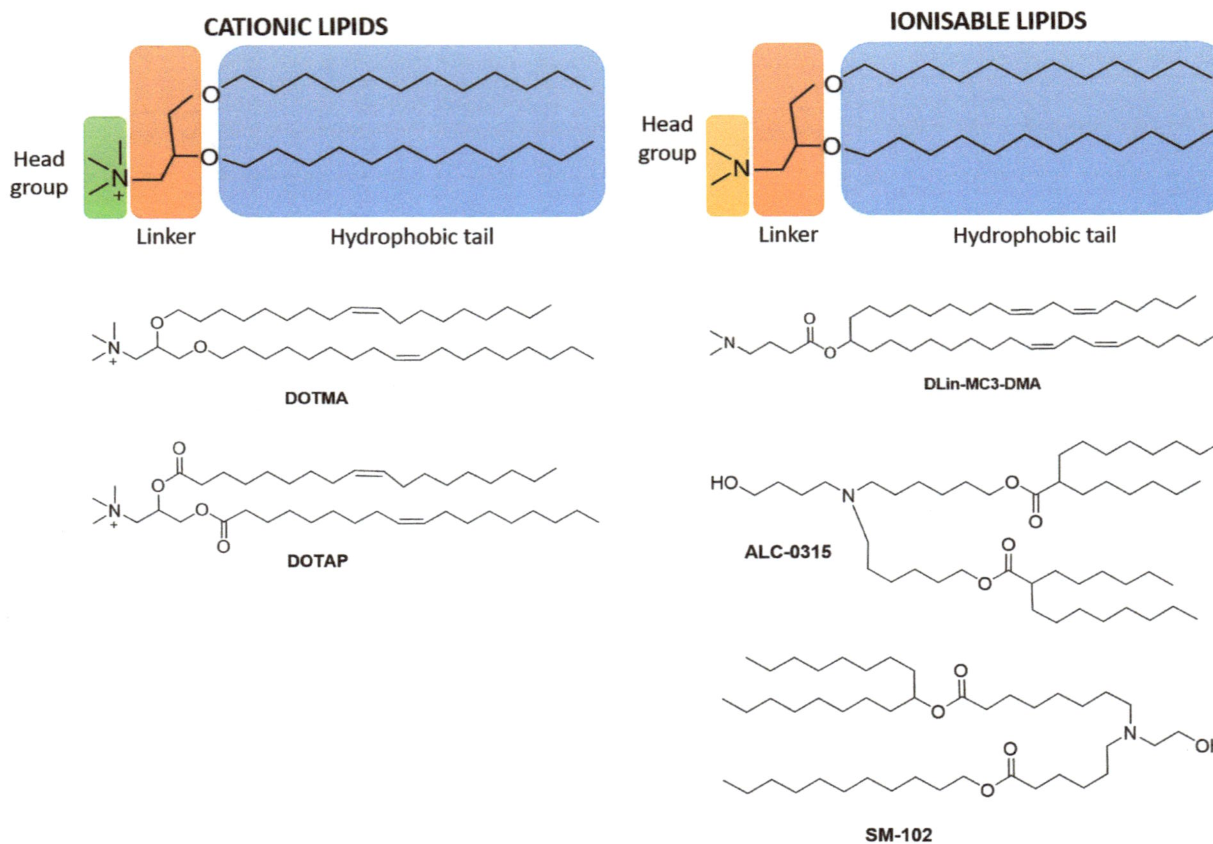

Fig. 4 Differences in the chemical structure of cationic and ionisable lipids. DOTMA: N-[1-(2,3-dioleyloxy)propyl]-*N*,*N*,*N*-trimethylammonium chloride; DOTAP: 1,2-dioleoyl-3-trimethylammonium-propane; DLin-MC3-DMA: (6Z,9Z,28Z,31Z)-heptatriaconta-6,9,28,31-tetraen-19-yl 4-(dimethylamino) butanoate; ALC-0315: 2-hexyl-decanoic acid, 1,1′-[[(4-hydroxybutyl)imino]di-6,1-hexanediyl] ester; SM-102: 8-[(2-hydroxyethyl)[6-oxo-6-(undecyloxy)hexyl]amino]-octanoic acid, 1-octylnonyl ester.

the endosomal escape process, one of the major key challenges in nucleic acid delivery. Recently, ionizable lipids have been developed as an alternative to traditional cationic lipids.

Ionizable lipids present advantageous features for nucleic acid delivery due to their structure. Ionizable lipids are made up by three different structures: a tertiary amine in the head group, a linker and a hydrophobic tail (Fig. 4). The tertiary amine can become charged at low pH because of the protonation of the amine at acidic pH environment. This feature results useful to form lipoplexes, because at low pH electrostatic interactions can be triggered and facilitate the assembly between the nucleic acid and the positive lipids. Later the buffer is exchanged to physiologic pH and the formulations show lower surface charge, which improves their biocompatibility after in vivo administration due to their lower capacity to interact with the negative membranes (Meng et al., 2021). However, the delivery systems composed of ionizable lipids become increasingly more positive when pH values fall below their pK_a. This pH-dependent feature is crucial and advantageous for the endosomal escape, since the acidic environment of the endosomes induces the protonation of the lipids, provokes the destabilization of membrane and, consequently, the release of the mRNA into the cytoplasm (Schlich et al., 2021).

Ionizable lipids, originally developed for DNA delivery, and later for short interfering RNA (siRNA) delivery, have also been studied for mRNA therapeutics (Hou et al., 2021). In this sense, the incorporation of the ionizable lipid (6Z,9Z,28Z,31Z)-heptatriaconta-6,9,28,31-tetraen-19-yl 4-(dimethylamino) butanoate (DLin-MC3-DMA; MC3) in Onpattro, the first siRNA drug approved by the FDA, has opened the door for its use in mRNA therapy. Different MC3-based LNPs have been studied in mRNA therapies, including protein replacement (Jain et al., 2018; Meng et al., 2017; Nabhan et al., 2016; Patel et al., 2019b; Ryals et al., 2020; Zhang et al., 2020a) and antiviral therapies (Zhang et al., 2020b).

It should be highlighted that Pfizer/BioNTech and Moderna mRNA vaccines developed to fight against COVID-19 pandemic, contain ionizable lipids. More specifically, Pfizer/BioNTech's vaccine is formulated with the ionizable lipid ALC-0315, whereas Moderna's is formulated with the ionizable lipid SM-102 (Andresen and Fenton, 2021). These vaccines were the first mRNA therapies to receive the Emergency Use Authorization from United States Food and Drug Administration (FDA) (U.S. Food and Drug Administration, 2021) and Conditional Approval by the European Medicines Agency (EMA) (European Medicines

Agency, 2021a). Additionally, they are the only mRNA vaccines against SARS-CoV-2 that have been in use in patients during 2021 (Andresen and Fenton, 2021).

Apart from cationic and ionizable lipids, lipid-like materials have been also tested for the delivery of mRNA in vivo. Lipid-like materials are made up by multiple lipidic tails and different hydrophilic groups. Some examples of them are: 1,1′-((2-(4-(2-((2-(bis(2-hydroxydodecyl) amino)ethyl) (2-hydroxydodecyl)amino)ethyl) piperazin-1-yl)ethyl)azanediyl) bis(dodecan-2-ol) (C12-200) (Love et al., 2010), tetrakis(8-methylnonyl) 3,3′,3″,3‴-(((methylazanediyl) bis(propane-3,1 diyl))bis (azanetriyl)) tetrapropionate (306Oi10) (Whitehead et al., 2014), 3,6-bis(4-(bis(2-hydroxydodecyl)amino)butyl)piperazine-2,5-dione (cKK-E12) (Dong et al., 2014), and N1,N3,N5-tris(3-(didodecylamino)propyl)benzene-1,3,5-tricarboxamide (TT3). Among them, TT3 has demonstrated capacity to deliver mRNA encoding the human factor IX and to recover the normal levels of the protein in IX-deficient mice (Li et al., 2015). Moreover, a biodegradable analogue of TT3, specifically hexa(octan-3-yl) 9,9′,9″,9‴,9⁗,9‴″-((((benzene-1,3,5-tricarbonyl)ris(azanediyl)) tris (propane-3,1-diyl)) tris(azanetriyl)) hexanonanoate (FTT5), further enhanced the efficacy of in vivo delivery of mRNA encoding human factor VIII (Xinfu et al., 2021).

Another type of lipids applied for mRNA delivery are zwitteronic ionizable lipids. These lipids change their conformation at acidic pH, transforming into an inverted hexagonal phase (HII), which destabilizes the endosome membrane and triggers the delivery of the nucleic acid into the cytoplasm (Hou et al., 2021). They have demonstrated effective protein expression and efficient genome editing in vivo (Liu et al., 2021a).

3.2.5.2 Lipid-based nanocarriers

- Cationic nanoemulsions

Among lipid-based colloidal systems, cationic nanoemulsions are formed by an oil phase dispersion stabilized with a surfactant in an aqueous solution with droplets size of 200 nm. DOTAP is the cationic lipid most frequently used to obtain cationic nanoemulsions (Wang et al., 2021). These systems have been mainly applied for mRNA vaccination. Concretely, Brito et al. developed a well-tolerated self-amplifying RNA (SAM) vaccine formulated in a cationic nanoemulsion, for synthesis of antigens of respiratory syncytial virus (RSV), human immunodeficiency virus (HIV), and human cytomegalovirus (CMV). This vaccine demonstrated induced immune responses in multiple animals at lower doses than required for pDNA vaccines (Brito et al., 2014). Moreover, a SAM vaccine encoding HIV type 1 envelope based on cationic nanoemulsion produced potent immune responses in rhesus macaques (Bogers et al., 2015). Currently, the first time-in-human study has entered to the clinic (NCT04062669) to evaluate the safety, reactogenicity and immunogenicity of an experimental rabies glycoprotein G SAM vaccine, formulated in a cationic emulsion, after intramuscular administration.

• Liposomes

Liposomes are spherical vesicles containing unilamellar or multilamellar phospholipid bilayers, enclosing an aqueous core. The interaction between hydrophilic and hydrophobic groups induces the formation of the vesicles (Hajj and Whitehead, 2017). Liposomes have been widely used because they are biodegradable, easy to formulate, present high efficacy and low toxicity (Aldosari et al., 2021). They are usually formed by a cationic lipid together with a helper lipid, cholesterol and a polyethylene glycol-lipid (PEG-lipid). The cationic or ionizable lipids interact with the negative charges of the nucleic acid, leading to structural changes and forming new structures. These structures can be formed by lamellar, hexagonal or even cubic stacks that intercalate nucleic acids (Fig. 5) (Ewert et al., 2021; Gaspar et al., 2020; Koltover et al., 1998; Ziller et al., 2018).

The incorporation of these components can improve the stability of formulation, the efficacy of the nucleic acid delivery, tolerability, biodistribution and transfection efficacy (Hajj and Whitehead, 2017; Kim et al., 2021; Meng et al., 2021). For example, DOPE is a phospholipid used to adopt an inverted hexagonal H(II) phase, and destabilize the endosomal membranes (Kauffman et al., 2015; Koltover et al., 1998). DSPC (1,2-distearoyl-sn-glycero-3-phosphocholine), and POPE (1-palmitoyl-2-oleoylsn-glycero-3-phosphoethanolamine) are also helper lipids commonly used in these systems to stabilize the lipid bilayer of the structure (Li et al., 2019). Cholesterol modulates the integrity and rigidity of the membrane, and thus, increases the stability of the particle (Hajj and Whitehead, 2017; Kim et al., 2021; Meng et al., 2021). PEG-lipids prevent particle aggregation and enhance its stability. In order to connect with specific ligands and reach the target tissue, PEG-lipids can be used as linkers. The effects of PEG-lipids depend on their properties (e.g., lipid tail length, lipid tail saturation, and lipid charge), and their proportion in the liposomes. It has been reported that PEG-lipids with shorter lipid tails like 1,2-dimyristoyl-sn-glycerol-3-methoxypolyethylene glycol (DMG) are preferred. The reason for this

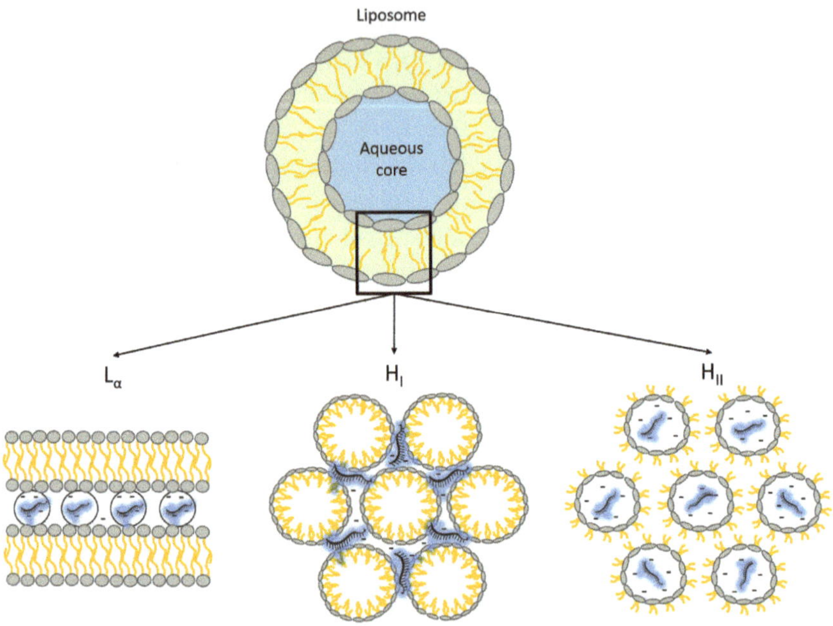

Fig. 5 Schematic representation of the spontaneous formation of liposome–mRNA complexes. Lα: lamellar structure, where the mRNA is sandwiched between the lipid bilayers; HI: hexagonal phase, with lipid elongated micelles arranged on a hexagonal lattice and mRNA inserted in the gaps with honeycomb symmetry; and HII: inverted hexagonal phase, comprised of mRNA inserted within inverse lipid tubules.

is that, in addition to reducing the immune response to PEG (Judge et al., 2006), shorter tails allow the PEG-lipids to dissociate more quickly from the LNPs after a period of time, resulting in a higher efficacy of transfection (MacLachlan and Cullis, 2005). However, the use of 1,2-distearoyl-sn-glycerol, methoxypolyethylene glycol (DSG) lipid, also known as persistent PEG lipid, could be interesting in cases in where larger circulation times are required. Indeed, in a recent study, the saturated alkyl chain of PEG2000–DMG and PEG2000-DSG affected the circulation time and efficacy of siRNA, resulting more effective the formulation containing PEG2000-DSG (Zhu et al., 2017).

The administration of liposome–mRNA complexes was reported almost 30 years ago (Lu et al., 1994). The transfection of tumor cells in vitro and in vivo employing cationic mRNA-liposomes demonstrated their potential utility as an alternative to pDNA immunotherapy. Currently, nucleoside-modified mRNA encoding the pre-membrane and envelope glycoproteins of a strain from Zika virus was encapsulated in liposomes formed with the ionizable lipid phosphatidylcoline, cholesterol, and a PEG-lipid. Potent and durable neutralizing antibody responses and protective immunity were observed in mice and non-human primates after intradermal administration of a 50 μg dose (Pardi et al., 2017). Additionally, the administration of synthetic SAM-liposomes encoding influenza H1HA antigen from N1H1 virus showed antibody responses comparable to the licensed vaccine (Hekele et al., 2013). In 1999, Zhou et al. administered for the first time mRNA-liposomes for cancer therapy in a mouse-melanoma model, and since then, its application has increased drastically (Zhou et al., 1999). In this study, hemagglutinating virus of Japan (HVJ)-based liposomes containing mRNA encoding human melanoma-associated antigen glycoprotein 100 (gp100) were administered by direct injection into the mouse spleen. Both antibody expression and cellular immune responses were induced.

Liposomes have also been applied for clustered regularly interspaced short palindromic repeats (CRISPR)-associated nuclease Cas9 (CRISPR/Cas9) delivery. Finn et al. (Finn et al., 2018) formulated LNPs composed by a biodegradable and ionizable lipid termed LP01, Spy Cas9 mRNA, and a single RNA guide (sgRNA). A single systemic administration to mice of the formulation produced significant editing of the mouse transthyretin (*Ttr*) gene in the liver, resulting in >97% reduction of TTR serum protein levels that persisted for at least 12 months. More recently, the safety, tolerability, pharmacokinetics and pharmacodynamics of single escalating doses of this system has been studied in a clinical trial (NCT04601051) in six patients with hereditary transthyretin amyloidosis with polyneuropathy. Although a small group of patients showed a few mild adverse effects, a reduction of the concentration of TTR protein in serum through targeted knockout of *Ttr* was produced (Gillmore et al., 2021). Currently, most clinical trials with liposomes complexed with mRNA are focused on cancer therapy, including ovarian cancer (NCT04163094), pediatric high-grade gliomas and adult glioblastoma (NCT00024648) and neoplasms (NCT04573140).

Recent advances in the development of novel vaccine strategies to address the COVID-19 pandemic have been made, thanks to the design of appropriate LNPs. Toward this end, LNPs for mRNA vaccines developed by Pfizer/BioNTech and Moderna have been clinically rolled out to hundreds of millions of patients around the world. These vaccines contain modified mRNA encoding the viral Spike (S) glycoprotein of SARS-CoV-2 encapsulated in LNPs. Both Pfizer/BioNTech's vaccine and

Moderna's vaccine consist of LNPs formulated by the combination of four chemical components: an ionizable cationic lipid, a neutral phospholipid, cholesterol and a lipid-anchored polymer. Specifically, Pfizer/BioNTech's vaccine is formulated with the ionizable lipid ALC-0315, cholesterol, DSPC, and PEG-lipid ALC-0159, whereas Moderna's is formulated with the ionizable lipid SM-102, cholesterol, DSPC, and PEG2000-DMG (Andresen and Fenton, 2021). Indeed, the LNPs of these vaccines present lipids in the core (Schoenmaker et al., 2021). Both vaccines have been well-tolerated and have demonstrated approximately 95% efficacy against COVID-19, with few adverse effects after intramuscular administration (Bajaj et al., 2021; Gediz Erturk et al., 2021).

- Solid Lipid Nanoparticles (SLNs)

SLNs are promising new lipid-based colloidal gene delivery systems with exciting potential for gene therapy (Solinís et al., 2015). SLNs are spherical nanometric particles formed by an aqueous dispersion stabilized by surfactants surrounding a solid lipid core matrix. They are usually made of well-tolerated physiological lipids (Chattopadhyay et al., 2007; Ramamoorth and Narvekar, 2015; Trucillo and Campardelli, 2019; Yadav et al., 2013). Initially, SLNs were developed with a view to address the efficacy and safety limitations that liposomes present (Müller et al., 2002). In this sense, SLNs show good stability and they can be sterilized and lyophilized. Moreover, the versatility of SLNs confers to them the capacity to be functionalized incorporating specific ligands, taking into account the features of the nucleic acid and the target cell (Apaolaza et al., 2015, 2014; Delgado et al., 2013). SLNs have demonstrated efficacy as nucleic acid delivery systems in vitro and in vivo, mainly for pDNA delivery in lysosomal storage disorders (Rodríguez-Castejón et al., 2021; Ruiz De Garibay et al., 2015), various types of cancer (De, 2021; del Pozo-Rodríguez et al., 2016; Granja et al., 2021; Ma et al., 2015) and ocular pathologies (Apaolaza et al., 2016; Torrecilla et al., 2018; Vicente-Pascual et al., 2018, 2020). More recently, SLN-based vectors have been assessed as delivery systems of mRNA. Gómez-Aguado et al. (Gómez-Aguado et al., 2020a) showed that formulation-related factors have a more significant impact on mRNA vectors than on pDNA vectors in terms of nucleic acid delivery, transfection efficacy and long-term stability during storage. In another study, SLN-based vectors bearing mRNA encoding intereleukin-10 (IL-10) were evaluated in vitro and in vivo (Gómez-Aguado et al., 2021). The study showed the capacity of SLNs to transfect human corneal epithelial (HCE-2) cells in vitro, and the capacity to transfect and produce IL-10 in the corneal epithelium of mice after topical administration as eye drops. Currently, there is no clinical trial comprising SLNs and mRNA under evaluation.

- Nanostructured lipid carriers (NLCs)

Nanostructured lipid carriers (NLCs) are a type of LNPs formed by a mixture of solid and liquid lipids, leading to a hybrid formulation between oil-in-water emulsions and SLNs (Aldosari et al., 2021). They present low cytotoxicity, their production is easy and economical and they can be sterilized (Beloqui et al., 2016). NLCs have been principally used to improve the oral bioavailability of poorly soluble drugs (Beloqui et al., 2017). However, NLCs are currently being evaluated for the delivery of mRNA vaccines. Concretely, the administration of NLCs containing a replicating viral mRNA encoding Zika virus antigens reported a complete protection of mice against a lethal Zika infection (Erasmus et al., 2018). For the time being, no NLC-based formulation has entered clinic evaluation.

- Exosomes

Exosomes are extracellular vesicles naturally secreted by numerous cells with a size range of 50–100 nm. They are emerging as promising natural carriers due to their role in the cell-cell communication process by exchanging proteins, lipids and nucleic acids to the recipient cells (Valadi et al., 2007). The interest of exosomes has increased as nucleic acid delivery systems due to their natural biocompatibility and minimal immune clearance (Valadi et al., 2007). Skog et al. demonstrated the capacity of exosomes to deliver mRNA to recipient cells and translate into a therapeutic protein. Exosomes were collected from glioblastoma cells and were transduced with mRNA encoding luciferase from Gaussia (Gluc). Cell-derived microvesicles with Gluc mRNA were used to treat human brain microvascular endothelial cells (HBMVEC) in vitro, resulting in an increase of Gluc activity (Skog et al., 2008). However, the use of these systems is hampered by the difficulties of large-scale production, isolation and purification (Ibba et al., 2021). Yang et al. proposed a cellular-nanoporation strategy, which consists of generating pores in membranes by using electrical nanochannels, to promote mRNA insertion. This method increased the exosome production by 50-fold and exosomal mRNA loading by 10^3-fold comparing with bulk electroporation (Yang et al., 2020). Additionally, the development of exosomes that expose specific targeting ligands on their surface is ongoing (Luan et al., 2017). More recently, mRNA-loaded exosomes have entered into clinical evaluation for the treatment of homozygous familial hypercholesterolemia (HoFH), providing a new therapeutic approach for patients suffering this disease. This first-in-human study is aimed to evaluate the safety and preliminary effectiveness of exosome-based LDL receptor mRNA nanoplatform in HoFH (NCT05043181).

- Hybrid systems

To conclude, the mixture of lipids with other materials, including polymers and peptides, among others, are defined as hybrid systems. As an advantage, these systems present all the benefits of each individual component in terms of stability, functionality and improved transfection efficacy (Ibba et al., 2021). Lipid-polymer hybrid nanoparticles, also known as lipopolyplexes, are formed by polymeric material surrounded by a lipid and/or lipid-PEG layer (Guevara et al., 2019). The efficacy of transfection of this nanosystem is influenced by the nature of the polymer (Zhao et al., 2018). Moreover, the combination of cationic lipid with peptides has been widely used for mRNA delivery, and significant suppression responses have been reported. For example, the systemic administration of lipid/protamine/synthetic mRNA encoding herpes simplex virus 1-thymidine kinase (HSV1-tk) to H460 xenograft-bearing nude mice produced higher efficacy and lower toxicity than the equivalent formulation with DNA (Wang et al., 2013). In another example, an intratumoral injection of a liposome–protamine complex containing IL-22 Binding Protein (IL-22BP) mRNA produced successful transfection into C26 colon carcinoma in vitro and in vivo with good safety profile (Zhang et al., 2018). Presently, there is no clinical trial under evaluation with this type of systems.

4. Clinical applications of mRNA

The attention placed over decades on mRNA design and the development of nucleic acid-based nanomedicines for therapeutic purposes is finally bearing fruit. mRNA therapeutics have dramatically changed the approach to the treatment of many diseases, with special interest in immunotherapy (treatment of infectious diseases and various types of cancer) and in protein replacement. Although most clinical trials using mRNA as active substance are focused on these fields, gene editing is a more recent clinical application of mRNA.

4.1 Gene editing

Gene editing is a novel therapeutic strategy that has increased in the recent years due to the different clinical conditions that can be applied for. This technology requires the action of engineered and programmable nucleases that generate a double stranded break (DSB) in the DNA by cleaving the double strand of DNA in a specific place of the genome. Two pathways can accomplish repair of DSBs: homolog-dependent repair (HDR) and non-homologous end joining (NHEJ). In the case of HDR, nucleases need a donor DNA that act as template containing a homologous sequence that will be introduced into DSB. This strategy is suitable when the repair of genomic mutations or the insertion of new therapeutic sequences are required. In contrast, NHEJ removes the specific sequence by joining DSBs in order to inactivate or correct an aberrant gene (Conway et al., 2019; Shim et al., 2017).

Zinc finger nucleases (ZFNs), transcription activator-like effector nucleases (TALENs) and CRISPR/Cas9 are the main editing technologies used. For gene editing, these nucleases can be delivered in protein, pDNA or mRNA forms (Zhang et al., 2019). mRNA delivery provides higher capability for therapeutic application compared to pDNA; since its expression is temporary, the effects of nuclease inside the cells are limited, and the possible risk of genome insertion is low. Moreover, the presence of nuclease is more consistent when it is expressed from mRNA than when the delivery of the nuclease is produced in protein form (Crudele and Chamberlain, 2018). Indeed, the administration of nuclease itself presents some limitations in vivo, such as, the activation of cellular and humoral response, challenging size and superficial charge for intravenous administration and protein purification process for large nucleases (Zhang et al., 2019).

As it observed in Table 2, clinical translation of mRNA-based gene editing is focused on ex vivo strategies that primarily use ZFN-mRNA. TALEN-mRNA has been evaluated only in one clinical trial. For in vivo application of CRISPR/Cas9-mRNA technology, intravenous administration of lipid nanosystems has also been evaluated. The corneal delivery of CRISPR/Cas9-mRNA in one clinical trial can be considered outstanding. There are hardly any clinical trials with nucleic acids (Gómez-Aguado et al., 2021; Rodríguez-Gascón et al., 2016) in cornea, and this strategy expands treatment options for a viral disease that severely limits the quality of life of patients suffering from it.

Regenerative medicine by cellular reprogramming and engineering with mRNA has just reached the clinical phase. Regenerative medicine consists of regrowing, repairing or replacing damaged or lost cells, organs or tissues, with the aim of recovering or restoring their natural activity (Jo et al., 2019). The transfection of somatic cells bearing mRNA encoding transcription factors is overcoming the conventional reprogramming and transdifferentiation strategies with somatic cells (Kwon et al., 2018). However, it is mainly

Table 2 Clinical trials of mRNA for gene editing therapy.

Disease	Biological active	Therapeutic mRNA	Target protein	Delivery system	Administration route	Clinical trial phase	Clinical trial identifier
Acquired immunodeficiency syndrome (AIDS) caused by HIV	SB-728mR	ZFN mRNA	CCR5	Ex vivo transfected autologous CD4 + T cells	Intravenous	Phase I	NCT02388594
	SB-728mR	ZFN mRNA	CCR5	Ex vivo transfected autologous CD4 + CAR-T cells	Intravenous	Phase I	NCT03617198
	SB-728mR-T	ZFN mRNA	CCR5	Ex vivo transfected autologous T cells	Intravenous	Phase I and I/II	NCT04201782, NCT02225665
	SB-728mR-HSPC	ZFN mRNA	CCR5	Ex vivo transfected autologous CD34 + hHSPCs	Intravenous	Phase I	NCT02500849
Sickle Cell Disease	BIVV003	ZFN mRNA	B-cell lymphoma/leukemia 11A (BCL11A)	Ex vivo transfected autologous CD34 + hematopoietic stem cells (HSPC)	Intravenous	Phase I/II	NCT03653247

Transfusion-Dependent Beta-thalassemia	ST-400	ZFN mRNA	B-cell lymphoma/leukemia 11A (BCL11A)	Ex vivo transfected autologous CD34+ hematopoietic stem/progenitor cells	Intravenous	Phase I/II	NCT03432364
B cell acute lymphoblastic leukemia	UCART19	TALEN mRNA	TCR and CD52	Ex vivo transfected allogenic T cells	Intravenous	Phase I	NCT02808442, NCT02746952, NCT02735083
B cell leukemia and B cell lymphoma	UCART019	CRISPR/Cas9 mRNA	TCR, B2M	Ex vivo transfected allogenic T cells	Intravenous	Phase I/II	NCT03166878
Herpetic Stromal Keratitis	BD111	CRISPR/Cas9 mRNA	Herpes simplex virus type I	NA	Corneal injection	Phase I/II	NCT04560790
Hereditary Transthyretin Amyloidosis	NTLA-2001	CRISPR/Cas9 mRNA	Transthyretin	Lipid nanosystem	Intravenous	Phase I	NCT04601051
Hereditary Angioedema	NTLA-2002	CRISPR/Cas9 mRNA	Kallikrein B1	Lipid nanosystem	Intravenous	Phase I/II	NCT05120830

B2M: β-2 microglobulin; CAR: chimeric antigen receptor; CCR5: CC chemokine receptor type 5; CRISPR/Cas9: clustered regularly interspaced short palindromic repeats/caspase 9; HIV: human immunodeficiency virus; NA: not available; TALEN: transcription activator-like effector nucleases; ZFN: Zinc finger nucleases.

limited to in vitro preclinical works, and only one clinical trial is ongoing in phase I/II (NCT02407470). Researchers have demonstrated that adipose tissue-derived mesenchymal stem cells (AD-MSCs) can be transdifferentiated into adipose tissue-derived hematopoietic stem cells (AD-HSCs) by transfection of small RNAs to the early region 1A (E1A)-like inhibitor of differentiation 1 (EID1) in the presence of specific cytokines. The purpose of the clinical trial is to evaluate the effectiveness and safety of a combination of patient's own AD-MSCs transdifferentiated into AD-HSCs and rabbit antithymoglobulin (polyclonal antibody) for the treatment of severe aplastic anemia.

MSCs engineering is another strategy of regenerative medicine, in which mRNA can be used to modulate the migratory properties of these cells, by expressing homing proteins in a brief, burst and temporary way. Once in the target tissue, MSCs have ability of self-renewal and differentiation into a wide range of cell lines, such as cartilage, adipocytes, and bone (Kwon et al., 2018; Park et al., 2015). Clinical application of mRNA-mediated engineered MSCs is already a reality. Indeed, a clinical trial (NCT04524962) has started recruiting patients to study the effect of RNA-engineered allogeneic MSCs (Descartes-30) in acute respiratory distress syndrome and COVID-19. Descartes-30 expresses a unique combination of DNases to eliminate extracellular networks of neutrophils, which are a key element of inflammation and coagulation in acute respiratory distress syndrome.

4.2 Immunotherapy

Triggering an immune response has been the main objective of mRNA-based therapies, resulting in a steady increase in clinical trials that evaluate immunotherapy with mRNA for infectious diseases and different types of cancer (Pardi et al., 2018). Besides those applications, a proof of concept by administering mRNA-transfected T cells against diabetogenic CD8+ T cells in mice offers a new approach for the treatment of type 1 diabetes (Fishman et al., 2017).

Recently, antibodies expression through mRNA for passive immunization has opened a wide range of opportunities for vaccination. The field of monoclonal antibodies (mAbs) is rapidly growing; however, the high cost of manufacturing and purification processes greatly limits their use. Given that situation, the pharmaceutical industry needs a more cost-effective synthesis and an appropriate delivery of mAbs. mRNA technology has come out as a highly attractive platform for any protein expression, being a suitable strategy for antibody production in vivo (Deal et al., 2021; Van Hoecke and Roose, 2019). In 2021, Moderna announced positive phase I results (NCT03829384) for the first systemic mRNA therapeutic (mRNA-1944) encoding an antibody (CHKV-24) with activity against chikungunya virus. Interim analysis of 38 healthy participants who received intravenous mRNA-1944, showed that the system was well tolerated with mild to moderate adverse effects, and that the vaccination produced CHKV-24 IgG with neutralizing activity at titres predicted to be therapeutically relevant (August et al., 2021). Although there have been few studies with mRNA encoding antibodies and further optimization is needed, this promising field is advancing rapidly.

mRNA possesses attractive features that make it an excellent choice for immunotherapy, for instance, innate immunogenicity, transient expression of the antigen of interest and versatility of application (prophylaxis, therapy and personalized vaccines). The safety profile and the simple, fast and cost-effective synthesis of mRNA vaccines make them advantageous over conventional subunit, killed and live attenuated virus vaccines and DNA-based vaccines. (Liu, 2019; Pardi et al., 2018).

In general, mRNA vaccines are classified as non-replicating or SAM. Fig. 6 outlines the differences between these two types of mRNA vaccines. Non-replicating mRNA vaccines only encode the immunogen of interest, avoiding the expression of other encoded proteins that could induce undesired immune responses (Kowalzik et al., 2021). SAM constructs, also known as replicons, encode not only the target antigen, but also an RNA-dependent RNA polymerase (RDRP) complex. The sequence that encodes de RDRP complex is derived from a positive single-stranded RNA virus genome, generally alphaviruses, where the genes that encode the replicative machinery of the virus are not modified, but structural genes are replaced by the sequence that encodes the immunogen. The ability of the replicons to self-amplify increases the efficacy and duration of the expression of the constructs and the consequent production of the encoded immunogen, and usually a lower dose is needed compared to non-replicating mRNA vaccines (Blakney et al., 2021). Additionally, SAM constructs allow the incorporation of multiple genes, enabling the expression of both the target antigen and immunomodulatory molecules to enhance potency (Maruggi et al., 2019). The transient self-replication generates double-stranded RNA and, thus, SAM tends to activate innate immune pathways. It is well known that SAMs activate IFN-I response via binding to TLRs 3, 7 and 8,

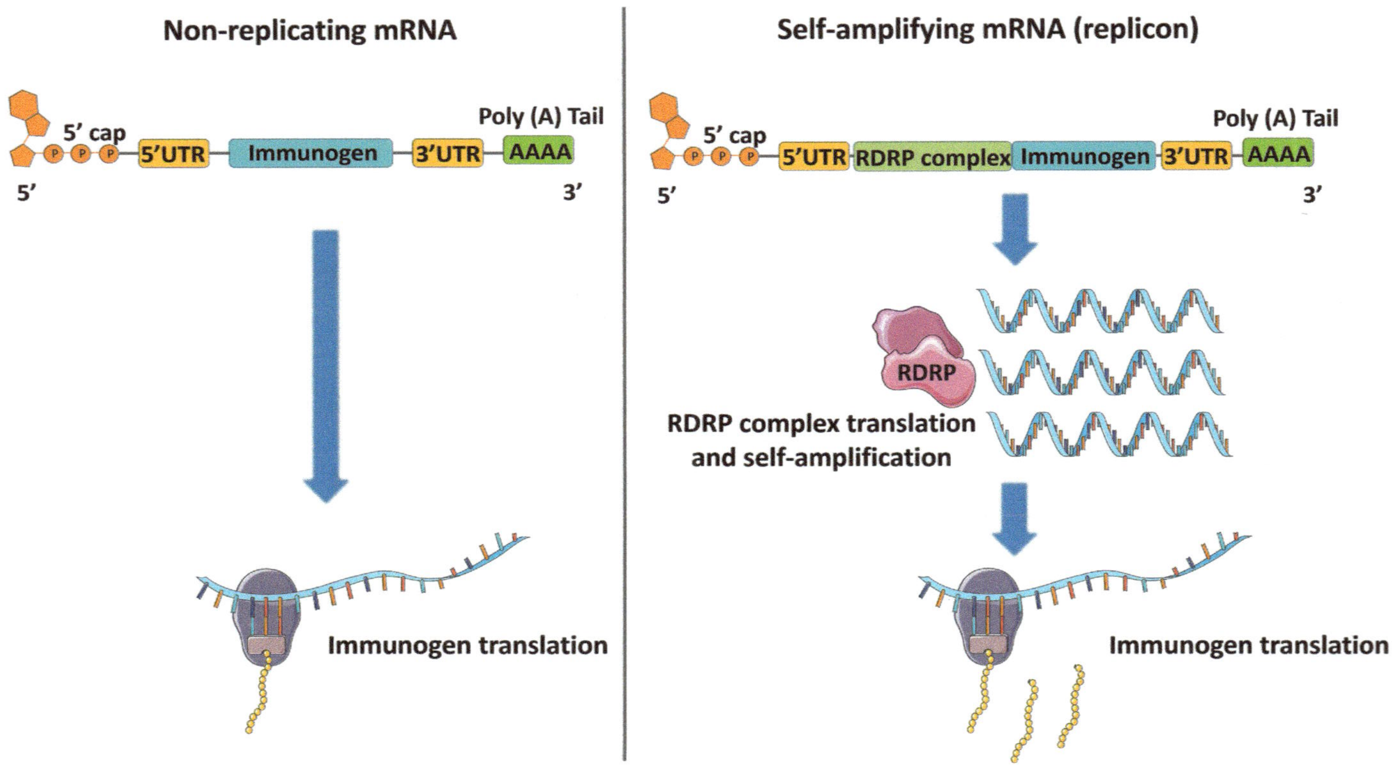

Fig. 6 Non-replicating and self-amplifying (replicon) mRNA constructs. Non-replicating mRNA constructs are composed of a coding sequence (immunogen) flanked by 5′ and 3′ untranslated regions (UTR), a 5′ Cap and a 3′ polyadenylated tail (Poly (A) Tail). Additionally, self-amplifying mRNA constructs encode a RNA-dependent RNA polymerase (RDRP) complex to amplify the immunogen sequence.

and via melanoma differentiation-associated protein 5 (MDA5), RIG-I, protein kinase R (PKR), and 2′-5′oligoadenylate synthetase (OAS). This contributes to the enhanced immunogenicity of SAM vaccines. However, IFN-I activation can lead to inhibition of translation and degradation of the mRNA compromising the potency of the vaccine (Blakney et al., 2021; Rodríguez-Gascón et al., 2014). In addition, SAM can bear only few modifications of synthetic nucleotides and sequence alterations without losing self-replicating activity (Kowalski et al., 2019; Pardi et al., 2018), and the production and stability is challenging since they are larger than non-replicating mRNA vaccines (≈9.3 vs 2.2 Kb) (Kowalzik et al., 2021).

For the development of a vaccine, there are two important steps to take in consideration. First, an antigen from the target pathogen must be identified, and second, a delivery system capable of producing a strong humoral and cellular immunity must be developed (Park et al., 2021). Stimulation of a specific or adaptive immune response by mRNA vaccines begins when the antigen is expressed by the ribosome in the cytosol of antigen-presenting cells (APCs), such as DCs. The expressed antigen is processed by the proteasome complex and presented on the cell surface to cytotoxic CD8 + T cells by major histocompatibility complex (MHC) class I proteins, activating cellular response. Additionally, secreted antigens can be taken up by cells, degraded inside endosomes and displayed to CD4+ helper T cells by MHC class II proteins. CD4+ helper T cells can stimulate B cells (humoral response) to produce neutralizing antibodies and can secrete cytokines to activate phagocytic macrophages (Chaudhary et al., 2021; Heine et al., 2021).

4.2.1 mRNA vaccines against infectious diseases

Vaccination is the most effective prophylactic measure to provide protection against a wide variety of infectious agents. There are currently about 24 mRNA-based vaccines under clinical evaluation for infectious diseases, excluding those developed for SARS-CoV-2. The majority of targeted infectious agents are viruses (e.g., rabies virus, Zika virus, CMV, human metapneumovirus and parainfluenza virus type 3 (hMPV/PIV3), influenza virus, chikungunya virus, RSV and HIV). mRNA-based vaccines under evaluation are mainly administered by intradermal, intramuscular or subcutaneous injection, being taken up by immune or non-immune cells and translated into antigens, which are presented to T and B cells. Furthermore, structure of the mRNA and the delivery platform greatly condition the immunogenicity and efficacy of the vaccine. Table 3 shows clinical trials of mRNA vaccines against infectious diseases apart from COVID-19.

Table 3 Clinical trials of mRNA vaccines against infectious diseases apart from COVID-19.

Infectious disease	Biological active	Encoded protein	Delivery system	Administration route	Clinical trial phase	Clinical trial identifier
Influenza	mRNA-1440 (VAL-506440)	Hemagglutinin from influenza A H10N8 strain	Lipid nanosystem	Intramuscular	Phase I	NCT03076385
	mRNA-1851 (VAL-339851)	Hemagglutinin from influenza A H7N9 strain	Lipid nanosystem	Intramuscular	Phase I	NCT03345043
	AVX502	H3 hemagglutinin from A/Wyoming/03/2003 H3N2 strain	Virus-like replicon particles	Intramuscular Subcutaneous	Phase I/II	NCT00440362, NCT00706732
	mRNA-1010	Influenza A (H1N1, H3N2), Influenza B (Yamagata lineage, Victoria lineage)	NA	Intramuscular	Phase I/II	NCT04956575
Rabies	CV7201	Rabies virus glycoprotein (RABV-G)	Polypeptide system (RNActive®)	Intradermal Intramuscular	Phase I	NCT02241135
	CV7202	Rabies virus glycoprotein (RABV-G)	Lipid nanosystem	Intramuscular	Phase I	NCT03713086
	GSK3903133A	Rabies glycoprotein G (RG)	Cationic nanoemulsion	Intramuscular	Phase I	NCT04062669

Respiratory syncytial virus (RSV)	mRNA-1345	Prefusion F glycoprotein	Lipid nanosystem	Intramuscular	Phase I and II/III	NCT04528719, NCT05127434
hMPV / PIV3	mRNA-1653	Fusion proteins of hMPV and PIV3	Lipid nanosystem	Intramuscular	Phase I	NCT03392389, NCT04144348
Zika Virus	mRNA-1325	Structural proteins of Zika virus	Lipid nanosystem	Intramuscular	Phase I	NCT03014089
	mRNA-1893	Structural proteins of Zika virus	Lipid nanosystem	Intramuscular	Phase I and II	NCT04064905, NCT04917861
Chikungunya	mRNA-1944	Chikungunya antibody (CHKV-24 IgG)	Lipid nanosystem	Intravenous	Phase I	NCT03829384
	mRNA-1388 (VAL-181388)	Full native structural polyprotein (C-E3-E2-6k-E1)	Lipid nanosystem	Intramuscular	Phase I	NCT03325075
Human immunodeficiency virus 1 (HIV-1)	Ex vivo mRNA electroporated autologous DCs	HIV-1 Gag and Nef	Autologous DCs	Intradermal	Phase I/II	NCT00833781
	iHIVARNA-01	HIVACAT T cell immunogen (HTI) - TriMix	Naked mRNA	Intranodal	Phase I and II	NCT02413645, NCT02888756
	AVX101	HIV-1C Gag antigen	Virus-like replicon particles	Subcutaneous	Phase I	NCT00097838, NCT00063778

Continued

Table 3 Clinical trials of mRNA vaccines against infectious diseases apart from COVID-19.—cont'd

Infectious disease	Biological active	Encoded protein	Delivery system	Administration route	Clinical trial phase	Clinical trial identifier
	Core-g28v2 60mer mRNA vaccine (mRNA-1644) and eOD-GT8 60mer mRNA vaccine (mRNA-1644v2-Core)	NA	NA	Intramuscular	Phase I	NCT05001373
Cytomegalovirus (CMV)	mRNA-1647	Pentamer complex and full-length membrane-bound glycoprotein B (gB) of CMV	Lipid nanosystem	Intramuscular	Phase I, II and III	NCT03382405, NCT05105048, NCT04232280, NCT04975893, NCT05085366
	mRNA-1443	pp65 T cell antigen of CMV	Lipid nanosystem	Intramuscular	Phase I	NCT03382405
	AVX601	gB, pp65 and IE1 proteins of CMV	Virus-like replicon particles	Intramuscular Subcutaneous	Phase I	NCT00439803
Tuberculosis	GSK 692342	Immunogenic fusion protein (M72) derived from *Mycobacterium tuberculosis*	Lipid nanosystem	Intramuscular	Phase I and II	NCT00730795, NCT01669096, NCT01424501, NCT00397943, NCT01755598, NCT00621322, NCT01262976, NCT00950612, NCT00600782, NCT01098474, NCT00707967

DCs, dendritic cells; hMPV/PIV3, human metapneumovirus and parainfluenza virus type 3; NA, not available.

One of the most attractive candidates for mRNA-based vaccination is influenza virus. This was the target of the first demonstration of immunization against an infectious disease by a non-replicating mRNA vaccine, reported in 2012 by Petsch et al. (Petsch et al., 2012). RNActive® mRNA encoding hemagglutinin from the PR8 H1N1 influenza virus strain was administered by intradermal injections to mice, ferret and pigs. The vaccine induced immunity against H1N1 and H5N1 strains. In ferrets and pigs, mRNA vaccines induced immunological correlates of protection and protective effects similar to those of a licensed influenza vaccine, and although complete protection from virus replication was not achieved in pigs, the virus was cleared more quickly as compared to the licensed vaccine. In addition, vaccination protected the mice against a lethal viral dose. Since then, different delivery systems, non-replicating and SAM mRNAs and target antigens have been evaluated against influenza virus (Brazzoli et al., 2016; Freyn et al., 2020; Magini et al., 2016; Vogel et al., 2018; Zhuang et al., 2020). After the successful results in mice, ferrets and non-human primates of two LNP-formulated mRNA vaccines encoding hemagglutinin proteins of H10N8 or H7N9 (Bahl et al., 2017), Moderna has completed two phase I clinical trials (NCT03076385, NCT03345043) in which the safety and immunogenicity were evaluated. The administration of two intramuscular doses of the vaccines was well tolerated and elicited robust humoral immune responses in healthy adults (Feldman et al., 2019).

CureVac has also applied its platform RNActive® to develop an unmodified mRNA vaccine (CV7201) against the rabies virus glycoprotein. Preclinical studies in mice and pigs showed that two intradermal doses of 80 μg of the vaccine administered with an interval of 21 days, induced high neutralizing antibody titres and elicited antigen-specific CD4+ and CD8+T cell responses (Schnee et al., 2016). In a phase I clinical trial (NCT02241135), the administration route (intradermal or intramuscular) and the administration device (needle-syringe or a needle-free device) were evaluated. The administration route did not affect the immune response, but only with the needle-free system, functional antibodies were produced. Nevertheless, the humoral response resulted to have a short-life and the incidence of adverse effects was high (Alberer et al., 2017). To overcome those issues, CureVac used Acuitas Therapeutics' LNPs as delivery system of the mRNA for a new rabies vaccine CV7202 (Lutz et al., 2017). Results of the phase I clinical trial (NCT03713086) showed that two intramuscular doses of 1 or 2 μg of the vaccine were well tolerated and produced high neutralizing antibody titres (Aldrich et al., 2021).

The development of an effective vaccine for the RSV has faced numerous challenges. For instance, in 1968, a formalin-inactivated RSV vaccine candidate caused vaccine-associated enhanced disease (VAED) in children, resulting in severe bronchiolitis or pneumonia cases and two fatalities (Kim et al., 1969). Currently, Moderna is evaluating three mRNA-LNP vaccines encoding the prefusion F protein (mRNA-1172, mRNA-1777 and mRNA-1345); the three candidates differ in the type of LNPs (mRNA-1172 uses Merck's proprietary LNPs and mRNA-177 and mRNA-1345 Moderna's LNPs) and mRNA structure. In a phase I clinical trial, mRNA-1777 elicited a robust humoral response with neutralizing antibodies, CD4+ T cell response to F protein and no serious adverse effects (Aliprantis et al., 2021). mRNA-1345 is being evaluated in a phase I study in RSV-seropositive patients (NCT04528719) and in a phase II/III clinical trial in adults that are at least 60 years old (NCT05127434). In addition, Moderna has also evaluated his candidate mRNA-1653 against hMPV/PIV3 in healthy adults (NCT03392389). An interim analysis demonstrated that the intramuscular administration of the mRNA-1653 vaccine was generally well-tolerated at all dose levels and a single dose of mRNA-1653 boosted serum neutralization titres against both hMPV and PIV3 (Shaw et al., 2019a). Consequently, a phase I clinical trial is recruiting participants to evaluate the same vaccine in healthy adults (18–49 years old) and children (12–59 months old) with serological evidence of prior exposure to hMPV and PIV3 (NCT04144348).

Successful preclinical results of immunization against Zika virus with mRNA vaccines prompted two phase I clinical trials (NCT03014089, NCT04064905) to evaluate the safety, tolerability and immunogenicity of two Zika mRNA-LNP vaccine candidates, mRNA-1325 and mRNA-1893. mRNA-1325 was designed to encode structural proteins of the Zika virus inside the cells to make VLPs that are later secreted. mRNA-1893 relies on the same strategy but it contains a sequence distinct from mRNA-1325 with improved immunogenicity and production of Zika VLPs (Moderna Inc., 2019). Interim results of the clinical trial with mRNA-1893 (NCT04064905) showed that the vaccine was well tolerated and induced seroconversion in 94% and 100% of seronegative participants in 10 and 30 μg dose groups, respectively (Moderna Inc, 2020). A phase II clinical trial is recruiting adult participants living in endemic and non-endemic flavivirus (virus genus to which the Zika virus belongs) areas to evaluate safety, tolerability and immunogenicity of two dose levels of mRNA-1893 (NCT04917861).

The same strategy has been employed for the development of a vaccine against chikungunya virus, mRNA-1388 (VAL-181388). mRNA-1388 vaccine consists of a single mRNA encapsulated in LNPs encoding the full native structural polyprotein (C-E3-E2-6k-E1) which assemble into VLPs and are released from transfected cells. The structural proteins are naturally processed into C and E proteins. The E proteins on these VLPs are the main target of neutralizing and protective antibodies. Meanwhile the C protein provides structure to the VLP and contains T cell epitopes that induce immune responses (Moderna Inc, 2019). mRNA-1388 has been evaluated in a phase I clinical trial (NCT03325075). Results showed that the vaccine was safe and well tolerated at all dose levels, and produced a 100% seroconversion in all subjects treated with two doses of 100 μg of mRNA-1388 (Shaw et al., 2019b).

Two clinical trials evaluated mRNA vaccines for HIV infection. In a phase I trial (NCT02413645), three doses of the naked mRNA (iHIVARNA-01) were injected by intranodal administration in HIV-infected individuals under stable combined antiretroviral therapy. iHIVARNA-01 was composed by an mRNA encoding HIVACAT T-cell Immunogen (HTI) in combination with TriMix. HTI is a novel HIV immunogen sequence that encodes HIV-1 target epitopes in Gag, Pol, Vif and Nef. TriMix is a combination of mRNAs encoding APCs activation signals CD40L (CD40 ligand), CD70 (costimulatory molecule ligand of CD27) and caTLR4 (constitutively active TLR4). iHIVARNA-01 was well tolerated and induced moderate HIV-specific T cells responses (Leal et al., 2018), promoting a phase II clinical trial (NCT02888756), which was terminated owing to a lack of immunogenicity (De Jong et al., 2019).

Although the majority of mRNA vaccines against infectious diseases use non-replicating mRNA constructs, SAM vaccines are under clinical evaluation for CMVs, influenza virus, rabies virus and HIV. AVX601, a bivalent alphavirus replicon vaccine (alphavaccine) expressing three CMV proteins (gB, pp65 and IE1), was a vaccine candidate against CMV evaluated in healthy volunteers in 2007 (NCT00439803). The vaccine was well tolerated and induced neutralizing antibodies and multifunctional T cell responses against the three CMV antigens (Bernstein et al., 2009). The safety and immunogenicity of another alphavaccine (AVX502), expressing the H3 hemagglutinin from the A/Wyoming H3N2 strain was evaluated in healthy volunteers of 18–40 (NCT00440362) or 65 years old or older (NCT00706732), respectively. The vaccine was safe and well tolerated in both trials, and, especially in the younger group, antibody and T cell responses were efficiently stimulated and persisted for the 4-month study (AlphaVax, n.d.). In other study, an alphavaccine (AVX101) in which the alphavirus structural proteins were replaced with HIV-1 subtype C gag was administered subcutaneously in healthy HIV-1-uninfected adults, for safety and immunogenicity evaluation (NCT00097838, NCT00063778). Only low levels of binding antibodies and T cell responses were observed with the highest doses (Wecker et al., 2012). The SAM vaccine candidates mentioned above use VLPs as delivery system. Alternatively, a SAM vaccine against rabies virus, formulated in a cationic nanoemulsion and administered by intramuscular route, is under evaluation in a phase I clinical trial (NCT04062669).

Although being viruses the main infectious pathogens targeted by mRNA vaccines, other pathogens, such as *mycobacterium tuberculosis* or *plasmodium falciparum* are also candidates to be tackled with this kind of vaccination. For example, an mRNA-LNP vaccine encoding the immunogenic fusion protein M72 derived from *mycobacterium tuberculosis* was evaluated in phase I and phase II clinical trials (see clinical trial identifiers in Table 3). In a phase II study (NCT01669096), two intramuscular injections of the vaccine showed an acceptable safety profile and resulted immunogenic in a study involving a small population of healthy adults (van den Berg et al., 2018).

mRNA-loaded DCs have been used as ex vivo therapeutic approaches for vaccination. DCs internalize naked mRNA by different endocytic ways, although electroporation improves efficacy. Ex vivo strategy is well described for cancer immunotherapy, and will be further discussed. However, for infectious diseases, it has only been used for HIV immunotherapy in a phase I/II clinical trial (NCT00833781). Intradermal administration to HIV-1-infected subjects with autologous DCs electroporated with mRNA encoding HIV-1 Gag and Nef antigens did not induce significant responses. The authors concluded that DC vaccination should be optimized to elicit stronger and long-lasting immune responses to be effective as an HIV-1 therapeutic vaccine. (Gandhi et al., 2016).

- mRNA Vaccines Against SARS-CoV-2

Since December 2019, the global pandemic caused by the SARS-CoV-2 has extremely limited life-quality and has caused significant loss of life worldwide. Against this backdrop, researchers and pharmaceutical industry have made a great effort to develop effective vaccines to combat the

SARS-CoV-2, shifting the focus from conventional inactivated virus- and protein-based vaccines to newer technologies such as DNA and mRNA.

Neutralizing antibodies against SARS-CoV-2 found in patients who have overcome the disease mainly target epitopes within the S protein, located on the surface of the virus (Du et al., 2009). Therefore, most vaccine candidates against SARS-CoV-2 induce an immune response against the S protein. In fact, the S protein plays a critical role for the cellular infection; the receptor-binding domain (RBD) of S1 subunit of S protein interacts with human angiotensin-converting enzyme 2 (ACE2) on the host cell surface, followed by membrane fusion mediated by the S2 subunit (Tang et al., 2020).

Seizing the moment and taking advantage of the need for a vaccine to evade SARS-CoV-2 and mitigate its effects in society, mRNA-based vaccines have found the time to take a giant leap forward. Actually, by 18 June 2021, 287 SARS-CoV-2 vaccine candidates were under evaluation, 185 in preclinical development and 102 in clinical trials. Among those in clinical trials, 19 were mRNA-based vaccines. Generally, the mRNA vaccines encode the full-length S or its RBD. Table 4 summarizes some of the mRNA-based vaccines against SARS-CoV-2 in clinical stages.

To date, two mRNA vaccines have been approved, BNT162b2 (Comirnaty®) and mRNA-1273 (Spikevax). BNT162b2, developed by Pfizer/BioNTech, received emergency authorization from the Medicines and Healthcare products Regulatory Agency (MHRA) in the United Kingdom (Medicines and Healthcare products Regulatory Agency, 2020), from the FDA in the United States (US Food and Drug Administration, 2021), and from the EMA in the European Union (European Medicines Agency, 2021b) on December 2020, becoming the first mRNA drug approved for human use. Pfizer and BioNTech collaborated to develop two promising mRNA vaccine candidates: BNT162b1 and BNT62b2. Both of them use LNPs formulated with the Acuitas Therapeutics' ionisable lipid ALC-0315 and a nucleoside-modified mRNA to enhance mRNA translation. BNT162b1 encodes a secreted trimerized version of the S antigen's RBD.

BNT62b2 encodes the full-length S antigen stabilized in its prefusion conformation. Both vaccines induced high antibody and neutralizing titres and a robust CD4+ and CD8+ response with mild to moderate adverse effects after two 30 μg intramuscular doses administered 21 days apart. However, only the BNT62b2 vaccine advanced to phase II/III trials owing

Table 4 Examples of clinical trials of mRNA vaccines against SARS-CoV-2.

Vaccine	Encoded protein	Delivery system	Administration route	Clinical trial phase	Clinical trial identifier
BNT162b2	Full-length transmembrane prefusion spike	Lipid nanosystem	Intramuscular	Phase III	NCT04800133, NCT04713553
				Phase IV	NCT05057182, NCT05057169
mRNA-1273	Full-length transmembrane prefusion spike	Lipid nanosystem	Intramuscular	Phase III	NCT04811664, NCT04470427
				Phase IV	NCT04952402, NCT04978038
mRNA-1273.351	Full-length transmembrane prefusion spike of the SARS-CoV-2 B.1.351 variant	Lipid nanosystem	Intramuscular	Phase I	NCT04785144
mRNA-1273.211	Proteins encoded by mRNA-1273 +mRNA-1271.351	Lipid nanosystem	Intramuscular	Phase II	NCT04405076
mRNA-1283	RBD and N-terminal domain of spike	Lipid nanosystem	Intramuscular	Phase II	NCT05137236
CVnCoV	Full-length transmembrane prefusion spike	Lipid nanosystem	Intramuscular	Phase II/III	NCT04652102
ARCoV	RBD of secreted spike	Lipid nanosystem	Intramuscular	Phase III	NCT04847102
LNP-nCoVsaRNA	Full-length transmembrane prefusion spike	Lipid nanosystem	Intramuscular	Phase I	ISRCTN17072692
ARCT-021	Full-length transmembrane prefusion spike	Lipid nanosystem	Intramuscular	Phase II	NCT04728347, NCT04668339

RBD, receptor-binding domain.

to its safer profile (Mulligan et al., 2020; Walsh et al., 2020). In phase III clinical trials, BNT62b2 showed 95% effectiveness at preventing COVID-19 disease (Polack et al., 2020).

One week after the approval of Pfizer/BioNTech's BNT162b2 vaccine, Moderna's mRNA-1273 vaccine was also authorized in the United States. mRNA-1273 consists of LNPs formulated with the ionizable lipid SM-102 encapsulating a nucleoside-modified mRNA that encodes the prefusion stabilized full-length S antigen. mRNA-1273 showed extraordinary efficacy and was well tolerated in clinical trials. Vaccination induced binding and neutralizing antibody titres equivalent to convalescent serum samples, and S protein-specific CD4+ T cell responses (Baden et al., 2021; Jackson et al., 2020). In phase III clinical trials, two intramuscular doses of 100 μg of the vaccine administered 28 days apart, were 94.1% effective in preventing symptomatic infection and 100% in preventing severe COVID-19 disease (Baden et al., 2021).

CureVac has also developed an mRNA vaccine against SARS-CoV-2, CVnCoV. This vaccine uses LNPs formulated with the Acuitas Therapeutics' ionizable lipid ALC-0315 to encapsulate an unmodified mRNA encoding the full-length S protein stabilized in a prefusion conformation. Although in a phase I clinical study (NCT04449276) (Kremsner et al., 2020) injections of 12 μg of the vaccine generated neutralizing antibodies similar to those in patients recovered from COVID-19, CVnCoV resulted in 47% efficacy in phase IIb/III trials (NCT04652102) (Kremsner et al., 2021), which was attributed to emerging SARS-CoV-2 variants (e.g., B.1.1.7, B.1.351, P1 and B.1.617.2). CureVac, in collaboration with GSK, is developing a second-generation mRNA vaccine candidate, CV2CoV. mRNA from CV2CoV has been optimized to enhance translation and immunogenicity relative to CVnCoV by using a 5′ UTR from the human hydroxysteroid 17-β-dehydrogenase 4 gene and a 3′ UTR from the human proteasome 20S subunit β3 gene. Preclinical results have shown that vaccination with CV2CoV induces high titres of neutralizing antibodies and significant cross-neutralization of the B.1.1.7 and B.1.351 variants (Roth et al., 2021).

Since the emergence of different SARS-CoV-2 variants, the efficacy of current vaccines against new variants has been of great concern. Pfizer/BioNTech and Moderna have sent a message of hope by assuring that their BNT62b2 and mRNA-1273 vaccines provide protection against B.1.1.7 and B.1.351 variants. Moreover, it is also reassuring to know that flexibility of mRNA vaccines enables the reformulation of the vaccines to better protect against other variants (Burki, 2021; Muik et al., 2021; Xie et al., 2021).

In this regard, Moderna has modified its mRNA-1273 vaccine to develop two other new vaccines: mRNA-1273.351 encoding the S protein of the B.1.351 variant, and mRNA-1273.211, a mix of mRNA-1273 and mRNA-1273.351. The safety and immunogenicity of a single booster dose of these vaccines have been evaluated in a phase II clinical trial (NCT04405076). An interim analysis showed that the vaccines were safe and well tolerated, and that all boosters increased neutralization titres against various variants, including B.1.351, P.1. and B.1.617.2 (Choi et al., 2021).

In November 2021, a new SARS-CoV-2 variant emerged, B.1.1.529 (Omicron), and it expanded rapidly throughout the world. This variant consists of over 30 mutations within the S protein, and efficacy of current mRNA vaccines targeting the S protein have been questioned. A study has recently demonstrated that patients treated with the current posology (two doses) of BNT62b2 and mRNA-1273 vaccines did not show neutralizing activity against the Omicron variant, but a third dose did. This finding suggested that a third dose is required to provide robust neutralizing antibody responses against the Omicron lineage (Edara et al., 2022). In fact, Pfizer and BioNTech have recently announced the necessity of three doses of the vaccine to neutralize the Omicron variant (Pfizer Inc. and BioNTech SE, 2021), and evaluation of reactogenicity and immune response of the third dose of BNT62b2 is ongoing in phase IV clinical trials (NCT05057182, NCT05057169). A third dose of mRNA-1273 is also being evaluated in clinical trials (NCT04978038).

SAM vaccines are also under development for SARS-CoV-2. The Imperial College London and Acuitas Therapeutics have developed a SAM vaccine based on LNPs, LNP-nCoVsaRNA. This vaccine encodes the full-length S protein and replication machinery from the genome of Venezuelan equine encephalitis virus (McKay et al., 2020). LNP-nCoVsaRNA is being evaluated in a phase I clinical trial (ISRCTN17072692) with a dose ranging from 0.1 to 1 μg, the lowest dose of all mRNA vaccine candidates. Moreover, two phase II clinical trials (NCT04728347, NCT04668339) are evaluating another SAM vaccine encapsulated in LNPs, ARCT-021 (LUNAR-COV-19) (de Alwis et al., 2021).

4.2.2 Cancer immunotherapy

Cancer immunotherapy with mRNA harnesses the ability of the nucleic acid to deliver genetic information and its innate immunostimulatory activity to achieve a therapeutic effect. mRNA in cancer immunotherapy can be

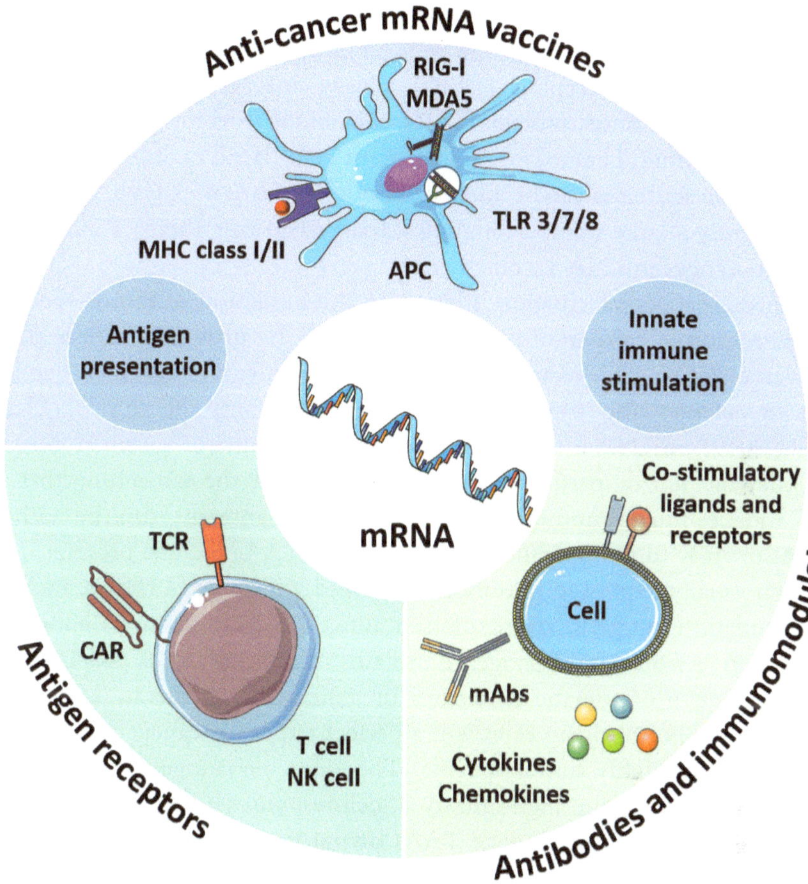

Fig. 7 Strategies for cancer immunotherapy with mRNA. mRNA can be employed for anti-cancer vaccination, by means of delivering cancer antigens to antigen-presenting cell (APCs) and stimulating innate immune system (top); to provide immune cells with antigen receptors (bottom left), and to express antibodies and immunomodulators in various cells (bottom right). MHC: major histocompatibility complex; CAR: chimeric antigen receptor; TCR: T-cell receptor; NK: natural killer; mAbs: monoclonal antibodies; TLR: Toll-like receptor; RIG-I: retinoic acid-inducible gene 1; MDA5: melanoma differentiation-associated protein 5.

employed with different purposes, which can be summarized in three strategies, as it is depicted in Fig. 7. On the one hand, mRNA can be used for anti-cancer vaccination to deliver cancer antigens to APCs, such as DCs, for the presentation on MHC class I and II, and to exert an adjuvant function by stimulating innate immune system through binding to pattern recognition receptors expressed by APCs (TLRs 3, 7 and 8, RIG-1, MDA5).

On the other hand, mRNA is used to engineer T cells and natural killer (NK) cells with antigen receptors such as chimeric antigen receptors (CARs) and T-cell receptors (TCRs). Finally, mRNA can express mAbs and immunomodulatory proteins (TLRs, chemokine receptors, co-stimulatory ligands, cytokines, chemokines) in different cells (Beck et al., 2021). Most of the above strategies can be addressed both in vivo and ex vivo, and there are currently a large number of clinical trials evaluating them (Table 5).

• Anti-cancer mRNA vaccines

In general, cancer vaccination is based on the induction of tumor-specific T-cell responses capable of fighting tumor cells, by providing cancer antigens in order to achieve an immunostimulating effect. Additionally, some cancer vaccines also trigger B-cell activation against the tumor (Zhao et al., 2019). mRNA is an attractive candidate for cancer vaccination as it encodes the full or partial sequence of antigens at the same time that it can induce innate immune-activation. Target antigens can be either TAAs or truly tumor-specific antigens (TSAs). TAAs include proteins that are abnormally expressed in tumors compared with normal tissues, such as differentiation antigens, overexpressed antigens or cancer/testis antigens. TSAs derive from viral oncogenic proteins or from proteins produced as a consequence of somatic mutations or gene rearrangements, known as neoantigens. Neoantigens represent a perfect target for cancer vaccination since they are highly tumor-specific. Therefore, neoantigens have higher probability of inducing high-affinity T-cell responses avoiding the central tolerance that can happen with TAAs owing to their self-antigen nature (Hollingsworth and Jansen, 2019; Liu et al., 2021b). Neoantigens can be identified by analyzing the entire somatic cancer mutations in an individual tumor (called mutatome), allowing the design of personalized neoepitope cancer vaccines (Blass and Ott, 2021).

Anti-cancer vaccination with mRNA can be performed by loading the mRNA encoding the desired antigen ex vivo on autologous DCs or by administering it in vivo. For the development of mRNA-based DC vaccines, early studies focused on ex vivo loading of mRNA into DCs to later reinfuse the transfected ones into patients by subcutaneous, intranodal or intravenous administration. DCs can be pulsed with bulk mRNA extracted from autologous tumors or with synthetic mRNA encoding defined TAAs (Filin et al., 2021). However, expressed antigen is presented on MHC class I rather than on class II molecules of the DCs, resulting in generally weak T-cell responses. Different strategies have been employed to improve MHC class II antigen presentation. For example, the fusion of the vaccine

Table 5 Examples of clinical trials of mRNA for cancer immunotherapy.

Vaccine type (Name)	Type of cancer	Administration route	Clinical trial phase	Clinical trial identifier
Anti-cancer mRNA vaccines				
WT1 mRNA DCs	AML, CML, MM	Intradermal	Phase II	NCT00965224, NCT01686334
hTERT/LAMP-1 mRNA DCs (GRNVAC1)	AML	Intradermal	Phase II	NCT00510133
HCMV pp65/LAMP mRNA DCs	Glioblastoma, astrocytoma, GCG, GM	Intradermal	Phase II	NCT02366728, NCT03688178
PSA, PAP, surviving, nTERT mRNA DCs	Prostatic neoplasm	Intradermal	Phase II	NCT01446731
HCMV pp65-shLAMP or pp65-flLAMP mRNA DCs	GM, glioblastoma, MG, astrocytoma, GBM	Under the skin	Phase II	NCT02465268
Tyrosinase, gp100 mRNA DCs	Melanoma	Intradermal, intravenous	Phase II	NCT02285413
Tumor mRNA DCs	Uveal melanoma	Intravenous	Phase III	NCT01983748
Total tumor mRNA and CD40L mRNA DCs (AGS-003/Rocapuldencel-T)	Renal cell carcinoma	Intradermal	Phase II and III	NCT00678119, NCT01582672
MAGE-A3, MAGE-C2, tyrosinase and gp100 mRNA and TriMix (CD70, CD40L, caTLR4) mRNA DCs	Melanoma	Intradermal, intravenous	Phase II	NCT01302496, NCT01676779
TAA mRNA and TriMix (CD70, CD40L, caTLR4) mRNA (ECI-006)	Melanoma	Intranodal	Phase I	NCT03394937
HPV16 antigens mRNA-LPX (BNT113)	HPV16+ and PD-L1+ HNSCC	Intravenous	Phase II	NCT04534205
NY-ESO-1, MAGE-A3, tyrosinase and TPTE mRNA-LPX (BNT111)	Melanoma	Intravenous	Phase II	NCT04526899
Tumor neoantigens mRNA-LPX (RO7198457)	Melanoma	Intravenous	Phase II	NCT03815058

Continued

Table 5 Examples of clinical trials of mRNA for cancer immunotherapy.—cont'd

Vaccine type (Name)	Type of cancer	Administration route	Clinical trial phase	Clinical trial identifier
Tumor neoantigens mRNA-LNP (mRNA-4157)	Melanoma	Intramuscular	Phase II	NCT03897881
Antigen receptors				
Anti-BCMA CAR-T cells (Descartes-08)	MM	Intravenous	Phase I/II	NCT03448978
Anti-mesothelin CAR-T cells	Malignant pleural mesothelioma, metastatic PDAC	Intravenous	Phase I	NCT01355965, NCT01897415
cMET redirected CAR-T cells	Metastatic breast cancer, TNBC	Intratumoral	Phase I	NCT01837602
NKG2D-ligand targeting CAR-NK cells	Solid tumors	Intravenous	Phase I	NCT03415100
Immunomodulators				
IL-12 mRNA-LNP (MEDI1191)	Solid tumors	Intratumoral	Phase I	NCT03946800
OX40L, IL-23 and IL-36**γ** mRNA-LNP (mRNA-2752)	Relapsed/refractory solid tumor malignancies or lymphoma, TNBC, HNSCC, Non-Hodgkin lymphoma, urothelial cancer	Intratumoral	Phase I	NCT03739931
scIL-12, IL-15sushi, IFN**α** and GM-CSF mRNA (SAR441000)	Metastatic neoplasms	Intratumoral	Phase I	NCT03871348
OX40L mRNA-LNP (mRNA-2416)	Relapsed/refractory solid tumor malignancies or lymphoma, ovarian cancer	Intratumoral	Phase I/II	NCT03323398
CD70, CD40L, caTLR4 mRNA (TriMix)	Breast cancer	Intratumoral	Phase I	NCT03788083

AML, acute myeloid leukemia; BMCA, B cell maturation antigen; CAR, chimeric antigen receptor; CD40L, CD40 ligand; cMET, protein kinase; CML, chronic myeloid leukemia; DCs, dendritic cells; GBM, glioblastoma multiforme of brain; GCG, giant cell glioblastoma; GM, glioblastoma multiforme; GM-CSF, granulocyte-macrophage colony stimulating factor; HCMV, human cytomegalovirus; HNSCC, head and neck squamous cell carcinoma; HPV, human papilloma virus; hTERT, telomerase reverse transcriptase; IFN, interferon; IL, interleukin; LAMP, lysosome-associated membrane glycoprotein; LNP, lipid nanoparticle; LPX, lipoplex; MAGE, melanoma-associated antigen; MG, malignant glioma; MM, multiple myeloma; NK, natural killer; NKG2D, Natural killer group 2D; NY-ESO-1, New York esophageal squamous cell carcinoma-1; OX40L, OX40 ligand; PAP, prostatic acid phosphatase; PDAC, pancreatic ductal adenocarcinoma; PD-L1, programmed cell death ligand 1; PSA, prostate-specific antigen; TAA, tumor-associated antigen; TLR, toll-like receptor; TNBC, triple-negative breast cancer; TPTE, transmembrane phosphatase with tensin homology; WT1, Wilm's tumor protein 1.

antigen to lysosome-associated membrane proteins (LAMPs) or to MHC class I cytoplasmic and transmembrane domain, that direct the antigen through lysosomal compartments to enhance MHC class II antigen presentation (Bonehill et al., 2004; Kreiter et al., 2008; Wu et al., 1995). In phase I clinical trials (NCT02529072, NCT00626483, NCT00639639), mRNA-transfected DC vaccines encoding carcinoembryogenic antigen (CEA) fused to LAMP-1 were evaluated in glioblastoma multiforme patients, and prolongation of overall survival was observed (Batich et al., 2020; Peters et al., 2019; Vlahovic et al., 2016). CEA/LAMP-1 mRNA DC vaccines are now being evaluated in glioblastoma patients in combination with varilumab, a fully human mAb that targets CD27 (NCT03688178), and routine chemotherapy (NCT02465268) in phase II clinical trials. Other strategies to enhance T-cell responses of mRNA-trasfected DC vaccines is the co-transfection of DCs with mRNA encoding immunostimulatory ligands, receptors or cytokines (Bonehill et al., 2008; Daneshmandi et al., 2018; Dannull et al., 2005; Tcherepanova et al., 2008). Rocapuldencel-T, consisting of autologous tumor mRNA-transfected DCs co-transfected with an mRNA encoding CD40L, is one of the most clinically advanced DC vaccines. However, in a phase III clinical trial (NCT01582672), Rocapuldencel-T did not improve the overall survival of patients with metastatic renal cell carcinoma treated with sunitinib (Figlin et al., 2020). In a phase II clinical study with 39 pretreated melanoma patients (NCT01302496), DCs vaccines transfected with mRNAs encoding four TAAs (melanoma-associated antigen (MAGE)-A3, MAGE-C2, tyrosinase or gp100) fused to an MHC class II targeting signal and TriMix were evaluated in combination with the mAb ipilimumab. The vaccine was well tolerated, in 12 out of 15 patients T-cell responses were detected, and complete and partial responses were observed after 36 months follow-up in eight and in seven patients, respectively (De Keersmaecker et al., 2020; Wilgenhof et al., 2016).

Direct injection of the mRNA represents the simplest way of delivering mRNA-based cancer vaccines. Injected mRNA is taken up by local cells, including APCs, and translated in the cytoplasm. Intradermal and intranodal administration of naked mRNA has shown to efficiently induce T-cell responses in humans (Arance Fernandez et al., 2019; Rittig et al., 2011; Sahin et al., 2017). Intranodal injection facilitates the delivery of the vaccine directly at the site of activated T-cells to be uptaken by residing APCs. The feasibility and safety of direct injection of mRNA vaccines against cancer were first demonstrated in a phase I/II clinical trial in 15 patients with metastatic melanoma, after the intradermal administration of naked autologous tumor mRNA extracted from each patient's growing metastasis, in combination with granulocyte macrophage colony-stimulating factor (GM-CSF) as adjuvant (Weide et al., 2008). Although an increase in antitumor humoral immune response was seen in some patients, clinical effectiveness was not shown. In another phase I/II clinical trial, the feasibility, safety, and immunological and clinical response of an intradermal mRNA-based vaccine using GM-CSF as adjuvant was assessed in 30 patients with stage IV renal cell cancer (Rittig et al., 2011). The mRNA vaccine administered encoded the TAAs mucin 1 (MUC1), CEA, human epidermal growth factor receptor 2 (Her-2/neu), telomerase, surviving and melanoma-associated antigen 1 (MAGE-A1). The vaccine was well tolerated with no severe side effects and induced clinical responses (stable disease in 15 patients and one partial response). In addition, induction of CD4 + and CD8 + T cell responses was shown for several TAAs in 12 out of 17 patients. The safety and tolerability of intranodal injection of mRNA was firstly evaluated in a phase I study of 29 advanced melanoma patients (NCT01684241). In 6 out of 13 patients, T-cell responses were observed against the encoded antigen (New York esophageal squamous cell carcinoma-1 and tyrosinase), and new lesions only appeared in 8 of the 29 participants (Beck et al., 2021). In another phase I clinical study including 20 patients with resected melanoma (NCT03394937), the safety and immunogenicity of ECI-006 vaccine administered intranodally was evaluated. ECI-006 vaccine contained a combination of TriMix and mRNAs encoding for melanoma-specific TAAs (tyrosinase, gp100, MAGE-A3, MAGE-C2, and PRAME). The vaccine was generally well tolerated and demonstrated to be immunogenic in a proportion of patients (Arance Fernandez et al., 2019). Nevertheless, the study was terminated due to the expiry of the study medication.

Formulation of mRNA vaccines with peptides, polymers or cationic/ionizable lipids to protect the genetic material and improve uptake by APCs have gained much attention in recent years. One of the most clinically advanced formulations in cancer mRNA vaccination is based on protamine. Protamine-mRNA vaccine was first evaluated in 21 metastatic melanoma patients after intradermal administration. The vaccine encoded MART-1, tyrosinase, gp100, MAGE-A1, MAGE-A3 and surviving with GM-CSF either alone or in combination with keyhole limpet hemocyanin as adjuvant (NCT00204607) (Weide et al., 2009). The study demonstrated the safety and feasibility of protamine-mRNA formulations for cancer vaccination. Since then, several clinical trials are underway or have been carried out with protamine-complexed mRNA vaccines (e.g., NCT01817738, NCT00923312, NCT03164772). The intravenous administration to mice of negative to neutral charged mRNA lipoplexes (mRNA-LPX) have shown to target to APCs resident in lymphoid organs, and the induction of strong local and systemic immune modulation. In addition, in mouse tumor models, RNA-LPX vaccines induced strong effector and memory T-cell responses against the encoded TSAs or TAAs, and mediated tumor growth control or rejection, and improvement of survival (Grunwitz et al., 2019; Kranz et al., 2016; Salomon et al., 2020). In a phase I clinical trial, 119 patients with melanoma were treated by intravenous administration of an RNA-LPX vaccine, which targets four non-mutated TAAs that are prevalent in melanoma (NCT02410733). The vaccine alone or in combination with checkpoint-inhibitor antagonistic anti-reprogrammed cell death protein-1 (PD-1) antibody, mediated durable objective responses (measurable responses). Vaccination also induced strong CD4+ and CD8+ T cell immunity against the vaccine antigens. The antigen-specific cytotoxic T-cell responses in some responders reached magnitudes typically reported for adoptive T-cell therapy (the transfer of T-cells to a patient) (Sahin et al., 2020). The same vaccine will be evaluated in a randomized phase II clinical trial in combination with the antagonistic anti-PD-1 antibody cemiplimab (NCT04526899). The same platform encoding other tumor type specific TAAs is being evaluated in various phase I and I/II clinical trials in patients with triple negative breast cancer (TNBC; NCT02316457), ovarian cancer (NCT04163094) and prostate cancer (NCT04382898).

Although SAM vaccines have been extensively studied for infectious diseases, the advance of this type of vaccines for cancer immunotherapy is more limited. Currently, there are two ongoing phase I/II clinical trials using SAM personalized neoantigen vaccines against non-small cell lung cancer, colorectal cancer, gastroesophageal adenocarcinoma, urothelial carcinoma, solid tumors, and pancreatic ductal adenocarcinoma (NCT03639714, NCT03953235), although results have not yet been published.

Personalized cancer vaccines based on mRNAs encoding neoantigens represent an important step forward in cancer immunotherapy. The fast production and easy and cost-effective manufacturing of mRNA allows the "on demand" design of vaccines. This approach was evaluated in a first-in-human trial (NCT02035956) in patients with metastatic melanoma, who were treated with intranodal injection of neoantigen-specific mRNA vaccines encoding 20 patient-unique mutations (IVAC MUTANOME).

Patients tolerated well the vaccine and showed CD4+ and CD8+ T-cell responses against neoepitopes. In addition, cumulative rate of recurrences was significantly reduced (Sahin et al., 2017). Since this first report, several clinical trials are testing mRNA-LNPs encoding TSAs. In a phase I clinical trial, mRNA-LNPs encoding specific neoantigens were administered intramuscularly to four gastrointestinal cancer patients (NCT03480152). It was seen that the vaccine is safe and induced mutation-specific T-cell responses. However, no objective clinical responses were observed in the four patients (Cafri et al., 2020). mRNA-4157, a lipid-encapsulated mRNA encoding neoantigens, has been evaluated alone in participants with resected solid tumors and in combination with pembrolizumab in participants with unresectable solid tumors (NCT03313778). Intramuscular vaccination was well tolerated and clinical responses were observed in 6 out of the 20 patients treated with the combination (Burris et al., 2019). Consequently, a phase II trial (NCT03897881) is ongoing to further evaluate the efficacy of the personalized vaccine mRNA-4157 and pembrolizumab in patients with high-risk melanoma. An individualized neoantigen-specific immunotherapy (iNeST) program, in which patients are treated with their "on demand" manufactured individualized cancer mutations, is testing in various tumor types (e.g., melanoma, non-small cell lung cancer, colorectal cancer) in phase I and II clinical trials (NCT03289962, NCT03815058, NCT04486378).

- Antigen receptors

Another immunotherapy strategy in cancer is the adoptive cell transfer (the transfer of cells to a patient) of engineered T cells with tumor-specific antigen receptors such as T cell receptors or CARs. This approach is of relevant interest since two DNA-based CAR-T cells products were marketed in 2017–2018, tisagenlecleucel (Kymriah) and axicabtagene ciloleucel (Yescarta), for acute B-cell lymphoblastic leukemia and large B-cell lymphoma, respectively (Maude et al., 2018; Neelapu et al., 2017). Generation of CAR-T cells with mRNA is mainly accomplished by electroporation, unlike classical CAR-T cells, which are transfected using viral vectors to deliver the DNA that encodes the CAR. Transient expression of T cells transduced by mRNA offers safety features and highly efficient recombinant protein translation, but repeated administrations are needed (Foster et al., 2019). Several phase I clinical trials are evaluating mRNA CAR-T cells against various antigens. Mesothelin-directed mRNA CAR-T cells were evaluated in two phase I clinical trials (NCT01355965, NCT01897415). However, repeated infusions of the mRNA CAR-T cells were needed, and one patient experienced anaphylactic shock after the third infusion. These results pointed out a safety issue owing to the potential immunogenicity of CARs derived from murine antibodies, especially when administering intermittently (Maus et al., 2013). CD19-directed mRNA CAR-T cells are the most advanced. Application of CD19 CAR-T cells in pediatric and adult patients with Hodgkin's lymphoma (NCT02624258, NCT02277522) was well tolerated and showed transient responses (Svoboda et al., 2018). Treatment of acute myeloid leukemia (AML) is being attempted by CD123 mRNA CAR-T cells (NCT02623582), but no anti-tumor effect has been demonstrated (Cummins et al., 2017). The intratumoral administration of protein kinase Met (cMet)-directed mRNA CAR-T cells to breast cancer patients have shown good tolerance and capacity to evoke an inflammatory response within tumors (NCT01837602) (Tchou et al., 2017). A phase I/II clinical trial is also evaluating mRNA CAR-T cells for the treatment of multiple myeloma (NCT03448978).

Apart from T cells, other type of cells can be modified with mRNA CARs, such as γδ T cells, NKT cells and NK cells (Ang et al., 2020; Boissel et al., 2012; Harrer et al., 2017; Shimasaki et al., 2012; Simon et al., 2018). In a phase I clinical trial (NCT03415100), intraperitoneal treatment with NKG2D-Ligand targeted CAR-NK cells resulted in clinical activity in patients with colorectal and liver metastatic tumor sites (Xiao et al., 2019).

- Antibodies and immunomodulators

Other approaches for cancer immunotherapy with mRNA comprize the expression of antibodies and immunomodulators (cytokines and co-stimulatory ligands and receptors). mAbs in cancer treatment are employed to target tumor cells and for the modulation of immune responses, and most of the approved recombinant mAbs are administered intravenously. One of the strategies for encoding anti-cancer antibodies with mRNA is the encapsulation of the mRNA in delivery systems targeting the liver to achieve a high protein production and to release the expressed mAbs into the circulation (Guevara et al., 2020). This approach is being evaluated in preclinical studies for breast cancer and different lymphomas, among others (Rybakova et al., 2019; Thran et al., 2017). Nevertheless, ex vivo transfection of DCs with mRNA encoding mAbs have taken a step forward compared to in vivo production of mAbs with mRNA. In fact, in a phase I clinical trial, subjects with metastatic melanoma were vaccinated with an intradermal injection of DCs transfected ex vivo with mRNA encoding melanoma TAAs (MART, MAGE-3, gp100 and tyrosinase), and co-administered with DCs transfected ex vivo with mRNA encoding soluble human GITR-L and/or anti-CTLA-4 mAb (NCT01216436). However, the clinical trial was terminated because the principal investigator retired.

Unlike mRNA-encoded mAbs, mRNA-based immunomodulators have successfully entered clinical testing. The immunomodulators expressed with mRNA for cancer immunotherapy include cytokines and stimulatory ligands and receptors. Several cytokines have been proposed for cancer immunotherapy (e.g., IL-2, IL-15, IL-12, IFNα, GM-CSF) with different immunomodulatory properties (Waldmann, 2018). Delivery of cytokines with mRNA is being tested in various clinical trials, by administering the mRNA encoding a single cytokine or a combination or mRNAs encoding different cytokines. The cytokines delivered via mRNA that are under clinical evaluation include IL-12 (MEDI1191) in patients with solid tumors (NCT03946800), a mixture of IL-36, IL-23 and OX40L (mRNA-2752) in patients with relapsed solid tumors or lymphoma (NCT03739931) (Patel et al., 2020), and a combination of IL-12 single chain, IL-15sushi, GM-CSF and IFNα (SAR441000) in metastatic neoplasm patients (NCT03871348). The three vaccines are administered intratumorally, and MEDI1191 and mRNA-2752 are formulated in LNPs.

mRNA can also be used to express stimulatory ligands and receptors on the surface of cells, providing a strong activation of immune signals. This strategy can be achieved by ex vivo modification of T cells, or by in vivo administration of mRNAs encoding for immune-stimulatory ligands or receptors, or a combination of both (Haabeth et al., 2019; Levin et al., 2018; Weinstein-Marom et al., 2019). In a phase I clinical trial, the intratumoral administration of TriMix is being evaluated in early breast cancer patients (NCT03788083) (Van Lint et al., 2016). Moreover, mRNA-LNP encoding OX40L (mRNA-2416) entered a phase I/II clinical trial to evaluate dose escalation and efficacy after intratumoral injection, alone and in combination with durvalumab, in participants with advanced malignancies, such as refractory solid tumors, lymphoma and ovarian cancer (NCT03323398). OX40L alone proved to be safe and showed pro-inflammatory activity in tumor lesions (Jimeno et al., 2020).

4.3 Protein replacement therapy

The main objective of protein replacement therapy with mRNA is to restore missing, down-expressed or non-functional proteins, responsible for the target disease. Due to the easy accessibility via the bloodstream, protein replacement therapies with mRNA have been targeted to express proteins mainly in the liver (Trepotec et al., 2019), lungs (Sahu et al., 2019) and heart (Magadum et al., 2019). Nevertheless, the skin (Lescan et al., 2018), the back of the eye (Patel et al., 2019b) and neuronal tissue through the nasal cavity (Baba et al., 2015) have been also proposed as tissues for protein synthesis. The first preclinical study of mRNA administration for protein replacement purposes was carried out in 1992, where the injection of mRNA encoding vasopressin in the hypothalamus of Brattleboro rats resulted in a reversal of diabetes insipidus for up to 5 days (Jirikowski et al., 1992). Since then, different genetic diseases have been proposed to be treated with mRNA-based protein replacement therapy: hemophilia B (DeRosa et al., 2016), Fabry disease (Zhu et al., 2019), methylmalonic acidemia (An et al., 2017), propionic acidemia (Jiang et al., 2020), acute intermittent porphyria (Jiang et al., 2018), ornithine transcarbamylase (OTC) deficiency (Prieve et al., 2018), cystic fibrosis (Robinson et al., 2018), William-Beuren syndrome (Lescan et al., 2018), glycogen storage disease type III (GSDIII) (Jauze et al., 2019) and familial hypercholesterolemia (Li et al., 2021). Nevertheless, it can also be employed to treat disease conditions without a genetic basis. In those cases, the therapy is based on the administration of an mRNA sequence encoding foreign proteins capable of activating or inhibiting cellular pathways. For instance, the expression of vascular endothelial growth factor-A (VEFG-A) with mRNA can induce the creation of new blood vessels in patients suffering ischemic cardiovascular disease, providing a regenerative effect due to the improved blood supply (Carlsson et al., 2018). Recently, it has been completed a phase II clinical trial evaluating safety of epicardial injection of VEGF-A mRNA (AZD8601) in patients with moderately decreased left ventricular function undergoing elective coronary artery bypass surgery (NCT03370887) (Anttila et al., 2020), but results are not yet available. The same treatment can be applied with healing purposes of diabetic wounds and other ischemic vascular diseases (Gan et al., 2019). In a phase I study (NCT02935712), safety and potential therapeutic effects of intradermal administration of AZD8601 was assessed. The study showed that intradermal AZD8601 was well tolerated and led to local functional VEGF-A protein expression and transient skin blood flow enhancement in men with type 2 diabetes mellitus, demonstrating the therapeutic potential for regenerative angiogenesis.

Table 6 summarizes the clinical trials with mRNA for protein replacement therapies. As it can be observed, mRNA is being evaluated in clinical

Table 6 Clinical trials of mRNA for protein replacement therapies.

Disease	Biological active	Encoded protein	Delivery system	Administration route	Clinical trial phase	Clinical trial identifier
Heart Failure	AZD8601	Vascular endotelial growth factor-A (VEGF-A)	Naked mRNA	Epicardial injection	Phase II	NCT03370887
Ulcers associated with type II diabetes	AZD8601	Vascular endotelial growth factor-A (VEGF-A)	Naked mRNA	Intradermal	Phase I	NCT02935712
Propionic Acidemia	mRNA-3927	α and β subunits of the mitochondrial enzyme propionyl-CoA carboxylase	Lipid nanosystem	Intravenous	Phase I/II	NCT04159103, NCT05130437
Isolated Methylmalonic Acidemia	mRNA-3704	Methylmalonyl-coenzyme A mutase (MUT)	Lipid nanosystem	Intravenous	Phase I/II	NCT03810690
	mRNA-3705	Methylmalonyl-coenzyme A mutase (MUT)	Lipid nanosystem	Intravenous	Phase I/II	NCT04899310
Ornithine Transcarbamylase Deficiency	MRT5201	Ornithine transcarbamylase	Lipid nanosystem	Intravenous	Phase I/II	NCT03767270
	ARCT-810	Ornithine transcarbamylase	Lipid nanosystem	Intravenous	Phase I	NCT04416126, NCT04442347
Cystic Fibrosis	MRT5005	Human cystic fibrosis transmembrane regulator protein (CFTR)	Lipid nanosystem	Respiratory tract (Nebulization)	Phase I/II	NCT03375047
Glycogen Storage Disease Type III	UX053	Glycogen debranching enzyme	Lipid nanosystem	Intravenous	Phase I/II	NCT04990388
Familial Hypercholesterolemia	Low Density Lipoprotein Receptor mRNA Exosomes	Low-density lipoprotein receptor	Exosome	Intravenous Intraperitoneal	Phase I	NCT05043181

trials for both, genetic diseases and illnesses without a genetic basis. Most of the biologicals under evaluation are formulated in lipidic carriers, including the first study with exosomes (NCT05043181).

Moderna has also developed mRNA-based therapies formulated in LNPs to be used in protein replacement therapies. The safety and clinical activity of intravenous doses of mRNA-3927 in LNPs is being evaluated in clinical trials for the treatment of propionic acidemia (NCT04159103, NCT05130437). This is an autosomal recessive inherited disease caused by mutations in the gene that encodes the enzyme propionyl-CoA carboxylase, leading to the accumulation of propionyl-CoA and other toxic compounds at mitochondrial level (García-Cazorla, 2010). Additionally, Moderna started a phase I and II clinical trial to evaluate an intravenous LNP formulation containing mRNA-3704 (NCT03810690) that encodes the enzyme deficient in patients with methylmalonic acidemia. This disease is caused by mutations in the gene encoding the vitamin B12-dependent enzyme methylmalonyl CoA mutase (Ohura, 1988). The clinical trial was withdrawn before the start of dosing due to a business decision and not due to safety or efficacy reasons; however, another clinical trial is recruiting participants to evaluate the pharmacologically optimized biological active mRNA-3705 (NCT04899310).

In the last few years, Translate Bio, another clinical-stage mRNA therapeutics company, has started two clinical trials for the treatment of ornithine transcarbamylase (OTC) deficiency and cystic fibrosis. OTC deficiency is characterized by the increase of ammonium levels in the blood caused by disorders in ammonium detoxification and deficiencies of mitochondrial OTC in the urea cycle. Currently, liver transplantation is the only therapeutic option for severe cases of the disease (Brassier et al., 2015). Translate Bio initiated a clinical trial (NCT03767270) to evaluate the safety, tolerability and pharmacokinetic/pharmacodynamics of a product (MRT5201) consisting of an mRNA encoding OTC in LNPs, after intravenous administration. Nevertheless, the clinical trial was discontinued because preclinical studies did not support the expected pharmacokinetic and safety profile, mainly related to non-optimal featured of the first-generation LNPs targeting the liver (Translate Bio Inc., 2019). Arcturus Therapeutics, Inc. is evaluating the safety, tolerability and pharmacokinetics of the intravenous administration of an mRNA-LNP formulation, ARCT-810, in healthy adults (NCT04416126) and in stable adults with OTC deficiency (NCT04442347).

Additionally, Translate Bio has developed a biological active, named MRT5005, for the treatment of cystic fibrosis, a monogenic disorder that affects approximately 70,000 people worldwide. It is caused by genetic variance within the coding region of the cystic fibrosis transmembrane conductance regulator (CFTR), an anion channel necessary for chloride efflux from secretory epithelial cells. Consequently, ion transport dysregulation happens leading to multisystem organ failure and death. Currently, around 300 different mutations have been described, and disease severity varies greatly from person to person which is mainly determined by the degree of the lung affectation (Robinson et al., 2018). MRT5005 is an inhaled medicinal product based on the delivery of an mRNA encoding CFTR by LNPs. Single- (8, 16 and 24 mg) and multiple-ascending doses (five once-weekly doses of 8, 12 and 16 mg) of MRT5005 are being evaluated in a phase I/II clinical trial (NCT03375047) in patients with cystic fibrosis. Interim data analysis showed that a single dose of the product was well tolerated at low and mid doses. In 4 out of 9 patients treated with MRT5005 the percent predicted forced expiratory volume in 1 s ($ppFEV_1$), a primary measure of lung function, were higher than expected based on known variability of $ppFEV_1$ (Zuckerman et al., 2019). A second interim analysis of the clinical trial has announced results of a new dose of the single-ascending dose group (20 mg), and results of the multiple-ascending dose group. In both cases, treatment-emergent adverse events were considered mild to moderate, and no marked increases in $ppFEV_1$ were observed. In the 20 mg single-ascending dose group just a serious adverse event, a pulmonary exacerbation 22 days after the dose, was observed. In the multiple-ascending dose group, three patients experienced transient, mild to moderate febrile reaction after the first dose; one of them discontinued due to the febrile reaction, but in the other two, the adverse effect did not recur with subsequent dosing (Translate Bio Inc, 2021).

GSDIII is caused by a deficiency of a glycogen debranching enzyme, resulting glycogen accumulation in the liver and muscle. Consequently, patients suffer hepatomegaly, hypoglycemia, hyperlipidemia, cirrhosis and myopathy (Jauze et al., 2019). Unfortunately, no therapies are approved for GSDIII. Protein replacement therapy with mRNA has been proposed to restore the deficient enzyme. An mRNA-LNP formulation (UX053) encoding glycogen debranching enzyme, demonstrated reductions in glycogen content in the liver in multiple models of the disease. Subsequently, a phase I/II clinical trial is evaluating the safety, tolerability and pharmacokinetics of intravenous administrations of UX053 in patients with GSDIII (NCT04990388).

5. Conclusions

mRNA has come a long way to finally demonstrate its great potential with the recent approval of two mRNA-based vaccines. Its multiple applications have enabled the development of therapies for a myriad of pathologies, with the number of clinical trials increasing dramatically in recent years. Nevertheless, mRNA chemical modifications together with the design of suitable vehicles adapted to its features and therapeutic purpose are key points to ensure clinical success. LPNs are the most advanced mRNA delivery systems, with liposomes being the main clinical translatable delivery technology. COVID-19 vaccines, as the first mRNA-based medicinal products marketed worldwide, and cancer immunotherapy are leading the way in the clinical use of mRNA. It represents a new therapeutic approach applicable to a wide variety of pathologies, and although it still has a long way to go, mRNA is here to stay.

Acknowledgments

I. Gómez-Aguado and M. Beraza-Millor thank the University of the Basque Country UPV/EHU (PIF17/067 and PIFG19/36, respectively) and J. Rodríguez-Castejón thanks the Basque Government (PRE_2020_2_0120) for their research grants. Figures were performed by using Smart Servier Medical Art (https://smart.servier.com (accessed on 28 December 2021)), licensed under a Creative Common Attribution 3.0 Unported License (https://creativecommons. org/licenses/by/3.0/ (accessed on 28 December 2021)).

References

Alberer, M., Gnad-Vogt, U., Hong, H.S., Mehr, K.T., Backert, L., Finak, G., Gottardo, R., Bica, M.A., Garofano, A., Koch, S.D., Fotin-Mleczek, M., Hoerr, I., Clemens, R., von Sonnenburg, F., 2017. Safety and immunogenicity of a mRNA rabies vaccine in healthy adults: an open-label, non-randomised, prospective, first-in-human phase 1 clinical trial. Lancet 390, 1511–1520. https://doi.org/10.1016/S0140-6736(17)31665-3.

Aldosari, B.N., Alfagih, I.M., Almurshedi, A.S., 2021. Lipid nanoparticles as delivery systems for RNA-based vaccines. Pharmaceutics 13, 1–29. https://doi.org/10.3390/pharmaceutics13020206.

Aldrich, C., Leroux-Roels, I., Huang, K.B., Bica, M.A., Loeliger, E., Schoenborn-Kellenberger, O., Walz, L., Leroux-Roels, G., von Sonnenburg, F., Oostvogels, L., 2021. Proof-of-concept of a low-dose unmodified mRNA-based rabies vaccine formulated with lipid nanoparticles in human volunteers: a phase 1 trial. Vaccine 39, 1310–1318. https://doi.org/10.1016/j.vaccine.2020.12.070.

Alexaki, A., Kames, J., Holcomb, D.D., Athey, J., Santana-Quintero, L.V., Lam, P.V.N., Hamasaki-Katagiri, N., Osipova, E., Simonyan, V., Bar, H., Komar, A.A., Kimchi-Sarfaty, C., 2019. Codon and codon-pair usage tables (CoCoPUTs): facilitating genetic variation analyses and recombinant gene design. J. Mol. Biol. 431, 2434–2441. https://doi.org/10.1016/j.jmb.2019.04.021.

Aliprantis, A.O., Shaw, C.A., Griffin, P., Farinola, N., Railkar, R.A., Cao, X., Liu, W., Sachs, J.R., Swenson, C.J., Lee, H., Cox, K.S., Spellman, D.S., Winstead, C.J., Smolenov, I., Lai, E., Zaks, T., Espeseth, A.S., Panther, L., 2021. A phase 1, randomized, placebo-controlled study to evaluate the safety and immunogenicity of an mRNA-based RSV prefusion F protein vaccine in healthy younger and older adults. Hum. Vaccin. Immunother. 17, 1248–1261. https://doi.org/10.1080/21645515.2020.1829899.

Alnasser, S.M., 2021. Review on mechanistic strategy of gene therapy in the treatment of disease. Gene 769, 145246. https://doi.org/10.1016/j.gene.2020.145246.

AlphaVax, n.d. AlphaVax Clinical Experience [WWW Document]. AlphaVax. URL https://www.alphavax.com/clinical-experience.html (accessed 12.8.21).

Al-Saif, M., Khabar, K.S.A., 2012. UU/UA dinucleotide frequency reduction in coding regions results in increased mRNA stability and protein expression. Mol. Ther. 20, 954–959. https://doi.org/10.1038/mt.2012.29.

An, D., Schneller, J.L., Frassetto, A., Liang, S., Zhu, X., Park, J.S., Theisen, M., Hong, S.J., Zhou, J., Rajendran, R., Levy, B., Howell, R., Besin, G., Presnyak, V., Sabnis, S., Murphy-Benenato, K.E., Kumarasinghe, E.S., Salerno, T., Mihai, C., Lukacs, C.M., Chandler, R.J., Guey, L.T., Venditti, C.P., Martini, P.G.V., 2017. Systemic messenger RNA therapy as a treatment for methylmalonic acidemia. Cell Rep. 21, 3548–3558. https://doi.org/10.1016/j.celrep.2017.11.081.

Andresen, J.L., Fenton, O.S., 2021. Nucleic acid delivery and nanoparticle design for COVID vaccines. MRS Bull. 46, 1–8. https://doi.org/10.1557/s43577-021-00169-2.

Andries, O., Mc Cafferty, S., De Smedt, S.C., Weiss, R., Sanders, N.N., Kitada, T., 2015. N1-methylpseudouridine-incorporated mRNA outperforms pseudouridine-incorporated mRNA by providing enhanced protein expression and reduced immunogenicity in mammalian cell lines and mice. J. Control. Release 217, 337–344. https://doi.org/10.1016/j.jconrel.2015.08.051.

Ang, W.X., Ng, Y.Y., Xiao, L., Chen, C., Li, Z., Chi, Z., Tay, J.C.K., Tan, W.K., Zeng, J., Toh, H.C., Wang, S., 2020. Electroporation of NKG2D RNA car improves Vγ9Vδ2 T cell responses against human solid tumor xenografts. Mol. Ther. Oncolytics 17, 421–430. https://doi.org/10.1016/j.omto.2020.04.013.

Anguela, X.M., High, K.A., 2019. Entering the modern era of gene therapy. Annu. Rev. Med. 70, 273–288. https://doi.org/10.1146/annurev-med-012017-043332.

Anttila, V., Saraste, A., Knuuti, J., Jaakkola, P., Hedman, M., Svedlund, S., Lagerström-Fermér, M., Kjaer, M., Jeppsson, A., Gan, L.M., 2020. Synthetic mRNA encoding VEGF-A in patients undergoing coronary artery bypass grafting: design of a phase 2a clinical trial. Mol. Ther. Methods Clin. Dev. 18, 464–472. https://doi.org/10.1016/j.omtm.2020.05.030.

Apaolaza, P.S., Delgado, D., Del Pozo-Rodríguez, A., Gascón, A.R., Solinís, M.Á., 2014. A novel gene therapy vector based on hyaluronic acid and solid lipid nanoparticles for ocular diseases. Int. J. Pharm. 465, 413–426. https://doi.org/10.1016/j.ijpharm.2014.02.038.

Apaolaza, P.S., del Pozo-Rodríguez, A., Torrecilla, J., Rodríguez-Gascón, A., Rodríguez, J.M., Friedrich, U., Weber, B.H.F., Solinís, M.A., 2015. Solid lipid nanoparticle-based vectors intended for the treatment of X-linked juvenile retinoschisis by gene therapy: In vivo approaches in Rs1h-deficient mouse model. J. Control. Release 217, 273–283. https://doi.org/10.1016/j.jconrel.2015.09.033.

Apaolaza, P.S., del Pozo-Rodríguez, A., Solinís, M.A., Rodríguez, J.M., Friedrich, U., Torrecilla, J., Weber, B.H.F., Rodríguez-Gascón, A., 2016. Structural recovery of the retina in a retinoschisin-deficient mouse after gene replacement therapy by solid lipid nanoparticles. Biomaterials 90, 40–49. https://doi.org/10.1016/j.biomaterials.2016.03.004.

Arance Fernandez, A.M., Baurain, J.-F., Vulsteke, C., Rutten, A., Soria, A., Carrasco, J., Neyns, B., De Keersmaecker, B., Van Assche, T., Lindmark, B., 2019. A phase I study (E011-MEL) of a TriMix-based mRNA immunotherapy (ECI-006) in resected melanoma patients: analysis of safety and immunogenicity. J. Clin. Oncol. 37, 2641. https://doi.org/10.1200/jco.2019.37.15_suppl.2641.

August, A., Attarwala, H.Z., Himansu, S., Kalidindi, S., Lu, S., Pajon, R., Han, S., Lecerf, J.-M., Tomassini, J.E., Hard, M., Ptaszek, L.M., Crowe, J.E., Zaks, T., 2021. A phase 1 trial of lipid-encapsulated mRNA encoding a monoclonal antibody with neutralizing activity against Chikungunya virus. Nat. Med. 27, 2224–2233. https://doi.org/10.1038/s41591-021-01573-6.

Baba, M., Itaka, K., Kondo, K., Yamasoba, T., Kataoka, K., 2015. Treatment of neurological disorders by introducing mRNA in vivo using polyplex nanomicelles. J. Control. Release 201, 41–48. https://doi.org/10.1016/j.jconrel.2015.01.017.

Baden, L.R., El Sahly, H.M., Essink, B., Kotloff, K., Frey, S., Novak, R., Diemert, D., Spector, S.A., Rouphael, N., Creech, C.B., McGettigan, J., Khetan, S., Segall, N., Solis, J., Brosz, A., Fierro, C., Schwartz, H., Neuzil, K., Corey, L., Gilbert, P., Janes, H., Follmann, D., Marovich, M., Mascola, J., Polakowski, L., Ledgerwood, J., Graham, B.S., Bennett, H., Pajon, R., Knightly, C., Leav, B., Deng, W., Zhou, H., Han, S., Ivarsson, M., Miller, J., Zaks, T., 2021. Efficacy and safety of the mRNA-1273 SARS-CoV-2 vaccine. N. Engl. J. Med. 384, 403–416. https://doi.org/10.1056/nejmoa2035389.

Bahl, K., Senn, J.J., Yuzhakov, O., Bulychev, A., Brito, L.A., Hassett, K.J., Laska, M.E., Smith, M., Almarsson, Ö., Thompson, J., Ribeiro, A.(.M.)., Watson, M., Zaks, T., Ciaramella, G., 2017. Preclinical and clinical demonstration of immunogenicity by mRNA vaccines against H10N8 and H7N9 influenza viruses. Mol. Ther. 25, 1316–1327. https://doi.org/10.1016/j.ymthe.2017.03.035.

Baiersdörfer, M., Boros, G., Muramatsu, H., Mahiny, A., Vlatkovic, I., Sahin, U., Karikó, K., 2019. A facile method for the removal of dsRNA contaminant from in vitro-transcribed mRNA. Mol. Ther. Nucleic Acids 15, 26–35. https://doi.org/10.1016/j.omtn.2019.02.018.

Bajaj, V., Gadi, N., Spihlman, A.P., Wu, S.C., Choi, C.H., Moulton, V.R., 2021. Aging, immunity, and COVID-19: how age influences the host immune response to coronavirus infections? Front. Physiol. 11, 1793. https://doi.org/10.3389/fphys.2020.571416.

Bali, V., Bebok, Z., 2015. Decoding mechanisms by which silent codon changes influence protein biogenesis and function. Int. J. Biochem. Cell Biol. 64, 58–74. https://doi.org/10.1016/j.biocel.2015.03.011.

Barrett, S.P., Salzman, J., 2016. Circular RNAs: analysis, expression and potential functions. Development 143, 1838–1847. https://doi.org/10.1242/dev.128074.

Batich, K.A., Mitchell, D.A., Healy, P., Herndon, J.E., Sampson, J.H., 2020. Once, twice, three times a finding: reproducibility of dendritic cell vaccine trials targeting cytomegalovirus in glioblastoma. Clin. Cancer Res. 26 (20), 5297–5303. https://doi.org/10.1158/1078-0432.CCR-20-1082.

Beck, J.D., Reidenbach, D., Salomon, N., Sahin, U., Türeci, Ö., Vormehr, M., Kranz, L.M., 2021. mRNA therapeutics in cancer immunotherapy. Mol. Cancer 20 (1), 69. https://doi.org/10.1186/s12943-021-01348-0.

Behr, J.-P., 1997. The proton sponge: a trick to enter cells the viruses did not exploit. Chimia (Aarau) 51, 34–36.

Beloqui, A., Solinís, M.Á., Rodríguez-Gascón, A., Almeida, A.J., Préat, V., 2016. Nanostructured lipid carriers: promising drug delivery systems for future clinics. Nanomedicine 12, 143–161. https://doi.org/10.1016/j.nano.2015.09.004.

Beloqui, A., del Pozo-Rodríguez, A., Isla, A., Rodríguez-Gascón, A., Solinís, M.Á., 2017. Nanostructured lipid carriers as oral delivery systems for poorly soluble drugs. J. Drug Deliv. Sci. Technol. 42, 144–154. https://doi.org/10.1016/j.jddst.2017.06.013.

Berg, K., Kristian Selbo, P., Prasmickaite, L., Tjelle, T.E., Sandvig, K., Moan, J., Gaudernack, G., Fodstad, Ø., Kjølsrud, S., Anholt, H., Rodal, G.H., Rodal, S.K., Høgset, A., 1999. Photochemical Internalization. Cancer Res. 59, 1180–1183.

Berkovits, B.D., Mayr, C., 2015. Alternative 3′ UTRs act as scaffolds to regulate membrane protein localization. Nature 522, 363–367. https://doi.org/10.1038/nature14321.

Bernstein, D.I., Reap, E.A., Katen, K., Watson, A., Smith, K., Norberg, P., Olmsted, R.A., Hoeper, A., Morris, J., Negri, S., Maughan, M.F., Chulay, J.D., 2009. Randomized, double-blind, Phase 1 trial of an alphavirus replicon vaccine for cytomegalovirus in CMV seronegative adult volunteers. Vaccine 28, 484–493. https://doi.org/10.1016/j.vaccine.2009.09.135.

Björklund, T., 2016. Non-Viral, Lipid-Mediated DNA and mRNA Gene Therapy of the Central Nervous System (CNS): Chemical-Based Transfection. pp. 41–56. https://doi.org/10.1007/978-1-4939-3271-9.

Blakney, A.K., Ip, S., Geall, A.J., 2021. An update on self-amplifying mRNA vaccine development. Vaccine 9, 1–26. https://doi.org/10.3390/vaccines9020097.

Blass, E., Ott, P.A., 2021. Advances in the development of personalized neoantigen-based therapeutic cancer vaccines. Nat. Rev. Clin. Oncol. 18 (4), 215–229. https://doi.org/10.1038/s41571-020-00460-2.

Bogers, W.M., Oostermeijer, H., Mooij, P., Koopman, G., Verschoor, E.J., Davis, D., Ulmer, J.B., Brito, L.A., Cu, Y., Banerjee, K., Otten, G.R., Burke, B., Dey, A., Heeney, J.L., Shen, X., Tomaras, G.D., Labranche, C., Montefiori, D.C., Liao, H.X., Haynes, B., Geall, A.J., Barnett, S.W., 2015. Potent immune responses in rhesus macaques induced by nonviral delivery of a self-amplifying RNA vaccine expressing HIV type 1 envelope with a cationic nanoemulsion. J Infect Dis 211, 947–955. https://doi.org/10.1093/infdis/jiu522.

Boissel, L., Betancur, M., Lu, W., Wels, W.S., Marino, T., Van Etten, R.A., Klingemann, H., 2012. Comparison of mRNA and lentiviral based transfection of natural killer cells with chimeric antigen receptors recognizing lymphoid antigens. Leuk. Lymphoma 53, 958–965. https://doi.org/10.3109/10428194.2011.634048.

Bonehill, A., Heirman, C., Tuyaerts, S., Michiels, A., Breckpot, K., Brasseur, F., Zhang, Y., van der Bruggen, P., Thielemans, K., 2004. Messenger RNA-electroporated dendritic cells presenting MAGE-A3 simultaneously in HLA class I and class II molecules. J. Immunol. 172, 6649–6657. https://doi.org/10.4049/jimmunol.172.11.6649.

Bonehill, A., Tuyaerts, S., Van Nuffel, A.M.T., Heirman, C., Bos, T.J., Fostier, K., Neyns, B., Thielemans, K., 2008. Enhancing the t-cell stimulatory capacity of human dendritic cells by co-electroporation with CD40L, CD70 and constitutively active TLR4 encoding mRNA. Mol. Ther. 16, 1170–1180. https://doi.org/10.1038/mt.2008.77.

Brassier, A., Gobin, S., Arnoux, J.B., Valayannopoulos, V., Habarou, F., Kossorotoff, M., Servais, A., Barbier, V., Dubois, S., Touati, G., Barouki, R., Lesage, F., Dupic, L., Bonnefont, J.P., Ottolenghi, C., De Lonlay, P., 2015. Long-term outcomes in ornithine transcarbamylase deficiency: a series of 90 patients. Orphanet J. Rare Dis. 10 (1), 58. https://doi.org/10.1186/s13023-015-0266-1.

Brazzoli, M., Magini, D., Bonci, A., Buccato, S., Giovani, C., Kratzer, R., Zurli, V., Mangiavacchi, S., Casini, D., Brito, L.M., De Gregorio, E., Mason, P.W., Ulmer, J.B., Geall, A.J., Bertholet, S., 2016. Induction of broad-based immunity and protective efficacy by self-amplifying mrna vaccines encoding influenza virus hemagglutinin. J. Virol. 90, 332–344. https://doi.org/10.1128/jvi.01786-15.

Brito, L.A., Chan, M., Shaw, C.A., Hekele, A., Carsillo, T., Schaefer, M., Archer, J., Seubert, A., Otten, G.R., Beard, C.W., Dey, A.K., Lilja, A., Valiante, N.M., Mason, P.W., Mandl, C.W., Barnett, S.W., Dormitzer, P.R., Ulmer, J.B., Singh, M., O'Hagan, D.T., Geall, A.J., 2014. A cationic nanoemulsion for the delivery of next-generation RNA vaccines. Mol. Ther. 22, 2118–2129. https://doi.org/10.1038/mt.2014.133.

Burki, T., 2021. Understanding variants of SARS-CoV-2. Lancet (London, England) 397, 462. https://doi.org/10.1016/S0140-6736(21)00298-1.

Burris, H.A., Patel, M.R., Cho, D.C., Clarke, J.M., Gutierrez, M., Zaks, T.Z., Frederick, J., Hopson, K., Mody, K., Binanti-Berube, A., Robert-Tissot, C., Goldstein, B., Breton, B., Sun, J., Zhong, S., Pruitt, S.K., Keating, K., Meehan, R.S., Gainor, J.F., 2019. A phase I multicenter study to assess the safety, tolerability, and immunogenicity of mRNA-4157 alone in patients with resected solid tumors and in combination with pembrolizumab in patients with unresectable solid tumors. J. Clin. Oncol. 37, 2523. https://doi.org/10.1200/jco.2019.37.15_suppl.2523.

Cafri, G., Gartner, J.J., Zaks, T., Hopson, K., Levin, N., Paria, B.C., Parkhurst, M.R., Yossef, R., Lowery, F.J., Jafferji, M.S., Prickett, T.D., Goff, S.L., McGowan, C.T., Seitter, S., Shindorf, M.L., Parikh, A., Chatani, P.D., Robbins, P.F., Rosenberg, S.A., 2020. mRNA vaccine–induced neoantigen-specific T cell immunity in patients with gastrointestinal cancer. J. Clin. Invest. 130, 5976–5988. https://doi.org/10.1172/JCI134915.

Campillo-Davo, D., De Laere, M., Roex, G., Versteven, M., Flumens, D., Berneman, Z.N., Van Tendeloo, V.F.I., Anguille, S., Lion, E., 2021. The ins and outs of messenger RNA electroporation for physical gene delivery in immune cell-based therapy. Pharmaceutics 13, 396. https://doi.org/10.3390/pharmaceutics13030396.

Carlsson, L., Clarke, J.C., Yen, C., Gregoire, F., Albery, T., Billger, M., Egnell, A.C., Gan, L.M., Jennbacken, K., Johansson, E., Linhardt, G., Martinsson, S., Sadiq, M.W., Witman, N., Wang, Q.D., Chen, C.H., Wang, Y.P., Lin, S., Ticho, B., Hsieh, P.C.H., Chien, K.R., Fritsche-Danielson, R., 2018. Biocompatible, purified VEGF-A mRNA improves cardiac function after intracardiac injection 1 week post-myocardial infarction in swine. Mol. Ther. Methods Clin. Dev. 9, 330–346. https://doi.org/10.1016/j.omtm.2018.04.003.

Chahal, J.S., Khan, O.F., Cooper, C.L., McPartlan, J.S., Tsosie, J.K., Tilley, L.D., Sidik, S.M., Lourido, S., Langer, R., Bavari, S., Ploegh, H.L., Anderson, D.G., 2016. Dendrimer-RNA nanoparticles generate protective immunity against lethal Ebola, H1N1 influenza, and Toxoplasma gondii challenges with a single dose. Proc. Natl. Acad. Sci. U. S. A. 113, E4133–E4142. https://doi.org/10.1073/pnas.1600299113.

Chahal, J.S., Fang, T., Woodham, A.W., Khan, O.F., Ling, J., Anderson, D.G., Ploegh, H.L., 2017. An RNA nanoparticle vaccine against Zika virus elicits antibody and CD8+ T cell responses in a mouse model. Sci. Rep. 7, 252. https://doi.org/10.1038/s41598-017-00193-w.

Chattopadhyay, P., Shekunov, B.Y., Yim, D., Cipolla, D., Boyd, B., Farr, S., 2007. Production of solid lipid nanoparticle suspensions using supercritical fluid extraction of emulsions (SFEE) for pulmonary delivery using the AERx system. Adv. Drug Deliv. Rev. 59, 444–453. https://doi.org/10.1016/j.addr.2007.04.010.

Chaudhary, N., Weissman, D., Whitehead, K.A., 2021. mRNA vaccines for infectious diseases: principles, delivery and clinical translation. Nat. Rev. Drug Discov. 20 (11), 817–838. https://doi.org/10.1038/s41573-021-00283-5.

Chiper, M., Tounsi, N., Kole, R., Kichler, A., Zuber, G., 2017. Self-aggregating 1.8 kDa polyethylenimines with dissolution switch at endosomal acidic pH are delivery carriers for plasmid DNA, mRNA, siRNA and exon-skipping oligonucleotides. J. Control. Release 246, 60–70. https://doi.org/10.1016/j.jconrel.2016.12.005.

Choi, A., Koch, M., Wu, K., Chu, L., Ma, L.Z., Hill, A., Nunna, N., Huang, W., Oestreicher, J., Colpitts, T., Bennett, H., Legault, H., Paila, Y., Nestorova, B., Ding, B., Montefiori, D., Pajon, R., Miller, J.M., Leav, B., Carfi, A., McPhee, R., Edwards, D.K., 2021. Safety and immunogenicity of SARS-CoV-2 variant mRNA vaccine boosters in healthy adults: an interim analysis. Nat. Med. 27, 2025–2031. https://doi.org/10.1038/s41591-021-01527-y.

Conway, A., Mendel, M., Kim, K., McGovern, K., Boyko, A., Zhang, L., Miller, J.C., DeKelver, R.C., Paschon, D.E., Mui, B.L., Lin, P.J.C., Tam, Y.K., Barbosa, C., Redelmeier, T., Holmes, M.C., Lee, G., 2019. Non-viral delivery of zinc finger nuclease mRNA enables highly efficient in vivo genome editing of multiple therapeutic gene targets. Mol. Ther. 27, 866–877. https://doi.org/10.1016/j.ymthe.2019.03.003.

Crudele, J.M., Chamberlain, J.S., 2018. Cas9 immunity creates challenges for CRISPR gene editing therapies. Nat. Commun. 9, 3497. https://doi.org/10.1038/s41467-018-05843-9.

Cummins, K.D., Frey, N., Nelson, A.M., Schmidt, A., Luger, S., Isaacs, R.E., Lacey, S.F., Hexner, E., Melenhorst, J.J., June, C.H., Porter, D.L., Gill, S.I., 2017. Treating relapsed/refractory (RR) AML with biodegradable anti-CD123 CAR modified T cells. Blood 130, 1359. https://doi.org/10.1182/blood.V130.Suppl_1.1359.1359.

Dammes, N., Peer, D., 2020. Paving the road for RNA therapeutics. Trends Pharmacol. Sci. 41, 755–775. https://doi.org/10.1016/j.tips.2020.08.004.

Daneshmandi, S., Pourfathollah, A.A., Forouzandeh-Moghaddam, M., 2018. Enhanced CD40 and ICOSL expression on dendritic cells surface improve anti-tumor immune responses; effectiveness of mRNA/chitosan nanoparticles. Immunopharmacol. Immunotoxicol. 40, 375–386. https://doi.org/10.1080/08923973.2018.1510959.

Dannull, J., Nair, S., Su, Z., Boczkowski, D., DeBeck, C., Yang, B., Gilboa, E., Vieweg, J., 2005. Enhancing the immunostimulatory function of dendritic cells by transfection with mRNA encoding OX40 ligand. Blood 105, 3206–3213. https://doi.org/10.1182/blood-2004-10-3944.

De, K., 2021. Decapeptide modified doxorubicin loaded solid lipid nanoparticles as targeted drug delivery system against prostate cancer. Langmuir 37, 13194–13207. https://doi.org/10.1021/acs.langmuir.1c01370.

de Alwis, R., Gan, E.S., Chen, S., Leong, Y.S., Tan, H.C., Zhang, S.L., Yau, C., Low, J.G.H., Kalimuddin, S., Matsuda, D., Allen, E.C., Hartman, P., Park, K.J.J., Alayyoubi, M., Bhaskaran, H., Dukanovic, A., Bao, Y., Clemente, B., Vega, J., Roberts, S., Gonzalez, J.A., Sablad, M., Yelin, R., Taylor, W., Tachikawa, K., Parker, S., Karmali, P., Davis, J., Sullivan, B.M., Sullivan, S.M., Hughes, S.G., Chivukula, P., Ooi, E.E., 2021. A single dose of self-transcribing and replicating RNA-based SARS-CoV-2 vaccine produces protective adaptive immunity in mice. Mol. Ther. 29, 1970–1983. https://doi.org/10.1016/j.ymthe.2021.04.001.

De Haes, W., Van Mol, G., Merlin, C., De Smedt, S.C., Vanham, G., Rejman, J., 2012. Internalization of mRNA lipoplexes by dendritic cells. Mol. Pharm. 9, 2942–2949. https://doi.org/10.1021/mp3003336.

De Jong, W., Aerts, J., Allard, S., Brander, C., Buyze, J., Florence, E., Van Gorp, E., Vanham, G., Leal, L., Mothe, B., Thielemans, K., Plana, M., Garcia, F., Gruters, R., 2019. IHIVARNA phase IIa, a randomized, placebo-controlled, double-blinded trial to evaluate the safety and immunogenicity of iHIVARNA-01 in chronically HIV-infected patients under stable combined antiretroviral therapy. Trials 20, 361. https://doi.org/10.1186/s13063-019-3409-1.

De Keersmaecker, B., Claerhout, S., Carrasco, J., Bar, I., Corthals, J., Wilgenhof, S., Neyns, B., Thielemans, K., 2020. TriMix and tumor antigen mRNA electroporated dendritic cell vaccination plus ipilimumab: link between T-cell activation and clinical responses in advanced melanoma. J. Immunother. Cancer 8, 329. https://doi.org/10.1136/jitc-2019-000329.

Deal, C.E., Carfi, A., Plante, O.J., 2021. Advancements in mRNA encoded antibodies for passive immunotherapy. Vaccine 9, 1–16. https://doi.org/10.3390/vaccines9020108.

del Pozo-Rodríguez, A., Solinís, M.Á., Rodríguez-Gascón, A., 2016. Applications of lipid nanoparticles in gene therapy. Eur. J. Pharm. Biopharm. 109, 184–193. https://doi.org/10.1016/j.ejpb.2016.10.016.

del Pozo-Rodríguez, A., Rodríguez-Gascón, A., Rodríguez-Castejón, J., Vicente-Pascual, M., Gómez-Aguado, I., Battaglia, L.S., Solinís, M.Á., 2019a. Gene therapy. In: Silva, A.C., Moreira, J.N., Lobo, J.M.S., Almeida, H. (Eds.), Current Applications of Pharmaceutical Biotechnology. Springer International Publishing, Cham, pp. 321–368. https://doi.org/10.1007/10_2019_109.

del Pozo-Rodríguez, A., Rodríguez-Gascón, A., Rodríguez-Castejón, J., Vicente-Pascual, M., Gómez-Aguado, I., Battaglia, L.S., Solinís, M.Á., 2019b. Gene therapy. In: Advances in Biochemical Engineering/Biotechnology, pp. 321–368. https://doi.org/10.1007/10.

Delgado, D., del Pozo-Rodríguez, A., Angeles Solinís, M., Bartkowiak, A., Rodríguez-Gascón, A., 2013. New gene delivery system based on oligochitosan and solid lipid nanoparticles: 'In vitro' and 'in vivo' evaluation. Eur. J. Pharm. Sci. 50, 484–491. https://doi.org/10.1016/j.ejps.2013.08.013.

DeRosa, F., Guild, B., Karve, S., Smith, L., Love, K., Dorkin, J.R., Kauffman, K.J., Zhang, J., Yahalom, B., Anderson, D.G., Heartlein, M.W., 2016. Therapeutic efficacy in a hemophilia B model using a biosynthetic mRNA liver depot system. Gene Ther. 23, 699–707. https://doi.org/10.1038/gt.2016.46.

Diebold, S.S., Massacrier, C., Akira, S., Paturel, C., Morel, Y., Reis e Sousa, C., 2006. Nucleic acid agonists for Toll-like receptor 7 are defined by the presence of uridine ribonucleotides. Eur. J. Immunol. 36, 3256–3267. https://doi.org/10.1002/eji.200636617.

Dong, Y., Love, K.T., Dorkin, J.R., Sirirungruang, S., Zhang, Y., Chen, D., Bogorad, R.L., Yin, H., Chen, Y., Vegas, A.J., Alabi, C.A., Sahay, G., Olejnik, K.T., Wang, W., Schroeder, A., Lytton-Jean, A.K.R., Siegwart, D.J., Akinc, A., Barnes, C., Barros, S.A., Carioto, M., Fitzgerald, K., Hettinger, J., Kumar, V., Novobrantseva, T.I., Qin, J., Querbes, W., Koteliansky, V., Langer, R., Anderson, D.G., 2014. Lipopeptide nanoparticles for potent and selective siRNA delivery in rodents and nonhuman primates. Proc. Natl. Acad. Sci. U. S. A. 111, 3955–3960. https://doi.org/10.1073/pnas.1322937111.

Du, L., He, Y., Zhou, Y., Liu, S., Zheng, B.J., Jiang, S., 2009. The spike protein of SARS-CoV—a target for vaccine and therapeutic development. Nat. Rev. Microbiol. 7 (3), 226–236. https://doi.org/10.1038/nrmicro2090.

Du, Z., Munye, M.M., Tagalakis, A.D., Manunta, M.D.I., Hart, S.L., 2014. The role of the helper lipid on the DNA transfection efficiency of lipopolyplex formulations. Sci. Rep. 4, 7107. https://doi.org/10.1038/srep07107.

Edara, V.-V., Manning, K.E., Ellis, M., Lai, L., Foster, S.L., Floyd, K., Davis-Gardner, M.E., Nyhoff, L.E., Bechnak, S., Alaaeddine, G., Naji, A., Lee, M., Bristow, L., Hussaini, L., Rose Ciric, C., Gagne, M., Roberts-Torres, J., Henry, A.R., Godbole, S., Sexton, M., Piantadosi, A., Waggoner, J.J., Douek, D.C., Anderson, J., Rouphael, N., Wrammert, J., Suthar, M.S., 2022. mRNA-1273 and BNT162b2 mRNA vaccines have reduced neutralizing 1 activity against the SARS-CoV-2 omicron variant. Cell Rep. Med. 3 (2), 100529. https://doi.org/10.1016/j.xcrm.2022.100529.

Erasmus, J.H., Khandhar, A.P., Guderian, J., Granger, B., Archer, J., Archer, M., Gage, E., Fuerte-Stone, J., Larson, E., Lin, S., Kramer, R., Coler, R.N., Fox, C.B., Stinchcomb, D.T., Reed, S.G., Van Hoeven, N., 2018. A nanostructured lipid carrier for delivery of a replicating viral RNA provides single, low-dose protection against zika. Mol. Ther. 26, 2507–2522. https://doi.org/10.1016/j.ymthe.2018.07.010.

European Medicines Agency, 2021a. COVID-19 vaccines [WWW Document]. Eur. Med. Agency. URL. https://www.ema.europa.eu/en/human-regulatory/overview/public-health-threats/coronavirus-disease-covid-19/treatments-vaccines/covid-19-vaccines. (accessed 12.30.21).

European Medicines Agency, 2021b. Comirnaty [WWW Document]. Eur. Med. Agency. URL https://www.ema.europa.eu/en/medicines/human/EPAR/comirnaty#authorisation-details-section. (accessed 1.2.22).

Ewert, K.K., Scodeller, P., Simón-Gracia, L., Steffes, V.M., Wonder, E.A., Teesalu, T., Safinya, C.R., 2021. Cationic liposomes as vectors for nucleic acid and hydrophobic drug therapeutics. Pharmaceutics 13 (9), 1365. https://doi.org/10.3390/pharmaceutics13091365.

Feldman, R.A., Fuhr, R., Smolenov, I., (Mick)Ribeiro, A., Panther, L., Watson, M., Senn, J.J., Smith, M., Almarsson, Ö., Pujar, H.S., Laska, M.E., Thompson, J., Zaks, T., Ciaramella, G., 2019. mRNA vaccines against H10N8 and H7N9 influenza viruses of pandemic potential are immunogenic and well tolerated in healthy adults in phase 1 randomized clinical trials. Vaccine 37, 3326–3334. https://doi.org/10.1016/j.vaccine.2019.04.074.

Figlin, R.A., Tannir, N.M., Uzzo, R.G., Tykodi, S.S., Chen, D.Y.T., Master, V., Kapoor, A., Vaena, D., Lowrance, W., Bratslavsky, G., DeBenedette, M., Gamble, A., Plachco, A., Norris, M.S., Horvatinovich, J., Tcherepanova, I.Y., Nicolette, C.A., Wood, C.G., 2020. Results of the ADAPT phase 3 study of rocapuldencel-T in combination with sunitinib as first-line therapy in patients with metastatic renal cell carcinoma. Clin. Cancer Res. 26, 2327–2335. https://doi.org/10.1158/1078-0432.CCR-19-2427.

Filin, I.Y., Kitaeva, K.V., Rutland, C.S., Rizvanov, A.A., Solovyeva, V.V., 2021. Recent advances in experimental dendritic cell vaccines for cancer. Front. Oncol. 11, 3863. https://doi.org/10.3389/fonc.2021.730824.

Finn, J.D., Smith, A.R., Patel, M.C., Shaw, L., Youniss, M.R., van Heteren, J., Dirstine, T., Ciullo, C., Lescarbeau, R., Seitzer, J., Shah, R.R., Shah, A., Ling, D., Growe, J., Pink, M., Rohde, E., Wood, K.M., Salomon, W.E., Harrington, W.F., Dombrowski, C., Strapps, W.R., Chang, Y., Morrissey, D.V., 2018. A single administration of CRISPR/Cas9 lipid nanoparticles achieves robust and persistent in vivo genome editing. Cell Rep. 22, 2227–2235. https://doi.org/10.1016/j.celrep.2018.02.014.

Fishman, S., Lewis, M.D., Siew, L.K., De Leenheer, E., Kakabadse, D., Davies, J., Ziv, D., Margalit, A., Karin, N., Gross, G., Wong, F.S., 2017. Adoptive transfer of mRNA-transfected T cells redirected against diabetogenic CD8 T cells can prevent diabetes. Mol. Ther. 25, 456–464. https://doi.org/10.1016/j.ymthe.2016.12.007.

Foster, J.B., Barrett, D.M., Karikó, K., 2019. The emerging role of in vitro-transcribed mRNA in adoptive T cell immunotherapy. Mol. Ther. 27 (4), 747–756. https://doi.org/10.1016/j.ymthe.2019.01.018.

Freyn, A.W., Ramos da Silva, J., Rosado, V.C., Bliss, C.M., Pine, M., Mui, B.L., Tam, Y.K., Madden, T.D., de Souza Ferreira, L.C., Weissman, D., Krammer, F., Coughlan, L., Palese, P., Pardi, N., Nachbagauer, R., 2020. A multi-targeting, nucleoside-modified mRNA influenza virus vaccine provides broad protection in mice. Mol. Ther. 28, 1569–1584. https://doi.org/10.1016/j.ymthe.2020.04.018.

Gan, L.-M., Lagerström-Fermér, M., Carlsson, L.G., Arfvidsson, C., Egnell, A.-C., Rudvik, A., Kjaer, M., Collén, A., Thompson, J.D., Joyal, J., Chialda, L., Koernicke, T., Fuhr, R., Chien, K.R., Fritsche-Danielson, R., 2019. Intradermal delivery of modified mRNA encoding VEGF-A in patients with type 2 diabetes. Nat. Commun. 10, 871. https://doi.org/10.1038/s41467-019-08852-4.

Gandhi, R.T., Kwon, D.S., Macklin, E.A., Shopis, J.R., McLean, A.P., McBrine, N., Flynn, T., Peter, L., Sbrolla, A., Kaufmann, D.E., Porichis, F., Walker, B.D., Bhardwaj, N., Barouch, D.H., Kavanagh, D.G., 2016. Immunization of HIV-1-infected persons with autologous dendritic cells transfected with mRNA encoding HIV-1 Gag and Nef: results of a randomized, placebo-controlled clinical trial. J. Acquir. Immune Defic. Syndr. 71, 246–253. https://doi.org/10.1097/QAI.0000000000000852.

Gao, M., Zhang, Q., Feng, X.H., Liu, J., 2021. Synthetic modified messenger RNA for therapeutic applications. Acta Biomater. 131, 1–15. https://doi.org/10.1016/j.actbio.2021.06.020.

García-Cazorla, A., 2010. Propionic Acidemia. Encycl. Mov. Disord., 485–488. https://doi.org/10.1016/B978-0-12-374105-9.00373-7.

Gaspar, R., Coelho, F., Silva, B.F.B., 2020. Lipid-nucleic acid complexes: physicochemical aspects and prospects for cancer treatment. Molecules 25, 5006. https://doi.org/10.3390/molecules25215006.

Gediz Erturk, A., Sahin, A., Ay, E.B., Pelit, E., Bagdatli, E., Kulu, I., Gul, M., Mesci, S., Eryilmaz, S., Yildirim, T., Amparo, M., Faustino, F., Kijjoa, A., Gezici, S., 2021. A multidisciplinary approach to coronavirus disease (COVID-19). Molecules 26 (12), 3526. https://doi.org/10.3390/molecules26123526.

Gergen, J., Petsch, B., 2020. mRNA-Based Vaccines and Mode of Action. https://doi.org/10.1007/82_2020_230.

Gillmore, J.D., Gane, E., Taubel, J., Kao, J., Fontana, M., Maitland, M.L., Seitzer, J., O'Connell, D., Walsh, K.R., Wood, K., Phillips, J., Xu, Y., Amaral, A., Boyd, A.P., Cehelsky, J.E., McKee, M.D., Schiermeier, A., Harari, O., Murphy, A., Kyratsous, C.A., Zambrowicz, B., Soltys, R., Gutstein, D.E., Leonard, J., Sepp-Lorenzino, L., Lebwohl, D., 2021. CRISPR-Cas9 in vivo gene editing for transthyretin amyloidosis. N. Engl. J. Med. 385, 493–502. https://doi.org/10.1056/nejmoa2107454.

Gómez-Aguado, I., Rodríguez-Castejón, J., Vicente-Pascual, M., Rodríguez-Gascón, A., Del Pozo-Rodríguez, A., Solinís Aspiazu, M.Á., 2020. Nucleic acid delivery by solid lipid nanoparticles containing switchable lipids: plasmid DNA vs. messenger RNA. Molecules 25 (24), 5995. https://doi.org/10.3390/molecules25245995.

Gómez-Aguado, I., Rodríguez-Castejón, J., Vicente-Pascual, M., Rodríguez-Gascón, A., Solinís, M.Á., del Pozo-Rodríguez, A., 2020b. Nanomedicines to deliver mRNA: state of the art and future perspectives. Nanomaterials 10, 364. https://doi.org/10.3390/nano10020364.

Gómez-Aguado, I., Rodríguez-Castejón, J., Beraza-Millor, M., Vicente-Pascual, M., Rodríguez-Gascón, A., Garelli, S., Battaglia, L., del Pozo-Rodríguez, A., Solinís, M. Á., 2021. mRNA-based nanomedicinal products to address corneal inflammation by interleukin-10 supplementation. Pharmaceutics 13, 1472. https://doi.org/10.3390/pharmaceutics13091472.

Granja, A., Lima-Sousa, R., Alves, C.G., de Melo-Diogo, D., Pinheiro, M., Sousa, C.T., Correia, I.J., Reis, S., 2021. Mitoxantrone-loaded lipid nanoparticles for breast cancer therapy—quality-by-design approach and efficacy assessment in 2D and 3D in vitro cancer models. Int. J. Pharm. 607, 121044. https://doi.org/10.1016/j.ijpharm.2021.121044.

Grier, A.E., Burleigh, S., Sahni, J., Clough, C.A., Cardot, V., Choe, D.C., Krutein, M.C., Rawlings, D.J., Jensen, M.C., Scharenberg, A.M., Jacoby, K., 2016. pEVL: a linear plasmid for generating mRNA IVT templates with extended encoded poly(A) sequences. Mol. Ther. Nucleic Acids 5, e306. https://doi.org/10.1038/mtna.2016.21.

Grunwitz, C., Salomon, N., Vascotto, F., Selmi, A., Bukur, T., Diken, M., Kreiter, S., Türeci, Ö., Sahin, U., 2019. HPV16 RNA-LPX vaccine mediates complete regression of aggressively growing HPV-positive mouse tumors and establishes protective T cell memory. Onco. Targets. Ther. 8, e1629259. https://doi.org/10.1080/2162402X.2019.1629259.

Guevara, M.L., Persano, S., Persano, F., 2019. Lipid-based vectors for therapeutic mRNA-based anti-cancer vaccines. Curr. Pharm. Des. 25 (13), 1443–1454.

Guevara, M.L., Persano, F., Persano, S., 2020. Advances in lipid nanoparticles for mRNA-based cancer immunotherapy. Front. Chem. 8, 1–17. https://doi.org/10.3389/fchem.2020.589959.

Haabeth, O.A.W., Blake, T.R., McKinlay, C.J., Tveita, A.A., Sallets, A., Waymouth, R.M., Wender, P.A., Levy, R., 2019. Local delivery of OX40L, CD80, and CD86 mRNA kindles global anticancer immunity. Cancer Res. 79, 1624–1634. https://doi.org/10.1158/0008-5472.CAN-18-2867.

Hajj, K.A., Whitehead, K.A., 2017. Tools for translation: non-viral materials for therapeutic mRNA delivery. Nat. Rev. Mater. 2, 1–17. https://doi.org/10.1038/natrevmats.2017.56.

Harrer, D.C., Simon, B., Fujii, S.-I., Shimizu, K., Uslu, U., Schuler, G., Gerer, K.F., Hoyer, S., Dörrie, J., Schaft, N., 2017. RNA-transfection of γ/δ T cells with a chimeric antigen receptor or an α/β T-cell receptor: a safer alternative to genetically engineered α/β T cells for the immunotherapy of melanoma. BMC Cancer 17, 551. https://doi.org/10.1186/s12885-017-3539-3.

Heine, A., Juranek, S., Brossart, P., 2021. Clinical and immunological effects of mRNA vaccines in malignant diseases. Mol. Cancer 20 (1), 52. https://doi.org/10.1186/s12943-021-01339-1.

Hekele, A., Bertholet, S., Archer, J., Gibson, D.G., Palladino, G., Brito, L.A., Otten, G.R., Brazzoli, M., Buccato, S., Bonci, A., Casini, D., Maione, D., Qi, Z.-Q., Gill, J.E., Caiazza, N.C., Urano, J., Hubby, B., Gao, G.F., Shu, Y., De Gregorio, E., Mandl, C.W., Mason, P.W., Settembre, E.C., Ulmer, J.B., Craig Venter, J., Dormitzer, P.R., Rappuoli, R., Geall, A.J., 2013. Rapidly produced SAM(®) vaccine against H7N9 influenza is immunogenic in mice. Emerg. Microbes Infect. 2, e52. https://doi.org/10.1038/emi.2013.54.

Hellen, C.U.T., Sarnow, P., 2001. Internal ribosome entry sites in eukaryotic mRNA molecules. Genes Dev. 15, 1593–1612. https://doi.org/10.1101/gad.891101.

Henderson, J.M., Ujita, A., Hill, E., Yousif-Rosales, S., Smith, C., Ko, N., McReynolds, T., Cabral, C.R., Escamilla-Powers, J.R., Houston, M.E., 2021. Cap 1 messenger RNA synthesis with co-transcriptional CleanCap® analog by in vitro transcription. Curr. Protoc. 1 (2), e39. https://doi.org/10.1002/cpz1.39.

Hinnebusch, A.G., Ivanov, I.P., Sonenberg, N., 2016. Translational control by 5′-untranslated regions of eukaryotic mRNAs. Science 352, 1413–1416.

Hollingsworth, R.E., Jansen, K., 2019. Turning the corner on therapeutic cancer vaccines. NPJ Vaccines 4 (1), 7. https://doi.org/10.1038/s41541-019-0103-y.

Holtkamp, S., Kreiter, S., Selmi, A., Simon, P., Koslowski, M., Huber, C., Türeci, Ö., Sahin, U., 2006. Modification of antigen-encoding RNA increases stability, translational efficacy, and T-cell stimulatory capacity of dendritic cells. Blood 108, 4009–4017. https://doi.org/10.1182/blood-2006-04-015024.

Hornung, V., Kato, H., Poeck, H., Akira, S., Conzelmann, K., Schlee, M., 2010. 5′-triphosphate RNA is the ligand for RIG-I. Science 994, 1–5.

Hou, X., Zaks, T., Langer, R., Dong, Y., 2021. Lipid nanoparticles for mRNA delivery. Nat. Rev. Mater. 6 (12), 1078–1094. 0123456789. https://doi.org/10.1038/s41578-021-00358-0.

Huang, H.W., Chen, F.-Y., Lee, M.-T., 2004. Molecular mechanism of peptide-induced pores in membranes. Phys. Rev. Lett. 92, 198304. https://doi.org/10.1103/PhysRevLett.92.198304.

Hunt, R.C., Simhadri, V.L., Iandoli, M., Sauna, Z.E., Kimchi-Sarfaty, C., 2014. Exposing synonymous mutations. Trends Genet. 30, 308–321. https://doi.org/10.1016/j.tig.2014.04.006.

Ibba, M.L., Ciccone, G., Esposito, C.L., Catuogno, S., Giangrande, P.H., 2021. Advances in mRNA non-viral delivery approaches. Adv. Drug Deliv. Rev. 177, 113930. https://doi.org/10.1016/J.ADDR.2021.113930.

Islam, M.A., Reesor, E.K.G., Xu, Y., Zope, H.R., Zetter, B.R., Shi, J., 2015. Biomaterials for mRNA delivery. Biomater. Sci. 3, 1519–1533. https://doi.org/10.1039/c5bm00198f.

Jackson, L.A., Anderson, E.J., Rouphael, N.G., Roberts, P.C., Makhene, M., Coler, R.N., McCullough, M.P., Chappell, J.D., Denison, M.R., Stevens, L.J., Pruijssers, A.J., McDermott, A., Flach, B., Doria-Rose, N.A., Corbett, K.S., Morabito, K.M., O'Dell, S., Schmidt, S.D., Swanson, P.A., Padilla, M., Mascola, J.R., Neuzil, K.M., Bennett, H., Sun, W., Peters, E., Makowski, M., Albert, J., Cross, K., Buchanan, W., Pikaart-Tautges, R., Ledgerwood, J.E., Graham, B.S., Beigel, J.H., 2020. An mRNA vaccine against SARS-CoV-2—preliminary report. N. Engl. J. Med. 383, 1920–1931. https://doi.org/10.1056/NEJMoa2022483.

Jain, R., Frederick, J.P., Huang, E.Y., Burke, K.E., Mauger, D.M., Andrianova, E.A., Farlow, S.J., Siddiqui, S., Pimentel, J., Cheung-Ong, K., McKinney, K.M., Köhrer, C., Moore, M.J., Chakraborty, T., 2018. MicroRNAs enable mRNA therapeutics to selectively program cancer cells to self-destruct. Nucleic Acid Ther. 28, 285–296. https://doi.org/10.1089/nat.2018.0734.

Jang, S.K., Paek, K.Y., 2016. Cap-dependent translation is mediated by 'RNA looping' rather than 'ribosome scanning'. RNA Biol. 13, 1–5. https://doi.org/10.1080/15476286.2015.1107700.

Jauze, L., Monteillet, L., Mithieux, G., Rajas, F., Ronzitti, G., 2019. Challenges of gene therapy for the treatment of glycogen storage diseases type I and type III. Hum. Gene Ther. 30, 1263–1273. https://doi.org/10.1089/hum.2019.102.

Jere, D., Jiang, H., Arote, R., Kim, Y., Choi, Y., Cho, M., Akaike, T., Cho, C., 2009. Degradable polyethylenimines as DNA and small interfering RNA carriers. Expert Opin. Drug Deliv. 6, 827–834. https://doi.org/10.1517/17425240903029183.

Jiang, Y., Xu, X.-S., Russell, J.E., 2006. A nucleolin-binding 3′ untranslated region element stabilizes β-globin mRNA in vivo. Mol. Cell. Biol. 26, 2419–2429. https://doi.org/10.1128/mcb.26.6.2419-2429.2006.

Jiang, L., Berraondo, P., Jericó, D., Guey, L.T., Sampedro, A., Frassetto, A., Benenato, K.E., Burke, K., Santamaría, E., Alegre, M., Pejenaute, Á., Kalariya, M., Butcher, W., Park, J.S., Zhu, X., Sabnis, S., Kumarasinghe, E.S., Salerno, T., Kenney, M., Lukacs, C.M., Ávila, M.A., Martini, P.G.V., Fontanellas, A., 2018. Systemic messenger RNA as an etiological treatment for acute intermittent porphyria. Nat. Med. 24, 1899–1909. https://doi.org/10.1038/s41591-018-0199-z.

Jiang, L., Park, J.S., Yin, L., Laureano, R., Jacquinet, E., Yang, J., Liang, S., Frassetto, A., Zhuo, J., Yan, X., Zhu, X., Fortucci, S., Hoar, K., Mihai, C., Tunkey, C., Presnyak, V., Benenato, K.E., Lukacs, C.M., Martini, P.G.V., Guey, L.T., 2020. Dual mRNA therapy restores metabolic function in long-term studies in mice with propionic acidemia. Nat. Commun. 11, 1–10. https://doi.org/10.1038/s41467-020-19156-3.

Jimeno, A., Gupta, S., Sullivan, R., Do, K.T., Akerley, W.L., Wang, D., Teoh, D., Schalper, K., Zacharek, S.J., Sun, J., Laino, A.S., Frederick, J., Zhou, H., Randolph, W., Pascarella, S., Johansen, L., Cohen, P.S., Meehan, R.S., Bauer, T.M., 2020. Abstract CT032: A Phase 1/2, Open-Label, Multicenter, Dose Escalation and Efficacy Study of mRNA-2416, A Lipid Nanoparticle Encapsulated mRNA Encoding Human OX40L, for Intratumoral Injection Alone or in Combination With Durvalumab for Patients With Advance. American Association for Cancer Research (AACR), p. CT032. https://doi.org/10.1158/1538-7445.am2020-ct032.

Jirikowski, G.F., Sanna, P.P., Maciejewski-Lenoir, D., Bloom, F.E., 1992. Reversal of diabetes insipidus in Brattleboro tats: intrahypothalamic injection of vasopressin mRNA. Science 255, 996–998. https://doi.org/10.1126/science.1546298.

Jo, J.-I., Gao, J.-Q., Tabata, Y., 2019. Biomaterial-based delivery systems of nucleic acid for regenerative research and regenerative therapy. Regen. Ther. 11, 123–130. https://doi.org/10.1016/j.reth.2019.06.007.

Johnson, A.G., Grosely, R., Petrov, A.N., Puglisi, J.D., Puglisi, J.D., 2017. Dynamics of IRES-mediated translation. Philos. Trans. R. Soc. Lond. B Biol. Sci. 372, 20160177.

Jones, V.A.S., Bucher, M., Hambleton, E.A., Guse, A., 2018. Microinjection to deliver protein, mRNA, and DNA into zygotes of the cnidarian endosymbiosis model Aiptasia sp. Sci. Rep. 8, 16437. https://doi.org/10.1038/s41598-018-34773-1.

Judge, A., McClintock, K., Phelps, J.R., MacLachlan, I., 2006. Hypersensitivity and loss of disease site targeting caused by antibody responses to PEGylated liposomes. Mol. Ther. 13, 328–337. https://doi.org/10.1016/j.ymthe.2005.09.014.

Kaczmarek, J.C., Patel, A.K., Kauffman, K.J., Fenton, O.S., Webber, M.J., Heartlein, M.W., DeRosa, F., Anderson, D.G., 2016. Polymer–lipid nanoparticles for systemic delivery of mRNA to the lungs. Angew. Chem. Int. Ed. 55, 13808–13812. https://doi.org/10.1002/anie.201608450.

Kallen, K.-J., Heidenreich, R., Schnee, M., Petsch, B., Schlake, T., Thess, A., Baumhof, P., Scheel, B., Koch, S.D., Fotin-Mleczek, M., 2013. A novel, disruptive vaccination technology. Hum. Vaccin. Immunother. 9, 2263–2276. https://doi.org/10.4161/hv.25181.

Karikó, K., Buckstein, M., Ni, H., Weissman, D., 2005. Suppression of RNA recognition by Toll-like receptors: the impact of nucleoside modification and the evolutionary origin of RNA. Immunity 23, 165–175. https://doi.org/10.1016/j.immuni.2005.06.008.

Karikó, K., Muramatsu, H., Welsh, F.A., Ludwig, J., Kato, H., Akira, S., Weissman, D., 2008. Incorporation of pseudouridine into mRNA yields superior nonimmunogenic vector with increased translational capacity and biological stability. Mol. Ther. 16, 1833–1840. https://doi.org/10.1038/mt.2008.200.

Karikó, K., Muramatsu, H., Ludwig, J., Weissman, D., 2011. Generating the optimal mRNA for therapy: HPLC purification eliminates immune activation and improves translation of nucleoside-modified, protein-encoding mRNA. Nucleic Acids Res. 39, 1–10. https://doi.org/10.1093/nar/gkr695.

Kauffman, K.J., Dorkin, J.R., Yang, J.H., Heartlein, M.W., DeRosa, F., Mir, F.F., Fenton, O.S., Anderson, D.G., 2015. Optimization of lipid nanoparticle formulations for mRNA delivery in vivo with fractional factorial and definitive screening designs. Nano Lett. 15, 7300–7306. https://doi.org/10.1021/acs.nanolett.5b02497.

Kim, H.W., Canchola, J.G., Brandt, C.D., Pyles, G., Chanock, R.M., Jensen, K., Parrott, R.H., 1969. Respiratory syncytial virus disease in infants despite prior administration of antigenic inactivated vaccine. Am. J. Epidemiol. 89, 422–434. https://doi.org/10.1093/oxfordjournals.aje.a120955.

Kim, J., Eygeris, Y., Gupta, M., Sahay, G., 2021. Self-assembled mRNA vaccines. Adv. Drug Deliv. Rev. 170, 83–112. https://doi.org/10.1016/j.addr.2020.12.014.

Koh, W.S., Porter, J.R., Batchelor, E., 2019. Tuning of mRNA stability through altering 3′-UTR sequences generates distinct output expression in a synthetic circuit driven by p53 oscillations. Sci. Rep. 9, 1–8. https://doi.org/10.1038/s41598-019-42509-y.

Koltover, I., Salditt, T., Rädler, J.O., Safinya, C.R., 1998. An inverted hexagonal phase of cationic liposome-DNA complexes related to DNA release and delivery. Science 281, 78–81. https://doi.org/10.1126/science.281.5373.78.

Kormann, M.S.D., Hasenpusch, G., Aneja, M.K., Nica, G., Flemmer, A.W., Herber-Jonat, S., Huppmann, M., Mays, L.E., Illenyi, M., Schams, A., Griese, M., Bittmann, I., Handgretinger, R., Hartl, D., Rosenecker, J., Rudolph, C., 2011. Expression of therapeutic proteins after delivery of chemically modified mRNA in mice. Nat. Biotechnol. 29, 154–159. https://doi.org/10.1038/nbt.1733.

Kowalska, J., Zuberek, J., Darzynkiewicz, Z.M., Lukaszewicz, M., Darzynkiewicz, E., Jemielity, J., 2008. The first examples of mRNA cap analogs bearing boranophosphate modification. Nucleic Acids Symp. Ser., 289–290. https://doi.org/10.1093/nass/nrn146.

Kowalski, P.S., Rudra, A., Miao, L., Anderson, D.G., 2019. Delivering the messenger: advances in technologies for therapeutic mRNA delivery. Mol. Ther. 27, 710–728. https://doi.org/10.1016/j.ymthe.2019.02.012.

Kowalzik, F., Schreiner, D., Jensen, C., Teschner, D., Gehring, S., Zepp, F., 2021. Mrna-based vaccines. Vaccine 9 (4), 390. https://doi.org/10.3390/vaccines9040390.

Kozak, M., 1984. Point mutations close to the AUG initiator codon affect the efficiency of translation of rat preproinsulin in vivo. Nature 308, 241–246. https://doi.org/10.1038/308241a0.

Kozak, M., 1987. At least six nucleotides preceding the AUG initiator codon enhance translation in mammalian cells. J. Mol. Biol. 196, 947–950. https://doi.org/10.1016/0022-2836(87)90418-9.

Kozak, M., 1991a. Structural features in eukaryotic mRNAs that modulate the initiation of translation. J. Biol. Chem. 266, 19867–19870. https://doi.org/10.1016/s0021-9258(18)54860-2.

Kozak, M., 1991b. An analysis of vertebrate mRNA sequences: Intimations of translational control. J. Cell Biol. 115, 887–903. https://doi.org/10.1083/jcb.115.4.887.

Kranz, L.M., Diken, M., Haas, H., Kreiter, S., Loquai, C., Reuter, K.C., Meng, M., Fritz, D., Vascotto, F., Hefesha, H., Grunwitz, C., Vormehr, M., Hüsemann, Y., Selmi, A., Kuhn, A.N., Buck, J., Derhovanessian, E., Rae, R., Attig, S., Diekmann, J., Jabulowsky, R.A., Heesch, S., Hassel, J., Langguth, P., Grabbe, S., Huber, C., Türeci, Ö., Sahin, U., 2016. Systemic RNA delivery to dendritic cells exploits antiviral defence for cancer immunotherapy. Nature 534, 396–401. https://doi.org/10.1038/nature18300.

Kreiter, S., Selmi, A., Diken, M., Sebastian, M., Osterloh, P., Schild, H., Huber, C., Türeci, Ö., Sahin, U., 2008. Increased antigen presentation efficiency by coupling antigens to MHC class I trafficking signals. J. Immunol. 180, 309–318. https://doi.org/10.4049/jimmunol.180.1.309.

Kremsner, P., Mann, P., Bosch, J., Fendel, R., Gabor, J.J., Kreidenweiss, A., Kroidl, A., Leroux-Roels, I., Leroux-Roels, G., Schindler, C., Schunk, M., Velavan, T.P., Fotin-Mleczek, M., Müller, S., Quintini, G., Schönborn-Kellenberger, O., Vahrenhorst, D., Verstraeten, T., Walz, L., Wolz, O.-O., Oostvogels, L., 2020. Phase 1 assessment of the safety and immunogenicity of an mRNA- lipid nanoparticle vaccine candidate against SARS-CoV-2 in human volunteers. medRxiv. 2020.11.09.20228551. https://doi.org/10.1101/2020.11.09.20228551.

Kremsner, P.G., Guerrero, A., Andrés, R., Arana-Arri, E., Aroca Martinez, G.J., Bonten, M., Chandler, R., Corral, G., De Block, E.J.L., Ecker, L., Gabor, J.J., Garcia Lopez, C.A., Gonzales, L., Granados González, M.A., Gorini, N., Grobusch, M.P., Hrabar, A.D., Junker, H., Kimura, A., Lanata, C.F., Lehmann, C., Leroux-Roels, I., Mann, P., Martinez-Reséndez, M.F., Ochoa, T.J., Poy, C.A., Reyes Fentanes, M.J., Rivera Mejia, L.M., Ruiz Herrera, V.V., Sáez-Llorens, X., Schönborn-Kellenberger, O., Schunk, M., Sierra Garcia, A., Vergara, I., Verstraeten, T., Vico, M., Oostvogels, L., Lovesio, L., Diez, F., Grazziani, F., Ganaha, M.C., Zalatnik, V.J., Dittrich, R.J., Espínola, L., Lambert, S., Longhi, A., Vecchio, C., Mastruzzo, M., Fernandez, A., Borchowiek, S., Potito, R., Guerrero, A., Andres, R., Guardiani, F.M., Castella, S., Foccoli, M., Pedernera, A., Braida, A., Durigan, V., Martella, C., Bobat, A., Boggia, B.E., Nemi, S.A., Tartaglione, J.G., Piedimonte, F.C., De Bie, J., Reynales Londoño, H., Rodríguez Ordoñez, P.A., García Cruz, J.M., Bautista Toloza, L., Ladino González, M.C., Zambrano Ochoa, A.P., Prieto Pradera, I., Torres Hernandez, D.,

Mazo Elorza, D.P., Collazos Lennis, M.F., Vanegas Dominguez, B., Solano Mosquera, L.M., Fendel, R., Fleischmann, W.A., Koehne, E., Kreidenweiss, A., Köhler, C., Esen, M., Horn, C., Eberts, S., Kroidl, A., Huber, K., Thiel, V., Mazara Rosario, S., Reyes, G., Rivera, L., Donastorg, Y., Lantigua, F., Torres Almanzar, D., Candelario, R., Peña Mendez, L., Rosario Gomez, N., Portolés-Pérez, A., Ascaso del Río, A., Laredo Velasco, L., Bustinduy Odriozola, M.J., Larrea Arranz, I., Martínez Alcorta, L.I., Durán Laviña, M.I., Imaz-Ayo, N., Meijide, S., García-de-Vicuña, A., Santorcuato, A., Gallego, M., Aguirre-García, G.M., Olmos Vega, J., González Limón, P., Vázquez Villar, A., Chávez Barón, J., Arredondo Saldaña, F., de Luján Palacios, J.D., Camacho Choza, L.J., Vázquez Saldaña, E.G., Ortega Dominguez, S.J., Vega Orozco, K.S., Torres Quiroz, I.A., Martinez Avendaño, A., Herrera Sanchez, J., Guzman, E., Castro Castrezana, L., Ruiz Palacios y Santos, G.M., de Winter, R.F.J., de Jonge, H.K., Schnyder, J.L., Boersma, W., Hessels, L., Djamin, R., van der Sar, S., DeAntonio, R., Peña, M., Rebollon, G., Rojas, M., Escobar, J., Hammerschlag Icaza, B., Wong, T.D.Y., Barrera Perigault, P., Ruiz, S., Chan, M., Arias Hoo, D.J., Gil, A.I., Celis, C.R., Balmaceda, M.P., Flores, O., Ochoa, M., Peña, B., de la Flor, C., Webb, C.M., Cornejo, E., Sanes, F., Mayorga, V., Valdiviezo, G., Ramírez Lamas, S.P., Grandez Castillo, G.A., Lama, J.R., Matta Aguirre, M.E., Arancibia Luna, L.A., Carbajal Paulet, Ó., Zambrano Ortiz, J., Camara, A., Guzman Quintanilla, F., Diaz-Parra, C., Morales-Oliva, J., Cornejo, R.E., Ricalde, S.A., Vidal, J., Rios Nogales, L., Cheatham-Seitz, D., Gregoraci, G., Brecx, A., Walz, L., Vahrenhorst, D., Seibel, T., Quintini, G., 2021. Efficacy and safety of the CVnCoV SARS-CoV-2 mRNA vaccine candidate in ten countries in Europe and Latin America (HERALD): a randomised, observer-blinded, placebo-controlled, phase 2b/3 trial. Lancet Infect. Dis. 22 (3), 329–340. https://doi.org/10.1016/S1473-3099(21)00677-0.

Kristensen, L.S., Andersen, M.S., Stagsted, L.V.W., Ebbesen, K.K., Hansen, T.B., Kjems, J., 2019. The biogenesis, biology and characterization of circular RNAs. Nat. Rev. Genet. 20, 675–691. https://doi.org/10.1038/s41576-019-0158-7.

Kübler, H., Scheel, B., Gnad-Vogt, U., Miller, K., Schultze-Seemann, W., Dorp, F., Parmiani, G., Hampel, C., Wedel, S., Trojan, L., Jocham, D., Maurer, T., Rippin, G., Fotin-Mleczek, M., Mülbe, F., Probst, J., Hoerr, I., Kallen, K.J., Lander, T., Stenzl, A., 2015. Self-adjuvanted mRNA vaccination in advanced prostate cancer patients: a first-in-man phase I/IIa study. J. Immunother. Cancer 3, 1–14. https://doi.org/10.1186/s40425-015-0068-y.

Kuhn, A.N., Diken, M., Kreiter, S., Selmi, A., Kowalska, J., Jemielity, J., Darzynkiewicz, E., Huber, C., Türeci, Ö., Sahin, U., 2010. Phosphorothioate cap analogs increase stability and translational efficiency of RNA vaccines in immature dendritic cells and induce superior immune responses in vivo. Gene Ther. 17, 961–971. https://doi.org/10.1038/gt.2010.52.

Kwon, H., Kim, M., Seo, Y., Moon, Y.S., Lee, H.J., Lee, K., Lee, H., 2018. Emergence of synthetic mRNA: in vitro synthesis of mRNA and its applications in regenerative medicine. Biomaterials 156, 172–193. https://doi.org/10.1016/j.biomaterials.2017.11.034.

Laczkó, D., Hogan, M.J., Toulmin, S.A., Hicks, P., Lederer, K., Gaudette, B.T., Castaño, D., Amanat, F., Muramatsu, H., Oguin, T.H., Ojha, A., Zhang, L., Mu, Z., Parks, R., Manzoni, T.B., Roper, B., Strohmeier, S., Tombácz, I., Arwood, L., Nachbagauer, R., Karikó, K., Greenhouse, J., Pessaint, L., Porto, M., Putman-Taylor, T., Strasbaugh, A., Campbell, T.A., Lin, P.J.C., Tam, Y.K., Sempowski, G.D., Farzan, M., Choe, H., Saunders, K.O., Haynes, B.F., Andersen, H., Eisenlohr, L.C., Weissman, D., Krammer, F., Bates, P., Allman, D., Locci, M., Pardi, N., 2020. A single immunization with nucleoside-modified mRNA vaccines elicits strong cellular and humoral immune responses against SARS-CoV-2 in mice. Immunity 53, 724–732.e7. https://doi.org/10.1016/j.immuni.2020.07.019.

Landesman-Milo, D., Peer, D., 2014. Toxicity profiling of several common RNAi-based nanomedicines: a comparative study. Drug Deliv. Transl. Res. 4, 96–103. https://doi.org/10.1007/s13346-013-0158-7.

Lane, M., Mis, E.K., Khokha, M.K., 2022. Microinjection of Xenopus tropicalis embryos. Cold Spring Harb. Protoc. 2022 (4). https://doi.org/10.1101/pdb.prot107644. pdb.prot107644.

Leal, L., Guardo, A.C., Morón-López, S., Salgado, M., Mothe, B., Heirman, C., Pannus, P., Vanham, G., van den Ham, H.J., Gruters, R., Andeweg, A., Van Meirvenne, S., Pich, J., Arnaiz, J.A., Gatell, J.M., Brander, C., Thielemans, K., Martínez-Picado, J., Plana, M., García, F., 2018. Phase I clinical trial of an intranodally administered mRNA-based therapeutic vaccine against HIV-1 infection. AIDS 32, 2533–2545. https://doi.org/10.1097/QAD.0000000000002026.

Lescan, M., Perl, R.M., Golombek, S., Pilz, M., Hann, L., Yasmin, M., Behring, A., Keller, T., Nolte, A., Gruhn, F., Kochba, E., Levin, Y., Schlensak, C., Wendel, H.P., Avci-Adali, M., 2018. De novo synthesis of elastin by exogenous delivery of synthetic modified mRNA into skin and elastin-deficient cells. Mol. Ther. Nucleic Acids 11, 475–484. https://doi.org/10.1016/j.omtn.2018.03.013.

Levin, N., Weinstein-Marom, H., Pato, A., Itzhaki, O., Besser, M.J., Eisenberg, G., Peretz, T., Lotem, M., Gross, G., 2018. Potent activation of human T cells by mRNA encoding constitutively active CD40. J. Immunol. 201, 2959–2968. https://doi.org/10.4049/jimmunol.1701725.

Li, B., Luo, X., Deng, B., Wang, J., McComb, D.W., Shi, Y., Gaensler, K.M.L., Tan, X., Dunn, A.L., Kerlin, B.A., Dong, Y., 2015. An orthogonal array optimization of lipid-like nanoparticles for mRNA delivery in vivo. Nano Lett. 15, 8099–8107. https://doi.org/10.1021/acs.nanolett.5b03528.

Li, B., Zhang, X., Dong, Y., 2019. Nanoscale platforms for messenger RNA delivery. Wiley Interdiscip. Rev. Nanomed. Nanobiotechnol. 11, 1–14. https://doi.org/10.1002/wnan.1530.

Li, Z., Zhao, P., Zhang, Y., Wang, J., Wang, C., Liu, Y., Yang, G., Yuan, L., 2021. Exosome-based Ldlr gene therapy for familial hypercholesterolemia in a mouse model. Theranostics 11, 2953–2965. https://doi.org/10.7150/THNO.49874.

Linares-Fernández, S., Lacroix, C., Exposito, J.Y., Verrier, B., 2020. Tailoring mRNA vaccine to balance innate/adaptive immune response. Trends Mol. Med. 26, 311–323. https://doi.org/10.1016/j.molmed.2019.10.002.

Liu, M.A., 2019. A comparison of plasmid DNA and mRNA as vaccine technologies. Vaccine 7 (2), 37. https://doi.org/10.3390/vaccines7020037.

Liu, S., Cheng, Q., Wei, T., Yu, X., Johnson, L.T., Farbiak, L., Siegwart, D.J., 2021a. Membrane-destabilizing ionizable phospholipids for organ-selective mRNA delivery and CRISPR–Cas gene editing. Nat. Mater. 20, 701–710. https://doi.org/10.1038/s41563-020-00886-0.

Liu, W., Tang, H., Li, L., Wang, X., Yu, Z., Li, J., 2021a. Peptide-based therapeutic cancer vaccine: current trends in clinical application. Cell Prolif. 54 (5), e13025. https://doi.org/10.1111/cpr.13025.

Lorenz, C., Fotin-Mleczek, M., Roth, G., Becker, C., Dam, T.C., Verdurmen, W.P.R., Brock, R., Probst, J., Schlake, T., 2011. Protein expression from exogenous mRNA: Uptake by receptor-mediated endocytosis and trafficking via the lysosomal pathway. RNA Biol. 8 (4), 627–636. https://doi.org/10.4161/rna.8.4.15394.

Love, K.T., Mahon, K.P., Levins, C.G., Whitehead, K.A., Querbes, W., Dorkin, J.R., Qin, J., Cantley, W., Qin, L.L., Racie, T., Frank-Kamenetsky, M., Yip, K.N., Alvarez, R., Sah, D.W.Y., de Fougerolles, A., Fitzgerald, K., Koteliansky, V., Akinc, A., Langer, R., Anderson, D.G., 2010. Lipid-like materials for low-dose, in vivo gene silencing. Proc. Natl. Acad. Sci. U. S. A. 107, 1864–1869. https://doi.org/10.1073/pnas.0910603106.

Lu, D., Benjamin, R., Kim, M., Conry, R.M., Curiel, D.T., 1994. Optimization of methods to achieve mRNA-mediated transfection of tumor cells in vitro and in vivo employing cationic liposome vectors. Cancer Gene Ther. 1 (4), 245–252.

Lu, J., Lu, G., Tan, S., Xia, J., Xiong, H., Yu, X., Qi, Q., Yu, X., Li, L., Yu, H., Xia, N., Zhang, T., Xu, Y., Lin, J., 2020. A COVID-19 mRNA vaccine encoding SARS-CoV-2 virus-like particles induces a strong antiviral-like immune response in mice. Cell Res. 30, 936–939. https://doi.org/10.1038/s41422-020-00392-7.

Luan, X., Sansanaphongpricha, K., Myers, I., Chen, H., Yuan, H., Sun, D., 2017. Engineering exosomes as refined biological nanoplatforms for drug delivery. Acta Pharmacol. Sin. 38, 754–763. https://doi.org/10.1038/aps.2017.12.

Lutz, J., Lazzaro, S., Habbeddine, M., Schmidt, K.E., Baumhof, P., Mui, B.L., Tam, Y.K., Madden, T.D., Hope, M.J., Heidenreich, R., Fotin-Mleczek, M., 2017. Unmodified mRNA in LNPs constitutes a competitive technology for prophylactic vaccines. NPJ Vaccines 2, 29. https://doi.org/10.1038/s41541-017-0032-6.

Ma, Z., Li, J., He, F., Wilson, A., Pitt, B., Li, S., 2005. Cationic lipids enhance siRNA-mediated interferon response in mice. Biochem. Biophys. Res. Commun. 330, 755–759. https://doi.org/10.1016/j.bbrc.2005.03.041.

Ma, Y., Zhang, X., Xu, X., Shen, L., Yao, Y., Yang, Z., Liu, P., 2015. STAT3 decoy oligodeoxynucleotides-loaded solid lipid nanoparticles induce cell death and inhibit invasion in ovarian cancer cells. PLoS One 10, e0124924.

MacLachlan, I., Cullis, P., 2005. Diffusible-PEG-lipid stabilized plasmid lipid particles. Adv. Genet. 53, 157–188. https://doi.org/10.1016/S0065-2660(05)53006-2.

Magadum, A., Kaur, K., Zangi, L., 2019. mRNA-based protein replacement therapy for the heart. Mol. Ther. 27, 785–793. https://doi.org/10.1016/j.ymthe.2018.11.018.

Magini, D., Giovani, C., Mangiavacchi, S., MacCari, S., Cecchi, R., Ulmer, J.B., De Gregorio, E., Geall, A.J., Brazzoli, M., Bertholet, S., 2016. Self-amplifying mRNA vaccines expressing multiple conserved influenza antigens confer protection against homologous and heterosubtypic viral challenge. PLoS One 11 (8), e0161193. https://doi.org/10.1371/journal.pone.0161193.

Malone, R.W., Felgner, P.L., Verma, I.M., 1989. Cationic liposome-mediated RNA transfection [cationic lipid vesicies/N-[1-(2,3-dioleyloxy)propyl]-NNN-timethylammonium chloride (DOTMA)/translationj]. Proc. Natl. Acad. Sci. U. S. A. 86, 6077–6081.

Martini, P.G.V., Guey, L.T., 2019. A new era for rare genetic diseases: messenger RNA therapy. Hum. Gene Ther. 30, 1180–1189. https://doi.org/10.1089/hum.2019.090.

Maruggi, G., Zhang, C., Li, J., Ulmer, J.B., Yu, D., 2019. mRNA as a transformative technology for vaccine development to control infectious diseases. Mol. Ther. 27, 757–772. https://doi.org/10.1016/j.ymthe.2019.01.020.

Matoulkova, E., Michalova, E., Vojtesek, B., Hrstka, R., 2012. The role of the 3′ untranslated region in post-transcriptional regulation of protein expression in mammalian cells. RNA Biol. 9, 563–576. https://doi.org/10.4161/rna.20231.

Maude, S.L., Laetsch, T.W., Buechner, J., Rives, S., Boyer, M., Bittencourt, H., Bader, P., Verneris, M.R., Stefanski, H.E., Myers, G.D., Qayed, M., De Moerloose, B., Hiramatsu, H., Schlis, K., Davis, K.L., Martin, P.L., Nemecek, E.R., Yanik, G.A., Peters, C., Baruchel, A., Boissel, N., Mechinaud, F., Balduzzi, A., Krueger, J., June, C.H., Levine, B.L., Wood, P., Taran, T., Leung, M., Mueller, K.T., Zhang, Y., Sen, K., Lebwohl, D., Pulsipher, M.A., Grupp, S.A., 2018. Tisagenlecleucel in children and young adults with B-cell lymphoblastic leukemia. N. Engl. J. Med. 378, 439–448. https://doi.org/10.1056/NEJMoa1709866.

Maus, M.V., Haas, A.R., Beatty, G.L., Albelda, S.M., Levine, B.L., Liu, X., Zhao, Y., Kalos, M., June, C.H., 2013. T cells expressing chimeric antigen receptors can cause anaphylaxis in humans. Cancer Immunol. Res. 1, 26–31. https://doi.org/10.1158/2326-6066.CIR-13-0006.

Mayr, C., 2017. Regulation by 3′-untranslated regions. Annu. Rev. Genet. 51, 171–194. https://doi.org/10.1146/annurev-genet-120116-024704.
McCarthy, C., Carrea, A., Diambra, L., 2017. Bicodon bias can determine the role of synonymous SNPs in human diseases. BMC Genomics 18, 1–11. https://doi.org/10.1186/s12864-017-3609-6.
McCracken, S., Fong, N., Rosonina, E., Yankulov, K., Brothers, G., Siderovski, D., Hessel, A., Foster, S., Shuman, S., Bentley, D.L., 1997. 5'-Capping enzymes are targeted to pre-mRNA by binding to the phosphorylated carboxy-terminal domain of RNA polymerase II. Genes Dev. 11, 3306–3318. https://doi.org/10.1101/gad.11.24.3306.
McCullough, K.C., Bassi, I., Milona, P., Suter, R., Thomann-Harwood, L., Englezou, P., Démoulins, T., Ruggli, N., 2014. Self-replicating replicon-RNA delivery to dendritic cells by chitosan-nanoparticles for translation in vitro and in vivo. Mol. Ther. Nucleic Acids 3, e173. https://doi.org/10.1038/mtna.2014.24.
McKay, P.F., Hu, K., Blakney, A.K., Samnuan, K., Brown, J.C., Penn, R., Zhou, J., Bouton, C.R., Rogers, P., Polra, K., Lin, P.J.C., Barbosa, C., Tam, Y.K., Barclay, W.S., Shattock, R.J., 2020. Self-amplifying RNA SARS-CoV-2 lipid nanoparticle vaccine candidate induces high neutralizing antibody titers in mice. Nat. Commun. 11, 1–7. https://doi.org/10.1038/s41467-020-17409-9.
McKinlay, C.J., Vargas, J.R., Blake, T.R., Hardy, J.W., Kanada, M., Contag, C.H., Wender, P.A., Waymouth, R.M., 2017. Charge-altering releasable transporters (CARTs) for the delivery and release of mRNA in living animals. Proc. Natl. Acad. Sci. U. S. A. 114, E448–E456. https://doi.org/10.1073/pnas.1614193114.
McKinlay, C.J., Benner, N.L., Haabeth, O.A., Waymouth, R.M., Wender, P.A., 2018. Enhanced mRNA delivery into lymphocytes enabled by lipid-varied libraries of charge-altering releasable transporters. Proc. Natl. Acad. Sci. U. S. A. 115, E5859–E5866. https://doi.org/10.1073/pnas.1805358115.
Medicines & Healthcare products Regulatory Agency, 2020. Regulatory approval of Pfizer/BioNTech vaccine for COVID-19 [WWW Document]. Med. Healthc. Prod. Regul. Agency. URL. https://www.gov.uk/government/publications/regulatory-approval-of-pfizer-biontech-vaccine-for-covid-19. (accessed 1.2.22).
Meng, Z., O'Keeffe-Ahern, J., Lyu, J., Pierucci, L., Zhou, D., Wang, W., 2017. A new developing class of gene delivery: messenger RNA-based therapeutics. Biomater. Sci. 5, 2381–2392. https://doi.org/10.1039/c7bm00712d.
Meng, C., Chen, Z., Li, G., Welte, T., Shen, H., 2021. Nanoplatforms for mRNA therapeutics. Adv. Ther. 4, 1–23. https://doi.org/10.1002/adtp.202000099.
Meyer, K.D., Patil, D.P., Zhou, J., Zinoviev, A., Skabkin, M.A., Elemento, O., Pestova, T.V., Qian, S.B., Jaffrey, S.R., 2015. 5′ UTR m6A promotes cap-independent translation. Cell 163, 999–1010. https://doi.org/10.1016/j.cell.2015.10.012.
Mirón-Barroso, S., Domènech, E.B., Trigueros, S., 2021. Nanotechnology-based strategies to overcome current barriers in gene delivery. Int. J. Mol. Sci. 22, 8537. https://doi.org/10.3390/ijms22168537.
Miyata, K., Oba, M., Nakanishi, M., Fukushima, S., Yamasaki, Y., Koyama, H., Nishiyama, N., Kataoka, K., 2008. Polyplexes from poly(aspartamide) bearing 1,2-diaminoethane side chains induce pH-selective, endosomal membrane destabilization with amplified transfection and negligible cytotoxicity. J. Am. Chem. Soc. 130, 16287–16294. https://doi.org/10.1021/ja804561g.
Mockey, M., Gonçalves, C., Dupuy, F.P., Lemoine, F.M., Pichon, C., Midoux, P., 2006. mRNA transfection of dendritic cells: synergistic effect of ARCA mRNA capping with poly(A) chains in cis and in trans for a high protein expression level. Biochem. Biophys. Res. Commun. 340, 1062–1068. https://doi.org/10.1016/j.bbrc.2005.12.105.
Moderna Inc, 2019. Item 16. Form 10-K Summary [WWW Document]. UNITED STATES Secur. Exch. Comm. URL. https://www.sec.gov/Archives/edgar/data/1682852/000168285219000009/moderna10-k12312018.htm. (accessed 1.5.22).

Moderna Inc, 2020. Moderna Highlights Opportunity of mRNA Vaccines at its First Vaccines Day [WWW Document]. Mod. Inc. URL. https://investors.modernatx.com/news-releases/news-release-details/moderna-highlights-opportunity-mrna-vaccines-its-first-vaccines. (accessed 1.3.22).

Mohammad, R., 2021. Key considerations in formulation development for gene therapy. Drug Discov. Today 27 (1), 292–303. https://doi.org/10.1016/j.drudis.2021.08.013.

Mu, X., Greenwald, E., Ahmad, S., Hur, S., 2018. An origin of the immunogenicity of in vitro transcribed RNA. Nucleic Acids Res. 46, 5239–5249. https://doi.org/10.1093/nar/gky177.

Muik, A., Wallisch, A.K., Sänger, B., Swanson, K.A., Mühl, J., Chen, W., Cai, H., Maurus, D., Sarkar, R., Türeci, Ö., Dormitzer, P.R., Şahin, U., 2021. Neutralization of SARS-CoV-2 lineage B.1.1.7 pseudovirus by BNT162b2 vaccine–elicited human sera. Science 371, 1152–1153. https://doi.org/10.1126/science.abg6105.

Müller, R.H., Radtke, M., Wissing, S.A., 2002. Solid lipid nanoparticles (SLN) and nanostructured lipid carriers (NLC) in cosmetic and dermatological preparations. Adv. Drug Deliv. Rev. 54, 131–155. https://doi.org/10.1016/S0169-409X(02)00118-7.

Mulligan, M.J., Lyke, K.E., Kitchin, N., Absalon, J., Gurtman, A., Lockhart, S., Neuzil, K., Raabe, V., Bailey, R., Swanson, K.A., Li, P., Koury, K., Kalina, W., Cooper, D., Fontes-Garfias, C., Shi, P.-Y., Türeci, Ö., Tompkins, K.R., Walsh, E.E., Frenck, R., Falsey, A.R., Dormitzer, P.R., Gruber, W.C., Şahin, U., Jansen, K.U., 2020. Phase I/II study of COVID-19 RNA vaccine BNT162b1 in adults. Nature 586, 589–593. https://doi.org/10.1038/s41586-020-2639-4.

Muttach, F., Muthmann, N., Rentmeister, A., 2017. Synthetic mRNA capping. Beilstein J. Org. Chem. 13, 2819–2832. https://doi.org/10.3762/bjoc.13.274.

Nabhan, J.F., Wood, K.M., Rao, V.P., Morin, J., Bhamidipaty, S., LaBranche, T.P., Gooch, R.L., Bozal, F., Bulawa, C.E., Guild, B.C., 2016. Intrathecal delivery of frataxin mRNA encapsulated in lipid nanoparticles to dorsal root ganglia as a potential therapeutic for Friedreich's ataxia. Sci. Rep. 6, 20019. https://doi.org/10.1038/srep20019.

Neelapu, S.S., Locke, F.L., Bartlett, N.L., Lekakis, L.J., Miklos, D.B., Jacobson, C.A., Braunschweig, I., Oluwole, O.O., Siddiqi, T., Lin, Y., Timmerman, J.M., Stiff, P.J., Friedberg, J.W., Flinn, I.W., Goy, A., Hill, B.T., Smith, M.R., Deol, A., Farooq, U., McSweeney, P., Munoz, J., Avivi, I., Castro, J.E., Westin, J.R., Chavez, J.C., Ghobadi, A., Komanduri, K.V., Levy, R., Jacobsen, E.D., Witzig, T.E., Reagan, P., Bot, A., Rossi, J., Navale, L., Jiang, Y., Aycock, J., Elias, M., Chang, D., Wiezorek, J., Go, W.Y., 2017. Axicabtagene ciloleucel CAR T-cell therapy in refractory large B-cell lymphoma. N. Engl. J. Med. 377, 2531–2544. https://doi.org/10.1056/NEJMoa1707447.

Oh, S., Kessler, J.A., 2018. Design, assembly, production, and transfection of synthetic modified mRNA. Methods 133, 29–43. https://doi.org/10.1016/j.ymeth.2017.10.008.

Ohura, T., 1988. Methylmalonic acidemia. Tanpakushitsu Kakusan Koso 33, 579–584.

Papachristofilou, A., Hipp, M.M., Klinkhardt, U., Früh, M., Sebastian, M., Weiss, C., Pless, M., Cathomas, R., Hilbe, W., Pall, G., Wehler, T., Alt, J., Bischoff, H., Geißler, M., Griesinger, F., Kallen, K.-J., Fotin-Mleczek, M., Schröder, A., Scheel, B., Muth, A., Seibel, T., Stosnach, C., Doener, F., Hong, H.S., Koch, S.D., Gnad-Vogt, U., Zippelius, A., 2019. Phase Ib evaluation of a self-adjuvanted protamine formulated mRNA-based active cancer immunotherapy, BI1361849 (CV9202), combined with local radiation treatment in patients with stage IV non-small cell lung cancer. J. Immunother. Cancer 7, 38. https://doi.org/10.1186/s40425-019-0520-5.

Pardi, N., Hogan, M.J., Pelc, R.S., Muramatsu, H., Andersen, H., DeMaso, C.R., Dowd, K.A., Sutherland, L.L., Scearce, R.M., Parks, R., Wagner, W., Granados, A., Greenhouse, J., Walker, M., Willis, E., Yu, J.S., McGee, C.E., Sempowski, G.D.,

Mui, B.L., Tam, Y.K., Huang, Y.J., Vanlandingham, D., Holmes, V.M., Balachandran, H., Sahu, S., Lifton, M., Higgs, S., Hensley, S.E., Madden, T.D., Hope, M.J., Karikó, K., Santra, S., Graham, B.S., Lewis, M.G., Pierson, T.C., Haynes, B.F., Weissman, D., 2017. Zika virus protection by a single low-dose nucleoside-modified mRNA vaccination. Nature 543, 248–251. https://doi.org/10.1038/nature21428.

Pardi, N., Hogan, M.J., Porter, F.W., Weissman, D., 2018. mRNA vaccines—a new era in vaccinology. Nat. Rev. Drug Discov. 17 (4), 261–279. https://doi.org/10.1038/nrd.2017.243.

Park, J.S., Suryaprakash, S., Lao, Y.H., Leong, K.W., 2015. Engineering mesenchymal stem cells for regenerative medicine and drug delivery. Methods 84, 3–16. https://doi.org/10.1016/j.ymeth.2015.03.002.

Park, K.S., Sun, X., Aikins, M.E., Moon, J.J., 2021. Non-viral COVID-19 vaccine delivery systems. Adv. Drug Deliv. Rev. 169, 137–151. https://doi.org/10.1016/j.addr.2020.12.008.

Parr, C.J.C., Wada, S., Kotake, K., Kameda, S., Matsuura, S., Sakashita, S., Park, S., Sugiyama, H., Kuang, Y., Saito, H., 2020. N1-Methylpseudouridine substitution enhances the performance of synthetic mRNA switches in cells. Nucleic Acids Res. 48, e35. https://doi.org/10.1093/nar/gkaa070.

Patel, A.K., Kaczmarek, J.C., Bose, S., Kauffman, K.J., Mir, F., Heartlein, M.W., DeRosa, F., Langer, R., Anderson, D.G., 2019. Inhaled nanoformulated mRNA polyplexes for protein production in lung epithelium. Adv. Mater. 31, 1805116. https://doi.org/10.1002/adma.201805116.

Patel, S., Athirasala, A., Menezes, P.P., Ashwanikumar, N., Zou, T., Sahay, G., Bertassoni, L.E., 2019a. Messenger RNA delivery for tissue engineering and regenerative medicine applications. Tissue Eng. Part A 25, 91–112. https://doi.org/10.1089/ten.tea.2017.0444.

Patel, S., Ryals, R.C., Weller, K.K., Pennesi, M.E., Sahay, G., 2019b. Lipid nanoparticles for delivery of messenger RNA to the back of the eye. J. Control. Release 303, 91–100. https://doi.org/10.1016/j.jconrel.2019.04.015.

Patel, M.R., Bauer, T.M., Jimeno, A., Wang, D., LoRusso, P., Do, K.T., Stemmer, S.M., Maurice-Dror, C., Geva, R., Zacharek, S., Laino, A.S., Sun, J., Frederick, J., Zhou, H., Randolph, W., Cohen, P.S., Meehan, R.S., Sullivan, R.J., 2020. A phase I study of mRNA-2752, a lipid nanoparticle encapsulating mRNAs encoding human OX40L, IL-23, and IL-36γ, for intratumoral (iTu) injection alone and in combination with durvalumab. J. Clin. Oncol. 38, 3092. https://doi.org/10.1200/jco.2020.38.15_suppl.3092.

Patel, P., Ibrahim, N.M., Cheng, K., 2021. The importance of apparent pKa in the development of nanoparticles encapsulating siRNA and mRNA. Trends Pharmacol. Sci. 42, 448–460. https://doi.org/10.1016/j.tips.2021.03.002.

Pelletier, J., Schmeing, T.M., Sonenberg, N., 2021. The multifaceted eukaryotic cap structure. Wiley Interdiscip. Rev. RNA 12, 1–24. https://doi.org/10.1002/wrna.1636.

Peng, J., Schoenberg, D.R., 2005. mRNA with a <20-nt poly(A) tail imparted by the poly(A)-limiting element is translated as efficiently in vivo as long poly(A) mRNA. RNA 11, 1131–1140. https://doi.org/10.1261/rna.2470905.

Peters, K.B., Archer, G.E., Norberg, P., Xie, W., Threatt, S., Lipp, E.S., Herndon, J.E., Healy, P., Congdon, K., Sanchez-Perez, L., Friedman, H.S., Desjardins, A., Vlahovic, G., Sampson, J.H., 2019. Safety of nivolumab in combination with dendritic cell vaccines in recurrent high-grade glioma. J. Clin. Oncol. 37, e13526. https://doi.org/10.1200/jco.2019.37.15_suppl.e13526.

Petsch, B., Schnee, M., Vogel, A.B., Lange, E., Hoffmann, B., Voss, D., Schlake, T., Thess, A., Kallen, K.J., Stitz, L., Kramps, T., 2012. Protective efficacy of in vitro synthesized, specific mRNA vaccines against influenza A virus infection. Nat. Biotechnol. 30, 1210–1216. https://doi.org/10.1038/nbt.2436.

Pfizer Inc, BioNTech SE, 2021. Pfizer and BioNTech Provide Update on Omicron Variant [WWW Document]. Pfizer Inc. URL. https://www.pfizer.com/news/press-release/press-release-detail/pfizer-and-biontech-provide-update-omicron-variant (accessed 12.8.21).

Platanias, L.C., 2005. Mechanisms of type-I- and type-II-interferon-mediated signalling. Nat. Rev. Immunol. 5, 375–386. https://doi.org/10.1038/nri1604.

Polack, F.P., Thomas, S.J., Kitchin, N., Absalon, J., Gurtman, A., Lockhart, S., Perez, J.L., Pérez Marc, G., Moreira, E.D., Zerbini, C., Bailey, R., Swanson, K.A., Roychoudhury, S., Koury, K., Li, P., Kalina, W.V., Cooper, D., Frenck, R.W., Hammitt, L.L., Türeci, Ö., Nell, H., Schaefer, A., Ünal, S., Tresnan, D.B., Mather, S., Dormitzer, P.R., Şahin, U., Jansen, K.U., Gruber, W.C., 2020. Safety and efficacy of the BNT162b2 mRNA covid-19 vaccine. N. Engl. J. Med. 383, 2603–2615. https://doi.org/10.1056/NEJMoa2034577.

Presnyak, V., Alhusaini, N., Chen, Y.H., Martin, S., Morris, N., Kline, N., Olson, S., Weinberg, D., Baker, K.E., Graveley, B.R., Coller, J., 2015. Codon optimality is a major determinant of mRNA stability. Cell 160, 1111–1124. https://doi.org/10.1016/j.cell.2015.02.029.

Prieve, M.G., Harvie, P., Monahan, S.D., Roy, D., Li, A.G., Blevins, T.L., Paschal, A.E., Waldheim, M., Bell, E.C., Galperin, A., Ella-Menye, J.R., Houston, M.E., 2018. Targeted mRNA therapy for ornithine transcarbamylase deficiency. Mol. Ther. 26, 801–813. https://doi.org/10.1016/j.ymthe.2017.12.024.

Ramamoorth, M., Narvekar, A., 2015. Non viral vectors in gene therapy—an overview. J. Clin. Diagn. Res. 9, GE01–GE06. https://doi.org/10.7860/JCDR/2015/10443.5394.

Ramanathan, A., Robb, G.B., Chan, S.H., 2016. mRNA capping: biological functions and applications. Nucleic Acids Res. 44, 7511–7526. https://doi.org/10.1093/nar/gkw551.

Rittig, S.M., Haentschel, M., Weimer, K.J., Heine, A., Muller, M.R., Brugger, W., Horger, M.S., Maksimovic, O., Stenzl, A., Hoerr, I., Rammensee, H.G., Holderried, T.A.W., Kanz, L., Pascolo, S., Brossart, P., 2011. Intradermal vaccinations with RNA coding for TAA generate CD8 and CD4 immune responses and induce clinical benefit in vaccinated patients. Mol. Ther. 19, 990–999. https://doi.org/10.1038/mt.2010.289.

Robinson, E., MacDonald, K.D., Slaughter, K., McKinney, M., Patel, S., Sun, C., Sahay, G., 2018. Lipid nanoparticle-delivered chemically modified mRNA restores chloride secretion in cystic fibrosis. Mol. Ther. 26, 2034–2046. https://doi.org/10.1016/j.ymthe.2018.05.014.

Rodríguez-Castejón, J., Alarcia-Lacalle, A., Gómez-Aguado, I., Vicente-Pascual, M., Solinís Aspiazu, M.Á., del Pozo-Rodríguez, A., Rodríguez-Gascón, A., 2021. α-Galactosidase A augmentation by non-viral gene therapy: evaluation in fabry disease mice. Pharmaceutics 13, 771. https://doi.org/10.3390/pharmaceutics13060771.

Rodríguez-Gascón, A., del Pozo-Rodríguez, A., Solinís, M.Á., 2014. Development of nucleic acid vaccines: use of self-amplifying RNA in lipid nanoparticles. Int. J. Nanomedicine 9, 1833. https://doi.org/10.2147/IJN.S39810.

Rodríguez-Gascón, A., Pozo-Rodríguez, A., Isla, A., Solinís, M.A., 2016. Gene therapy in the cornea. In: ELS. Wiley, pp. 1–9. https://doi.org/10.1002/9780470015902.a0024274.

Ross, J., Sullivan, T.D., 1985. Half-lives of beta and gamma globin messenger RNAs and of protein synthetic capacity in cultured human reticulocytes. Blood 66, 1149–1154. https://doi.org/10.1182/blood.v66.5.1149.1149.

Roth, N., Schön, J., Hoffmann, D., Thran, M., Thess, A., Mueller, S.O., Petsch, B., Rauch, S., 2021. CV2CoV, an enhanced mRNA-based SARS-CoV-2 vaccine candidate, supports higher protein expression and improved immunogenicity in rats. bioRxiv. 2021.05.13.443734. https://doi.org/10.1101/2021.05.13.443734.

Ruiz De Garibay, A.P., Solinís, M.A., Del Pozo-Rodríguez, A., Apaolaza, P.S., Shen, J.S., Rodríguez-Gascón, A., 2015. Solid lipid nanoparticles as non-viral vectors for gene transfection in a cell model of fabry disease. J. Biomed. Nanotechnol. 11, 500–511. https://doi.org/10.1166/jbn.2015.1968.

Ryals, R.C., Patel, S., Acosta, C., McKinney, M., Pennesi, M.E., Sahay, G., 2020. The effects of PEGylation on LNP based mRNA delivery to the eye. PLoS One 15, 1–17. https://doi.org/10.1371/journal.pone.0241006.

Rybakova, Y., Kowalski, P.S., Huang, Y., Gonzalez, J.T., Heartlein, M.W., DeRosa, F., Delcassian, D., Anderson, D.G., 2019. mRNA delivery for therapeutic anti-HER2 antibody expression in vivo. Mol. Ther. 27, 1415–1423. https://doi.org/10.1016/j.ymthe.2019.05.012.

Rydzik, A.M., Kulis, M., Lukaszewicz, M., Kowalska, J., Zuberek, J., Darzynkiewicz, Z.M., Darzynkiewicz, E., Jemielity, J., 2012. Synthesis and properties of mRNA cap analogs containing imidodiphosphate moiety—fairly mimicking natural cap structure, yet resistant to enzymatic hydrolysis. Bioorg. Med. Chem. 20, 1699–1710. https://doi.org/10.1016/j.bmc.2012.01.013.

Sahin, U., Karikó, K., Türeci, Ö., 2014. MRNA-based therapeutics-developing a new class of drugs. Nat. Rev. Drug Discov. 13, 759–780. https://doi.org/10.1038/nrd4278.

Sahin, U., Derhovanessian, E., Miller, M., Kloke, B.P., Simon, P., Löwer, M., Bukur, V., Tadmor, A.D., Luxemburger, U., Schrörs, B., Omokoko, T., Vormehr, M., Albrecht, C., Paruzynski, A., Kuhn, A.N., Buck, J., Heesch, S., Schreeb, K.H., Müller, F., Ortseifer, I., Vogler, I., Godehardt, E., Attig, S., Rae, R., Breitkreuz, A., Tolliver, C., Suchan, M., Martic, G., Hohberger, A., Sorn, P., Diekmann, J., Ciesla, J., Waksmann, O., Brück, A.K., Witt, M., Zillgen, M., Rothermel, A., Kasemann, B., Langer, D., Bolte, S., Diken, M., Kreiter, S., Nemecek, R., Gebhardt, C., Grabbe, S., Höller, C., Utikal, J., Huber, C., Loquai, C., Türeci, Ö., 2017. Personalized RNA mutanome vaccines mobilize poly-specific therapeutic immunity against cancer. Nature 547, 222–226. https://doi.org/10.1038/nature23003.

Sahin, U., Oehm, P., Derhovanessian, E., Jabulowsky, R.A., Vormehr, M., Gold, M., Maurus, D., Schwarck-Kokarakis, D., Kuhn, A.N., Omokoko, T., Kranz, L.M., Diken, M., Kreiter, S., Haas, H., Attig, S., Rae, R., Cuk, K., Kemmer-Brück, A., Breitkreuz, A., Tolliver, C., Caspar, J., Quinkhardt, J., Hebich, L., Stein, M., Hohberger, A., Vogler, I., Liebig, I., Renken, S., Sikorski, J., Leierer, M., Müller, V., Mitzel-Rink, H., Miederer, M., Huber, C., Grabbe, S., Utikal, J., Pinter, A., Kaufmann, R., Hassel, J.C., Loquai, C., Türeci, Ö., 2020. An RNA vaccine drives immunity in checkpoint-inhibitor-treated melanoma. Nature 585, 107–112. https://doi.org/10.1038/s41586-020-2537-9.

Sahu, I., Haque, A.K.M.A., Weidensee, B., Weinmann, P., Kormann, M.S.D., 2019. Recent developments in mRNA-based protein supplementation therapy to target lung diseases. Mol. Ther. 27, 803–823. https://doi.org/10.1016/j.ymthe.2019.02.019.

Salomon, N., Vascotto, F., Selmi, A., Vormehr, M., Quinkhardt, J., Bukur, T., Schrörs, B., Löewer, M., Diken, M., Türeci, Ö., Sahin, U., Kreiter, S., 2020. A liposomal RNA vaccine inducing neoantigen-specific CD4+ T cells augments the antitumor activity of local radiotherapy in mice. Onco. Targets. Ther. 9, 1771925. https://doi.org/10.1080/2162402X.2020.1771925.

Scheel, B., Teufel, R., Probst, J., Carralot, J.P., Geginat, J., Radsak, M., Jarrossay, D., Wagner, H., Jung, G., Rammensee, H.G., Hoerr, I., Pascolo, S., 2005. Toll-like receptor-dependent activation of several human blood cell types by protamine-condensed mRNA. Eur. J. Immunol. 35, 1557–1566. https://doi.org/10.1002/eji.200425656.

Schlake, T., Thess, A., Fotin-Mleczek, M., Kallen, K.J., 2012. Developing mRNA-vaccine technologies. RNA Biol. 9, 1319–1330. https://doi.org/10.4161/rna.22269.

Schlake, T., Thess, A., Thran, M., Jordan, I., 2019. mRNA as novel technology for passive immunotherapy. Cell. Mol. Life Sci. 76, 301–328. https://doi.org/10.1007/s00018-018-2935-4.

Schlich, M., Palomba, R., Costabile, G., Mizrahy, S., Pannuzzo, M., Peer, D., Decuzzi, P., 2021. Cytosolic delivery of nucleic acids: The case of ionizable lipid nanoparticles. Bioeng. Transl. Med. 6, 1–16. https://doi.org/10.1002/btm2.10213.

Schnee, M., Vogel, A.B., Voss, D., Petsch, B., Baumhof, P., Kramps, T., Stitz, L., 2016. An mRNA vaccine encoding rabies virus glycoprotein induces protection against lethal infection in mice and correlates of protection in adult and newborn pigs. PLoS Negl. Trop. Dis. 10, e0004746. https://doi.org/10.1371/journal.pntd.0004746.

Schoenmaker, L., Witzigmann, D., Kulkarni, J.A., Verbeke, R., Kersten, G., Jiskoot, W., Crommelin, D.J.A., 2021. mRNA-lipid nanoparticle COVID-19 vaccines: structure and stability. Int. J. Pharm. 601, 120586. https://doi.org/10.1016/j.ijpharm.2021.120586.

Selby, L.I., Cortez-Jugo, C.M., Such, G.K., Johnston, A.P.R., 2017. Nanoescapology: progress toward understanding the endosomal escape of polymeric nanoparticles. Wiley Interdiscip. Rev. Nanomed. Nanobiotechnol. 9, e1452. https://doi.org/10.1002/wnan.1452.

Sharma, D., Arora, S., Singh, J., Layek, B., 2021. A review of the tortuous path of nonviral gene delivery and recent progress. Int. J. Biol. Macromol. 183, 2055–2073. https://doi.org/10.1016/j.ijbiomac.2021.05.192.

Shaw, C., Lee, H., Knightly, C., Kalidindi, S., Zaks, T., Smolenov, I., Panther, L., 2019a. 2754. Phase 1 trial of an mRNA-based combination vaccine against hMPV and PIV3. Open Forum Infect. Dis. 6, S970. https://doi.org/10.1093/ofid/ofz360.2431.

Shaw, C., Panther, L., August, A., Zaks, T., Smolenov, I., Bart, S., Watson, M., 2019b. Safety and immunogenicity of a mRNA-based chikungunya vaccine in a phase 1 dose-ranging trial. Int. J. Infect. Dis. 79, 17. https://doi.org/10.1016/j.ijid.2018.11.058.

Te Chang, C., Bercovich, N., Loh, B., Jonas, S., Izaurralde, E., 2014. The activation of the decapping enzyme DCP2 by DCP1 occurs on the EDC4 scaffold and involves a conserved loop in DCP1. Nucleic Acids Res. 42, 5217–5233. https://doi.org/10.1093/nar/gku129.

Shim, G., Kim, D., Park, G.T., Jin, H., Suh, S.K., Oh, Y.K., 2017. Therapeutic gene editing: delivery and regulatory perspectives. Acta Pharmacol. Sin. 38 (6), 738–753. https://doi.org/10.1038/aps.2017.2.

Shimasaki, N., Fujisaki, H., Cho, D., Masselli, M., Lockey, T., Eldridge, P., Leung, W., Campana, D., 2012. A clinically adaptable method to enhance the cytotoxicity of natural killer cells against B-cell malignancies. Cytotherapy 14, 830–840. https://doi.org/10.3109/14653249.2012.671519.

Shuai, Q., Zhu, F., Zhao, M., Yan, Y., 2021. mRNA delivery via non-viral carriers for biomedical applications. Int. J. Pharm. 607, 121020. https://doi.org/10.1016/j.ijpharm.2021.121020.

Simon, B., Wiesinger, M., März, J., Wistuba-Hamprecht, K., Weide, B., Schuler-Thurner, B., Schuler, G., Dörrie, J., Uslu, U., 2018. The generation of CAR-transfected natural killer T cells for the immunotherapy of melanoma. Int. J. Mol. Sci. 19, 2365. https://doi.org/10.3390/ijms19082365.

Skog, J., Würdinger, T., van Rijn, S., Meijer, D.H., Gainche, L., Curry, W.T., Carter, B.S., Krichevsky, A.M., Breakefield, X.O., 2008. Glioblastoma microvesicles transport RNA and proteins that promote tumour growth and provide diagnostic biomarkers. Nat. Cell Biol. 10, 1470–1476. https://doi.org/10.1038/ncb1800.

Soliní́s, M.Á., del Pozo-Rodríguez, A., Apaolaza, P.S., Rodríguez-Gascón, A., 2015. Treatment of ocular disorders by gene therapy. Eur. J. Pharm. Biopharm. 95, 331–342. https://doi.org/10.1016/j.ejpb.2014.12.022.

Steinle, H., Behring, A., Schlensak, C., Wendel, H.P., Avci-Adali, M., 2017. Application of in vitro transcribed messenger RNA for cellular engineering and reprogramming: progress and challenges. Stem Cells 35, 68–79. https://doi.org/10.1002/stem.2402.

Strenkowska, M., Kowalska, J., Lukaszewicz, M., Zuberek, J., Su, W., Rhoads, R.E., Darzynkiewicz, E., Jemielity, J., 2010. Towards mRNA with superior translational activity: synthesis and properties of ARCA tetraphosphates with single phosphorothioate modifications. New J. Chem. 34, 993–1007. https://doi.org/10.1039/b9nj00644c.

Svitkin, Y.V., Cheng, Y.M., Chakraborty, T., Presnyak, V., John, M., Sonenberg, N., 2017. N1-methyl-pseudouridine in mRNA enhances translation through eIF2α-dependent and independent mechanisms by increasing ribosome density. Nucleic Acids Res. 45, 6023–6036. https://doi.org/10.1093/nar/gkx135.

Svoboda, J., Rheingold, S.R., Gill, S.I., Grupp, S.A., Lacey, S.F., Kulikovskaya, I., Suhoski, M.M., Joseph Melenhorst, J., Loudon, B., Mato, A.R., Nasta, S.D., Landsburg, D.J., Youngman, M.R., Levine, B.L., Porter, D.L., June, C.H., Schuster, S.J., 2018. Nonviral RNA chimeric antigen receptor–modified T cells in patients with Hodgkin lymphoma. Blood 132, 1022–1026. https://doi.org/10.1182/blood-2018-03-837609.

Tang, T., Bidon, M., Jaimes, J.A., Whittaker, G.R., Daniel, S., 2020. Coronavirus membrane fusion mechanism offers a potential target for antiviral development. Antiviral Res. 178, 104792. https://doi.org/10.1016/j.antiviral.2020.104792.

Tcherepanova, I.Y., Adams, M.D., Feng, X., Hinohara, A., Horvatinovich, J., Calderhead, D., Healey, D., Nicolette, C.A., 2008. Ectopic expression of a truncated CD40L protein from synthetic post-transcriptionally capped RNA in dendritic cells induces high levels of IL-12 secretion. BMC Mol. Biol. 9, 90. https://doi.org/10.1186/1471-2199-9-90.

Tchou, J., Zhao, Y., Levine, B.L., Zhang, P.J., Davis, M.M., Melenhorst, J.J., Kulikovskaya, I., Brennan, A.L., Liu, X., Lacey, S.F., Posey, A.D., Williams, A.D., So, A., Conejo-Garcia, J.R., Plesa, G., Young, R.M., McGettigan, S., Campbell, J., Pierce, R.H., Matro, J.M., DeMichele, A.M., Clark, A.S., Cooper, L.J., Schuchter, L.M., Vonderheide, R.H., June, C.H., 2017. Safety and efficacy of intratumoral injections of chimeric antigen receptor (CAR) T cells in metastatic breast cancer. Cancer Immunol. Res. 5, 1152–1161. https://doi.org/10.1158/2326-6066.CIR-17-0189.

Thorne, B., Takeya, R., Vitelli, F., Swanson, X., 2018. Gene therapy. In: Kiss, B., Gottschalk, U., Pohlscheidt, M. (Eds.), New Bioprocessing Strategies: Development and Manufacturing of Recombinant Antibodies and Proteins. Springer International Publishing, Cham, pp. 351–399. https://doi.org/10.1007/10_2016_53.

Thran, M., Mukherjee, J., Pönisch, M., Fiedler, K., Thess, A., Mui, B.L., Hope, M.J., Tam, Y.K., Horscroft, N., Heidenreich, R., Fotin-Mleczek, M., Shoemaker, C.B., Schlake, T., 2017. mRNA mediates passive vaccination against infectious agents, toxins, and tumors. EMBO Mol. Med. 9, 1434–1447. https://doi.org/10.15252/emmm.201707678.

Torrecilla, J., del Pozo-Rodríguez, A., Vicente-Pascual, M., Solinís, M.Á., Rodríguez-Gascón, A., 2018. Targeting corneal inflammation by gene therapy: Emerging strategies for keratitis. Exp. Eye Res. 176, 130–140. https://doi.org/10.1016/j.exer.2018.07.006.
Translate Bio Inc, 2021. Results from Second Interim Data Analysis from Ongoing Phase 1/2 Clinical Trial of MRT5005 in Patients with Cystic Fibrosis (CF) was Generally Consistent with That Previously Reported for the 8, 16 and 24 mg Dose Groups. [WWW Document]. Transl. Bio, Inc. URL. https://investors.translate.bio/node/7886. (accessed 1.3.22).
Translate Bio Inc, 2019. Translate Bio Announces Pipeline Program Update [WWW Document]. Transl. Bio, Inc. URL. https://investors.translate.bio/node/7121/. (accessed 1.3.22).
Trepotec, Z., Lichtenegger, E., Plank, C., Aneja, M.K., Rudolph, C., 2019. Delivery of mRNA therapeutics for the treatment of hepatic diseases. Mol. Ther. 27 (4), 794–802. https://doi.org/10.1016/j.ymthe.2018.12.012.
TriLink BioTechnologies, 2020. CleanCap Technology: Leading the Way in mRNATM [WWW Document]. TriLink Biotechnol. URL. https://www.trilinkbiotech.com/cleancap. (accessed 12.30.21).
Trucillo, P., Campardelli, R., 2019. Production of solid lipid nanoparticles with a supercritical fluid assisted process. J. Supercrit. Fluids 143, 16–23. https://doi.org/10.1016/j.supflu.2018.08.001.
U.S. Food and Drug Administration, 2021. COVID-19 Vaccines [WWW Document]. U.S. Food Drug Adm. URL. https://www.fda.gov/emergency-preparedness-and-response/coronavirus-disease-2019-covid-19/covid-19-vaccines (accessed 12.30.21).
Uchida, H., Miyata, K., Oba, M., Ishii, T., Suma, T., Itaka, K., Nishiyama, N., Kataoka, K., 2011. Odd–even effect of repeating aminoethylene units in the side chain of N-substituted polyaspartamides on gene transfection profiles. J. Am. Chem. Soc. 133, 15524–15532. https://doi.org/10.1021/ja204466y.
Udhayakumar, V.K., De Beuckelaer, A., McCaffrey, J., McCrudden, C.M., Kirschman, J.L., Vanover, D., Van Hoecke, L., Roose, K., Deswarte, K., De Geest, B.G., Lienenklaus, S., Santangelo, P.J., Grooten, J., McCarthy, H.O., De Koker, S., 2017. Arginine-rich peptide-based mRNA nanocomplexes efficiently instigate cytotoxic T cell immunity dependent on the amphipathic organization of the peptide. Adv. Healthc. Mater. 6, 1–13. https://doi.org/10.1002/adhm.201601412.
Ulkoski, D., Bak, A., Wilson, J.T., Krishnamurthy, V.R., 2019. Recent advances in polymeric materials for the delivery of RNA therapeutics. Expert Opin. Drug Deliv. 16, 1149–1167. https://doi.org/10.1080/17425247.2019.1663822.
US Food & Drug Administration, 2021. FDA Approves First COVID-19 Vaccine [WWW Document]. US Food Drug Adm. URL. https://www.fda.gov/news-events/press-announcements/fda-approves-first-covid-19-vaccine (accessed 1.2.22).
Vaidyanathan, S., Azizian, K.T., Haque, A.K.M.A., Henderson, J.M., Hendel, A., Shore, S., Antony, J.S., Hogrefe, R.I., Kormann, M.S.D., Porteus, M.H., McCaffrey, A.P., 2018. Uridine depletion and chemical modification increase Cas9 mRNA activity and reduce immunogenicity without HPLC purification. Mol. Ther. Nucleic Acids 12, 530–542. https://doi.org/10.1016/j.omtn.2018.06.010.
Valadi, H., Ekström, K., Bossios, A., Sjöstrand, M., Lee, J.J., Lötvall, J.O., 2007. Exosome-mediated transfer of mRNAs and microRNAs is a novel mechanism of genetic exchange between cells. Nat. Cell Biol. 9, 654–659. https://doi.org/10.1038/ncb1596.
van den Berg, R.A., De Mot, L., Leroux-Roels, G., Bechtold, V., Clement, F., Coccia, M., Jongert, E., Evans, T.G., Gillard, P., van der Most, R.G., 2018. Adjuvant-associated peripheral blood mRNA profiles and kinetics induced by the adjuvanted recombinant protein candidate tuberculosis vaccine M72/AS01 in Bacillus Calmette-Guérin-vaccinated adults. Front. Immunol. 9, 1–15. https://doi.org/10.3389/fimmu.2018.00564.

Van Hoecke, L., Roose, K., 2019. How mRNA therapeutics are entering the monoclonal antibody field. J. Transl. Med. 17 (1), 54. https://doi.org/10.1186/s12967-019-1804-8.

Van Lint, S., Renmans, D., Broos, K., Goethals, L., Maenhout, S., Benteyn, D., Goyvaerts, C., Du Four, S., Van Der Jeught, K., Bialkowski, L., Flamand, V., Heirman, C., Thielemans, K., Breckpot, K., 2016. Intratumoral delivery of TriMix mRNA results in T-cell activation by cross-presenting dendritic cells. Cancer Immunol. Res. 4, 146–156. https://doi.org/10.1158/2326-6066.CIR-15-0163.

Varkouhi, A.K., Scholte, M., Storm, G., Haisma, H.J., 2011. Endosomal escape pathways for delivery of biologicals. J. Control. Release 151, 220–228. https://doi.org/10.1016/j.jconrel.2010.11.004.

Vicente-Pascual, M., Albano, A., Solinís, M., Serpe, L., Rodríguez-Gascón, A., Foglietta, F., Muntoni, E., Torrecilla, J., Del Pozo-Rodríguez, A., Battaglia, L., 2018. Gene delivery in the cornea: in vitro & ex vivo evaluation of solid lipid nanoparticle-based vectors. Nanomedicine 13, 1847–1864. https://doi.org/10.2217/nnm-2018-0112.

Vicente-Pascual, M., Gómez-Aguado, I., Rodríguez-Castejón, J., Rodríguez-Gascón, A., Muntoni, E., Battaglia, L., Del Pozo-Rodríguez, A., Aspiazu, M.Á.S., 2020. Topical administration of sln-based gene therapy for the treatment of corneal inflammation by de novo il-10 production. Pharmaceutics 12, 1–23. https://doi.org/10.3390/pharmaceutics12060584.

Vivinus, S., Baulande, S., Van Zanten, M., Campbell, F., Topley, P., Ellis, J.H., Dessen, P., 2001. An element within the 5′ untranslated region of human Hsp70 mRNA which acts as a general enhancer of mRNA translation. Eur. J. Biochem. 268, 1908–1917.

Vlahovic, G., Archer, G.E., Reap, E., Desjardins, A., Peters, K.B., Randazzo, D., Healy, P., Herndon, J.E., Friedman, A.H., Friedman, H.S., Bigner, D.D., Sampson, J.H., 2016. Phase I trial of combination of antitumor immunotherapy targeted against cytomegalovirus (CMV) plus regulatory T-cell inhibition in patients with newly-diagnosed glioblastoma multiforme (GBM). J. Clin. Oncol. 34, e13518. https://doi.org/10.1200/jco.2016.34.15_suppl.e13518.

Vogel, A.B., Lambert, L., Kinnear, E., Busse, D., Erbar, S., Reuter, K.C., Wicke, L., Perkovic, M., Beissert, T., Haas, H., Reece, S.T., Sahin, U., Tregoning, J.S., 2018. Self-amplifying RNA vaccines give equivalent protection against influenza to mRNA vaccines but at much lower doses. Mol. Ther. 26, 446–455. https://doi.org/10.1016/j.ymthe.2017.11.017.

Volloch, V., Housman, D., 1981. Stability of globin mRNA in terminally differentiating murine erythroleukemia cells. Cell 23, 509–514. https://doi.org/10.1016/0092-8674(81)90146-X.

Wadhwa, A., Aljabbari, A., Lokras, A., Foged, C., Thakur, A., 2020. Opportunities and challenges in the delivery of mrna-based vaccines. Pharmaceutics 12 (2), 102. https://doi.org/10.3390/pharmaceutics12020102.

Wahane, A., Waghmode, A., Kapphahn, A., Dhuri, K., Gupta, A., Bahal, R., 2020. Role of lipid-based and polymer-based non-viral vectors in nucleic acid delivery for next-generation gene therapy. Molecules 25 (12), 2866. https://doi.org/10.3390/molecules25122866.

Waldmann, T.A., 2018. Cytokines in cancer immunotherapy. Cold Spring Harb. Perspect. Biol. 10, a028472. https://doi.org/10.1101/cshperspect.a028472.

Walsh, E.E., Frenck, R.W., Falsey, A.R., Kitchin, N., Absalon, J., Gurtman, A., Lockhart, S., Neuzil, K., Mulligan, M.J., Bailey, R., Swanson, K.A., Li, P., Koury, K., Kalina, W., Cooper, D., Fontes-Garfias, C., Shi, P.-Y., Türeci, Ö., Tompkins, K.R., Lyke, K.E., Raabe, V., Dormitzer, P.R., Jansen, K.U., Şahin, U., Gruber, W.C., 2020. Safety and immunogenicity of two RNA-based covid-19 vaccine candidates. N. Engl. J. Med. 383, 2439–2450. https://doi.org/10.1056/nejmoa2027906.

Wang, Y., Su, H.H., Yang, Y., Hu, Y., Zhang, L., Blancafort, P., Huang, L., 2013. Systemic delivery of modified mRNA encoding herpes simplex virus 1 thymidine kinase for targeted cancer gene therapy. Mol. Ther. 21, 358–367. https://doi.org/10.1038/mt.2012.250.

Wang, Y., Zhang, Z., Luo, J., Han, X., Wei, Y., Wei, X., 2021. mRNA vaccine: a potential therapeutic strategy. Mol. Cancer 20, 1–23. https://doi.org/10.1186/s12943-021-01311-z.

Warminski, M., Kowalska, J., Buck, J., Zuberek, J., Lukaszewicz, M., Nicola, C., Kuhn, A.N., Sahin, U., Darzynkiewicz, E., Jemielity, J., 2013. The synthesis of isopropylidene mRNA cap analogs modified with phosphorothioate moiety and their evaluation as promoters of mRNA translation. Bioorg. Med. Chem. Lett. 23, 3753–3758. https://doi.org/10.1016/j.bmcl.2013.05.001.

Warminski, M., Kowalska, J., Nowak, E., Kubacka, D., Tibble, R., Kasprzyk, R., Sikorski, P.J., Gross, J.D., Nowotny, M., Jemielity, J., 2021. Structural insights into the interaction of clinically relevant phosphorothioate mRNA cap analogs with translation initiation factor 4E reveal stabilization via electrostatic thio-effect. ACS Chem. Biol. 16, 334–343. https://doi.org/10.1021/acschembio.0c00864.

Warren, L., Manos, P.D., Ahfeldt, T., Loh, Y.H., Li, H., Lau, F., Ebina, W., Mandal, P.K., Smith, Z.D., Meissner, A., Daley, G.Q., Brack, A.S., Collins, J.J., Cowan, C., Schlaeger, T.M., Rossi, D.J., 2010. Highly efficient reprogramming to pluripotency and directed differentiation of human cells with synthetic modified mRNA. Cell Stem Cell 7, 618–630. https://doi.org/10.1016/j.stem.2010.08.012.

Wecker, M., Gilbert, P., Russell, N., Hural, J., Allen, M., Pensiero, M., Chulay, J., Chiu, Y.L., Karim, S.S.A., Burke, D.S., 2012. Phase I safety and immunogenicity evaluations of an alphavirus replicon HIV-1 subtype C gag vaccine in healthy HIV-1-uninfected adults. Clin. Vaccine Immunol. 19, 1651–1660. https://doi.org/10.1128/CVI.00258-12.

Weide, B., Carralot, J.-P., Reese, A., Scheel, B., Eigentler, T.K., Hoerr, I., Rammensee, H.-G., Garbe, C., Pascolo, S., 2008. Results of the first phase I/II clinical vaccination trial with direct injection of mRNA. J. Immunother. 31, 180–188. https://doi.org/10.1097/CJI.0b013e31815ce501.

Weide, B., Pascolo, S., Scheel, B., Derhovanessian, E., Pflugfelder, A., Eigentler, T.K., Pawelec, G., Hoerr, I., Rammensee, H.-G., Garbe, C., 2009. Direct injection of protamine-protected mRNA: results of a phase 1/2 vaccination trial in metastatic melanoma patients. J. Immunother. 32, 498–507. https://doi.org/10.1097/CJI.0b013e3181a00068.

Weinstein-Marom, H., Levin, N., Pato, A., Shmuel, N., Sharabi-Nov, A., Peretz, T., Eisenberg, G., Lotem, M., Itzhaki, O., Besser, M.J., Gross, G., 2019. Combined expression of genetic adjuvants via mRNA electroporation exerts multiple immunostimulatory effects on antitumor T cells. J. Immunother. 42, 43–50. https://doi.org/10.1097/CJI.0000000000000252.

Weissman, D., 2014. mRNA transcript therapy. Expert Rev. Vaccines 14, 265–281. https://doi.org/10.1586/14760584.2015.973859.

Weng, Y., Li, C., Yang, T., Hu, B., Zhang, M., Guo, S., Xiao, H., Liang, X.J., Huang, Y., 2020. The challenge and prospect of mRNA therapeutics landscape. Biotechnol. Adv. 40, 107534. https://doi.org/10.1016/j.biotechadv.2020.107534.

Wesselhoeft, R.A., Kowalski, P.S., Anderson, D.G., 2018. Engineering circular RNA for potent and stable translation in eukaryotic cells. Nat. Commun. 9, 2629. https://doi.org/10.1038/s41467-018-05096-6.

Wesselhoeft, R.A., Kowalski, P.S., Parker-Hale, F.C., Huang, Y., Bisaria, N., Anderson, D.G., 2019. RNA circularization diminishes immunogenicity and can extend translation duration in vivo. Mol. Cell 74, 508–520.e4. https://doi.org/10.1016/j.molcel.2019.02.015.

Whitehead, K.A., Dorkin, J.R., Vegas, A.J., Chang, P.H., Veiseh, O., Matthews, J., Fenton, O.S., Zhang, Y., Olejnik, K.T., Yesilyurt, V., Chen, D., Barros, S., Klebanov, B., Novobrantseva, T., Langer, R., Anderson, D.G., 2014. Degradable lipid nanoparticles with predictable in vivo siRNA delivery activity. Nat. Commun. 5, 4277. https://doi.org/10.1038/ncomms5277.
Wilgenhof, S., Corthals, J., Heirman, C., Van Baren, N., Lucas, S., Kvistborg, P., Thielemans, K., Neyns, B., 2016. Phase II study of autologous monocyte-derived mRNA electroporated dendritic cells (TriMixDC-MEL) plus ipilimumab in patientswith pretreated advanced melanoma. J. Clin. Oncol. 34, 1330–1338. https://doi.org/10.1200/JCO.2015.63.4121.
Wolff, J.A., Malone, R.W., Williams, P., Chong, W., Acsadi, G., Jani, A., Felgner, P.L., 1990. Direct gene transfer into mouse muscle in vivo. Science). 247, 1465–1468. https://doi.org/10.1126/science.1690918.
Wu, G.Y., Wu, C.H., 1987. Receptor-mediated in vitro gene transformation by a soluble DNA carrier system. J. Biol. Chem. 262, 4429–4432. https://doi.org/10.1016/S0021-9258(18)61209-8.
Wu, T.C., Guarnieri, F.G., Staveley-O'Carroll, K.F., Viscidi, R.P., Levitsky, H.I., Hedrick, L., Cho, K.R., August, J.T., Pardoll, D.M., 1995. Engineering an intracellular pathway for major histocompatibility complex class II presentation of antigens. Proc. Natl. Acad. Sci. U. S. A. 92, 11671–11675. https://doi.org/10.1073/pnas.92.25.11671.
Xiao, L., Cen, D., Gan, H., Sun, Y., Huang, N., Xiong, H., Jin, Q., Su, L., Liu, X., Wang, K., Yan, G., Dong, T., Wu, S., Zhou, P., Zhang, J., Liang, W., Ren, J., Teng, Y., Chen, C., Xu, X.H., 2019. Adoptive transfer of NKG2D CAR mRNA-engineered natural killer cells in colorectal cancer patients. Mol. Ther. 27, 1114–1125. https://doi.org/10.1016/j.ymthe.2019.03.011.
Xie, X., Liu, Y., Liu, J., Zhang, X., Zou, J., Fontes-Garfias, C.R., Xia, H., Swanson, K.A., Cutler, M., Cooper, D., Menachery, V.D., Weaver, S.C., Dormitzer, P.R., Shi, P.Y., 2021. Neutralization of SARS-CoV-2 spike 69/70 deletion, E484K and N501Y variants by BNT162b2 vaccine-elicited sera. Nat. Med. 27, 620–621. https://doi.org/10.1038/s41591-021-01270-4.
Xinfu, Z., Weiyu, Z., Nguyen, N.G., Chengxiang, Z., Chunxi, Z., Jingyue, Y., Shi, D., Xucheng, H., Wenqing, L., Justin, J., Binbin, D., David, W.M., Robert, D., Aalok, S., Luis, B., Francine, G., Manmohan, S., Delai, C., Denise, E.S., Yizhou, D., 2021. Functionalized lipid-like nanoparticles for in vivo mRNA delivery and base editing. Sci. Adv. 6, eabc2315. https://doi.org/10.1126/sciadv.abc2315.
Xiong, Q., Lee, G.Y., Ding, J., Li, W., Shi, J., 2018. Biomedical applications of mRNA nanomedicine. Nano Res. 11, 5281–5309. https://doi.org/10.1007/s12274-018-2146-1.
Xu, S., Yang, K., Li, R., Zhang, L., 2020. Mrna vaccine era—mechanisms, drug platform and clinical prospection. Int. J. Mol. Sci. 21 (18), 6582. https://doi.org/10.3390/ijms21186582.
Yadav, N., Khatak, S., Singh Sara, U.V., 2013. Solid lipid nanoparticles—a review. Int. J. Appl. Pharm. 5, 8–18. https://doi.org/10.9790/3013-26103444.
Yang, Y., Wang, Z., 2019. IRES-mediated cap-independent translation, a path leading to hidden proteome. J. Mol. Cell Biol. 11, 911–919. https://doi.org/10.1093/jmcb/mjz091.
Yang, Y., Fan, X., Mao, M., Song, X., Wu, P., Zhang, Y., Jin, Y., Yang, Y., Chen, L.-L., Wang, Y., Wong, C.C., Xiao, X., Wang, Z., 2017. Extensive translation of circular RNAs driven by N 6-methyladenosine. Cell Res. 27, 626–641. https://doi.org/10.1038/cr.2017.31.
Yang, Z., Shi, J., Xie, J., Wang, Y., Sun, J., Liu, T., Zhao, Y., Zhao, X., Wang, X., Ma, Y., Malkoc, V., Chiang, C., Deng, W., Chen, Y., Fu, Y., Kwak, K.J., Fan, Y., Kang, C., Yin, C., Rhee, J., Bertani, P., Otero, J., Lu, W., Yun, K., Lee, A.S., Jiang, W., Teng, L.,

Kim, B.Y.S., Lee, L.J., 2020. Large-scale generation of functional mRNA-encapsulating exosomes via cellular nanoporation. Nat. Biomed. Eng. 4, 69–83. https://doi.org/10.1038/s41551-019-0485-1.

Yeom, J.H., Ryou, S.M., Won, M., Park, M., Bae, J., Lee, K., 2013. Inhibition of xenograft tumor growth by gold nanoparticle-DNA oligonucleotide conjugates-assisted delivery of BAX mRNA. PLoS One 8 (9), 1–10. https://doi.org/10.1371/journal.pone.0075369.

Zarghampoor, F., Azarpira, N., Khatami, S.R., Behzad-Behbahani, A., Foroughmand, A.M., 2019. Improved translation efficiency of therapeutic mRNA. Gene 707, 231–238. https://doi.org/10.1016/j.gene.2019.05.008.

Zhang, H., Leal, J., Soto, M.R., Smyth, H.D.C., Ghosh, D., 2020. Aerosolizable lipid nanoparticles for pulmonary delivery of mRNA through design of experiments. Pharmaceutics 12 (11), 1042. https://doi.org/10.3390/pharmaceutics12111042.

Zhang, H.X., Zhang, Y., Yin, H., 2019. Genome editing with mRNA encoding ZFN, TALEN, and Cas9. Mol. Ther. 27, 735–746. https://doi.org/10.1016/j.ymthe.2019.01.014.

Zhang, M., Sun, J., Li, M., Jin, X., 2020b. Modified mRNA-LNP vaccines confer protection against experimental DENV-2 infection in mice. Mol. Ther. Methods Clin. Dev. 18, 702–712. https://doi.org/10.1016/j.omtm.2020.07.013.

Zhang, R., Men, K., Zhang, X., Huang, R., Tian, Y., Zhou, B., Yu, C., Wang, Y., Ji, X., Hu, Q., Yang, L., 2018. Delivery of a modified mRNA encoding IL-22 binding protein (IL-22BP) for colon cancer gene therapy. J. Biomed. Nanotechnol. 14, 1239–1251. https://doi.org/10.1166/jbn.2018.2577.

Zhao, W., Zhang, C., Li, B., Zhang, X., Luo, X., Zeng, C., Li, W., Gao, M., Dong, Y., 2018. Lipid polymer hybrid nanomaterials for mRNA delivery. Cell. Mol. Bioeng. 11, 397–406. https://doi.org/10.1007/s12195-018-0536-9.

Zhao, Z., Zheng, L., Chen, W., Weng, W., Song, J., Ji, J., 2019. Delivery strategies of cancer immunotherapy: recent advances and future perspectives. J. Hematol. Oncol. 12 (1), 126. https://doi.org/10.1186/s13045-019-0817-3.

Zhong, Z., Mc Cafferty, S., Combes, F., Huysmans, H., De Temmerman, J., Gitsels, A., Vanrompay, D., Portela Catani, J., Sanders, N.N., 2018. mRNA therapeutics deliver a hopeful message. Nano Today 23, 16–39. https://doi.org/10.1016/j.nantod.2018.10.005.

Zhou, W.Z., Hoon, D.S.B., Huang, S.K.S., Fujii, S., Hashimoto, K., Morishita, R., Kaneda, Y., 1999. RNA melanoma vaccine: Induction of antitumor immunity by human glycoprotein 100 mRNA immunization. Hum. Gene Ther. 10, 2719–2724. https://doi.org/10.1089/10430349950016762.

Zhu, X., Tao, W., Liu, D., Wu, J., Guo, Z., Ji, X., Bharwani, Z., Zhao, L., Zhao, X., Farokhzad, O.C., Shi, J., 2017. Surface de-PEGylation controls nanoparticle-mediated siRNA delivery in vitro and in vivo. Theranostics 7, 1990–2002. https://doi.org/10.7150/thno.18136.

Zhu, X., Yin, L., Theisen, M., Zhuo, J., Siddiqui, S., Levy, B., Presnyak, V., Frassetto, A., Milton, J., Salerno, T., Benenato, K.E., Milano, J., Lynn, A., Sabnis, S., Burke, K., Besin, G., Lukacs, C.M., Guey, L.T., Finn, P.F., Martini, P.G.V., 2019. Systemic mRNA therapy for the treatment of fabry disease: preclinical studies in wild-type mice, fabry mouse model, and wild-type non-human primates. Am. J. Hum. Genet. 104, 625–637. https://doi.org/10.1016/j.ajhg.2019.02.003.

Zhuang, X., Qi, Y., Wang, M., Yu, N., Nan, F., Zhang, H., Tian, M., Li, C., Lu, H., Jin, N., 2020. MRNA vaccines encoding the HA protein of influenza a H1N1 virus delivered by cationic lipid nanoparticles induce protective immune responses in mice. Vaccine 8. https://doi.org/10.3390/vaccines8010123.

Ziemniak, M., Kowalska, J., Lukaszewicz, M., Zuberek, J., Wnek, K., Darzynkiewicz, E., Jemielity, J., 2015. Phosphate-modified analogues of m7GTP and m7Gppppm7G—synthesis and biochemical properties. Bioorg. Med. Chem. 23, 5369–5381. https://doi.org/10.1016/j.bmc.2015.07.052.

Ziller, A., Nogueira, S.S., Hühn, E., Funari, S.S., Brezesinski, G., Hartmann, H., Sahin, U., Haas, H., Langguth, P., 2018. Incorporation of mRNA in lamellar lipid matrices for parenteral administration. Mol. Pharm. 15, 642–651. https://doi.org/10.1021/acs.molpharmaceut.7b01022.

Zuckerman, J., McCoy, K., Schechter, M., Dorgan, D., Jain, M., MacDonald, K., Callison, C., Walker, S., Bodie, S., Barbier, A., Rowe, S., 2019. Safety and tolerability of a single dose of MRT5005, a nebulized CFTR mRNA therapeutic, in adult CF patients. Pediatr. Pulmonol. 54, 350.

Zytek, M., Kowalska, J., Lukaszewicz, M., Wojtczak, B.A., Zuberek, J., Ferenc-Mrozek, A., Darzynkiewicz, E., Niedzwiecka, A., Jemielity, J., 2014. Towards novel efficient and stable nuclear import signals: synthesis and properties of trimethylguanosine cap analogs modified within the 5′,5′-triphosphate bridge. Org. Biomol. Chem. 12, 9184–9199. https://doi.org/10.1039/c4ob01579g.

CHAPTER SEVEN

Advances in mRNA vaccines

Mengyun Li, Zining Wang, Chunyuan Xie, and Xiaojun Xia*
State Key Laboratory of Oncology in South China, Collaborative Innovation Center for Cancer Medicine, Sun Yat-sen University Cancer Center, Guangzhou, China
*Corresponding author: e-mail address: xiaxj@sysucc.org.cn

Contents

Abstract

mRNA vaccines have been increasingly recognized as a powerful vaccine platform since the FDA approval of two COVID-19 mRNA vaccines, which demonstrated outstanding prevention efficacy as well as great safety profile. Notably, nucleoside modification and lipid nanoparticle-facilitated delivery has greatly improved the immunogenicity, stability, and translation efficiency of mRNA molecule. Here we review the recent progress in mRNA vaccine development, including nucleoside modification, in vitro synthesis and product purification, and lipid nanoparticle vectors for in vivo delivery and efficient translation. We also briefly introduce the clinical application of mRNA vaccine in preventing infectious diseases and treating inflammatory diseases including cancer.

Messenger RNA (mRNA) was identified as the intermediate carrying genetic information form DNA to ribosomes for protein synthesis in 1961 (Brenner et al., 1961). After the in vitro transcription (IVT) method for RNA production was developed, scientists started to use mRNA to express desired protein in mouse (Krieg and Melton, 1984; Wolff et al., 1990). In theory, mRNA molecules inside the host cells would use the host

International Review of Cell and Molecular Biology, Volume 372
ISSN 1937-6448
https://doi.org/10.1016/bs.ircmb.2022.04.011
Copyright © 2022 Elsevier Inc.
All rights reserved.

translation machinery to produce protein and execute the desired physiological function. However, initially application of mRNA to express protein in vivo was hampered by the instability of mRNA in vivo and inherited inflammatory nature of mRNA with unmodified nucleotides. As the delivery technology quickly advanced in the recent years, and the development of nucleoside modification to avoid innate immune recognition, the applications of mRNA technology in biomedical field have been quickly expanding (Sahin et al., 2014). Vaccine is probably the most well-known field among the many applications, thanks to the excellent performance of Coronavirus Disease 2019 (COVID-19) mRNA vaccines recently approved by U.S. Food and Drug Administration (FDA).

Vaccine is nearly the most powerful weapon for human beings to battle bacteria- or virus-caused infectious diseases. Traditional vaccines are often made with live-attenuated or inactivated pathogens, or subunit antigens derived from pathogens. Traditional vaccines have been successful in preventing infectious diseases such as small pox, polio, etc., but the development and large-scale production could be time-consuming, which may not applicable for sudden outbreak of infectious diseases such as Severe Acute Respiratory Syndrome (SARS) and COVID-19. By contrast, mRNA can be quickly designed and synthesized, and the sequences can be modified according to the need for controlling viral variants. In addition, mRNA doesn't integrate into genome, thereby avoid the potential mutagenic risk. Using lipid nanoparticle-delivered mRNA encoding SPIKE antigen derived from SARS-COV-2 virus, two mRNA vaccines developed by BioNtech/Pfizer and Moderna Inc. have shown excellent prevention efficacy with good biosafety profile, which lead to unprecedented fast approval for human use by FDA (Chaudhary et al., 2021). In this chapter, we briefly summarize the recent progress in mRNA vaccine development, including mRNA design, delivery tools, and clinical applications.

1. Messenger RNA synthesis and modification for vaccine

mRNA vaccine is composed of targeted sequence encoding antigen and delivery system for intracellular release. The synthetic mRNA produced by in vitro transcription has a similar structure of natural mRNA, including 5′ cap, 5′UTR (untranslated region), nucleoside-modified ORF (open reading frame), 3′UTR, and poly(A) tail (Fig. 1), to overcome the problem in stability, translational efficacy as well as unwanted immunogenicity.

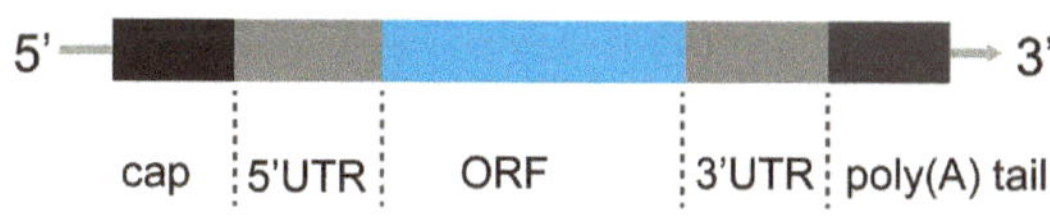

Fig. 1 The structure components of IVT mRNA. In-vitro transcribed (IVT) mRNA contains five structural elements: 5′cap, flanking 5′ and 3′ untranslated regions (UTRs), open reading frame (ORF) encoding the antigen and a poly(A) tail.

1.1 5′Cap structure

In eukaryotes transcription, 7-methylguanosine (m^7G) binds to the 5′ end of mRNA via 5′-5′ triphosphate to form a cap structure. The cap makes eukaryotic transcription initiation factor 4e (eLF4e) binding to initiate transcription, and at the same time prevents mRNA degradation induced by exonuclease Xrn1p. Besides, 5′cap works synergistically with the following mentioned "poly(A) tail" to enhance RNA stability (Mockey et al., 2006). There are two major strategies for capping in an IVT reaction, one is enzymatic capping post transcription; another one is capping simultaneously by additional cap analog in transcription mixture. Because of the easy reverse orientation of m^7G to 5′ nucleotides of mRNA via 3′-5′ phosphodiester, which reduces translation efficiency, anti-reverse cap analogues (ARCA) were developed in the past decades, such like P(1)-3′-deoxy-7-methyguanosine-5′ P3-guanosine-5′ triphosphate and P(1)-3′-O,7-dimethylguanosine-5′ P3-guanosine-5′ triphosphate (Stepinski et al., 2001). ARCA are made by chemical modification on m^7G including the replacement of 3′-OH into the 3′O-Me or the 3′deoxy substitution of oxygen molecule of the triphosphate bridge into sulfur, borane or selenium (Grudzien-Nogalska et al., 2007). Those ARCA are also resistant to decapping enzyme hydrolysis (Grudzien-Nogalska et al., 2007; Kowalska et al., 2008; Kuhn et al., 2010), benefiting mRNA stability and protein production.

1.2 UTRs

Untranslated regions (UTRs) flanking the ORF can bind with various regulatory protein and thus affect encoded protein translational efficacy. The sequence, length and secondary structure of UTR all affect mRNA translation efficacy. The presence of encephalomyocarditis virus internal ribosome entry site (IRES) within the 5′UTR even facilitate a 5′cap-independent protein expression pathway (Elroy-Stein et al., 1989; Tan and Wan, 2008). UTRs used in the synthetic RNA is usually from α/β-globin, human heat shock protein or virus. However, UTR performance may vary between cell types, and customized design of specific UTR for targeted cell type is necessary. Differ from 5′UTR, 3′UTR is more inclined to maintain the stability of mRNA and indirectly enhance gene expression.

1.3 Poly(A) tail

Polyadenylation naturally occurrs at the 3′ end of majority mRNAs in a co-transcriptional fashion and plays a critical role in transcript function and cell biology. On the one hand, poly(A) tail mediates the cytoplasm translocation of mature mRNA (Natalizio and Wente, 2013). On the other hand, poly(A) tail regulates translation efficiency and controls mRNA degradation (Bresson and Conrad, 2013; Wu et al., 2013). Poly(A) tailing of IVT mRNA can be accomplished either by enzymatic polyadenylation after IVT mixture or by cloning a poly(T) tail into the DNA template. The length of the poly(A) tail affects the translation efficiency of mRNA. Generally, longer poly(A) tail exhibits a higher promotion effect in protein translation. It was reported that antigen-coding sequence with a poly(A) tail (not longer than 120 bases) increased intracellular antigen protein expression and yield higher protein translation efficacy, thus improved T cell stimulatory capacity of dendritic cells (Holtkamp et al., 2006; Linares-Fernandez et al., 2021; Zhang et al., 2021). However, longer length is not always better, and recent findings suggest that 30-neucleotides length is likely the minimum length needed for cytoplasmic poly(A) binding protein binding and efficient translation (Gohin et al., 2014; Lima et al., 2017; Passmore and Coller, 2022). Further work is needed to decipher the detailed mechanism of poly(A) tail in mRNA stability and translation regulation for better mRNA design for efficient protein expression.

1.4 Nucleoside modification

The core of mRNA vaccine is the coding region expressing target protein antigen. Sequence modification within ORF is also required to elicit strong immune response. A strategy to enhance protein expression efficiency is codon optimization. It is known that different organisms prefers different codons (Gouy and Gautier, 1982). Protein expression outside host might be very difficult for the rarely used codons in the desired host. Therefore, replacing rare codon with frequently used codons in desired host enhance protein expression (Gustafsson et al., 2004).

The host immune system recognizes unmodified single-stranded RNA and switch on anti-virus responses once exogenous RNA entry into cell, and exerts anti-mRNA effect such as causing the instability of mRNA and downregulating translation efficacy (Pollard et al., 2013). Chemical modification on nucleosides improves the bottleneck of mRNA therapy by down-regulating RNA-mediated innate immune activation and maintaining mRNA stability. For example, Unmodified RNA stimulated pattern recognition receptors (PRRs) including Toll-like receptor 3 (TLR3), TLR7, TLR8 , leading to the induction of potent IFNα/γ as well as IL-12 (Kariko et al., 2005). Strikingly, mRNA bearing modified nucleosides such as 2-thiouridine and 5-methyl-cytidine modification prevented immune response compared with unmodified RNA and dedicated a higher protein expression level in vivo (Kormann et al., 2011).

Virus invasion leads to double- stranded RNA-dependent protein kinase (PKR) activation, translation initiation factor eIF2α subunit phosphorylation and production of 2′-5′ oligonucleotide synthase (OAS) and adenosinease (Pichlmair and Reis e Sousa, 2007). These downstream signals prevent virus replication, translation and promote exogenous nucleoside degradation. Based on virus depletion mechanism, modifications that help avoid enzymatic hydrolysis and translation inhibition are discovered. It is reported that 2′-*O*-methyl-U retain nuclease resistance, and incorporation of pseudouridine into mRNA can increase translation capacity while avoiding PRR recognition and PKR activation (Kariko et al., 2008). Furthermore, purification of IVT mRNA by Fast Protein Liquid Chromatography (FPLC) or High Pressure Liquid Chromatography (HPLC) would remove contaminated double-stranded RNA and greatly reduced PKR activation (Baiersdorfer et al., 2019; Weissman et al., 2013). Although some studies reported no discernible advantage of such nucleotide modification in mRNA translation (Kauffman et al., 2016; Thess et al., 2015), recent study confirmed that uridine modification and dsRNA impurity removal are the two most important factors determining the immunostimulating nature of synthesized mRNA (Nelson et al., 2020).

1.5 Self-amplifying RNA

There are two major types of RNA tested in vaccine application including self-amplifying RNA (saRNA) and non-replicating RNA (nrRNA). For traditional non-self-amplified mRNA, the expression level of encoded antigens depends on the number of copies of the delivered mRNA. In order to achieve stronger immune effect, it is necessary to increase the injection dose or frequency. Compared with nrRNA, saRNA has additional virus replication elemental between its 5′UTR and ORF. The viral replicase used in saRNA could be from alphaviruses, flaviviruses, measles viruses, and rhabdoviruses. Therefore, saRNA could produce a large amount of antigen protein and induce a strong immune response at a relatively small dosage. For example, Beissert et al. developed a trans-amplifying RNA vaccine, which in a nanogram was sufficient for inducing potent immune response (Beissert et al., 2020). However, due to the longer length, saRNA delivery is more challenging comparing to that of nrRNA (Lundstrom, 2018).

2. Carrier-mediated mRNA vaccine delivery

mRNA vaccine has shown therapeutic potential in preventing viral infection and treating inflammatory diseases including cancer. For in vivo administration, the most common injection routes for mRNA vaccines are intranodal (IN), intramuscular (IM), subcutaneous (SC), and intravenous (IV) administration (Chaudhary et al., 2021; Hou et al., 2021; Zeng et al., 2020). Despite the different administration routes, the common targets are the lymphoid organs and immune cells such as antigen-presenting cells (APCs). Many methods have been investigated for delivering mRNA vaccines, such as carrier-mediated and dendritic cell-based. The dendritic cell-based mRNA vaccine is a kind of cell therapy involving mRNA transfection into in vitro cultured dendritic cells and infusion of cells back to patient body, and requires strict in vitro culture process. In this section, we focus on the advanced strategies for carrier-mediated mRNA vaccines delivery.

The carrier-formulated mRNA can be easily and efficiently engulfed and translated by APCs when injected in vivo. The most commonly used carriers are lipid-based, polymer-based, peptide-based and pseudovirus-based particles (Fig. 2).

2.1 Lipid-based delivery

Lipids have been applied for exogenous mRNA delivery owing to several advantages, such as protecting mRNA from enzymatic degradation and high delivery efficiency into cell cytosol (Midoux and Pichon, 2015; Richner et al., 2017a; Sabnis et al., 2018). The lipids usually contain three essential domains: a polar head group, a linker and a hydrophobic tail region, while the polar charged head was for electrostatic interaction with mRNA and the hydrophobic tail was for nanoparticle formation.

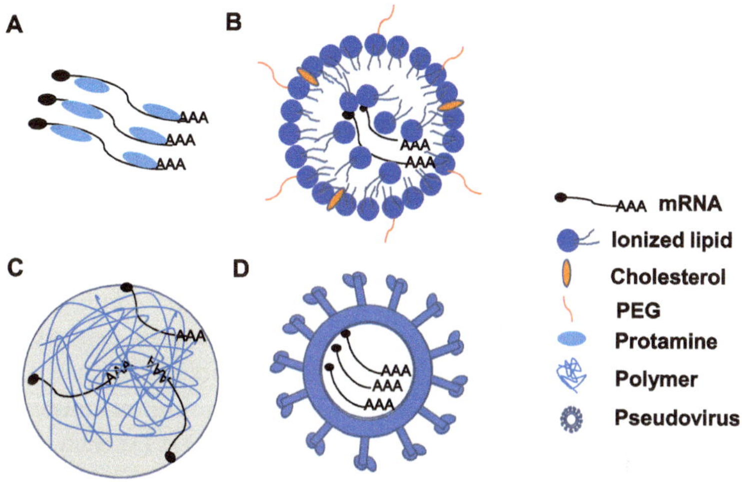

Fig. 2 Cartoon illustration of different mRNA delivery carriers. (A) Protamine-based carrier; (B) Lipid nanoparticle-based carrier; (C) Polymer-based carrier; (D) Pseudovirus-based carrier.

The 1,2-di-*O*-octadecenyl-3-trimethylammonium-propane (DOTMA), 1,2-dioleoyl-3-trimethylammonium-propane (DOTAP), 2,3 dioleyloxy-*N*-[2-(sperminecarboxamido) ethyl]-*N*,*N*-dimethyl-1-propanaminium trifluoroacetate (DOSPA) and Ethylphosphatidyl choline (ePC) are commonly used cationic lipids serving for mRNA delivery (Brito et al., 2014; Kormann et al., 2011; Malone et al., 1989; Persano et al., 2017). DOTMA-mRNA lipoplexes can target spleen specifically and serve as a systemic cancer vaccine, which can be used for melanoma and colon cancer therapy (Kranz et al., 2016a). DOTAP-based nanoemulsions and DOTAP-polymer hybrid nanoparticle can deliver mRNA for the treatment of colon cancer and protection of multiple types of virus infection, such as respiratory syncytial virus (RSV), human cytomegalovirus (hCMV) and human immunodeficiency virus (HIV) infection (Brito et al., 2014; Sahin et al., 2020a). DOSPA, which contains quaternary ammonium and spermine, can not only form mRNA complex but also stimulate innate immune responses, thereby serving as an immune adjuvant (Kormann et al., 2011). As another example, the engineered lipid-like material (synthesized through the ring opening of 1,2-epoxytetradecane by generation 0 of poly(amidoamine) (PAMAM) dendrimers with a different length of carbon tails) can deliver mRNA vaccine and potentiate antitumor efficacy through Toll-like receptor 4 (TLR4) signaling. Such lipid-like material-packaged mRNA nanoparticles showed efficient mRNA-encoded antigen expression and presentation, and induced the expression of inflammatory cytokines such as IL-12 via stimulating TLR4 signal pathway in dendritic cells. Moreover, the nanovaccine exhibited significant antitumor efficacy in both tumor prevention and therapeutic vaccine settings (Zhang et al., 2021).

Ionizable lipids are positively charged at low pH, and they can improve the biocompatibility of lipid nanoparticles due to the less interactions with the anionic members of blood cells, and facilitate endosome escape. The MC3 and MC3-based engineered lipids such as L319 (used for RNA interference (RNAi) delivery), SM-102(used for intramuscular administration of mRNA vaccines) and ALC-0315 (used for nucleic acids delivery) show great mRNA delivery efficacy and pharmacokinetics (Maier et al., 2013). More importantly, the SM-102 and ALC-0315 are the critical ionizable lipid components for mRNA-1273 and BNT162b COVID-19 vaccines, respectively (Baden et al., 2021; Vogel et al., 2021). Furthermore, the engineered ionizable lipid with a heterocyclic amine as head group can not only deliver the mRNA, but also active the STING (an critical innate immune sensor) signaling pathway in dendritic cells, thereby triggering the anti-tumor immune response (Miao et al., 2019).

In addition, the lipid helpers, such as DSPC, DOPE, cholesterol and polyethylene glycol (PEG) were often combined with ionizable lipids for mRNA delivery with different functions respectively. The DSPC could stabilize the structure of lipid nanoparticles (Koltover et al., 1998), which has been used in the mRNA-1273 and BNT162b COVID-19 vaccines (Baden et al., 2021; Polack et al., 2020). DOPE could destabilize the endosomal members and facilitate mRNA release into cytosol and was used for mRNA and siRNA delivery in vivo (Kauffman et al., 2015). Cholesterol can not only enhance the particle stability, but also affect the delivery efficacy and biodistribution of in vivo mRNA administration. For example, its analogues C-24 alkyl phytosterols increased the delivery efficacy of lipid nanoparticle (LNP) mRNA in vivo (Eygeris et al., 2020). PEG affects the lipid nanoparticle in multiple aspects, such as the nanoparticle formation, avoiding nanoparticle aggregation and prolonging the blood circulation of particles via providing a hydrophilic outer layer, and has been commonly applied for nanoparticle-mediated siRNA delivery in vitro and in vivo (Knop et al., 2010; Zhu et al., 2017).

Overall, lipid-based carriers could overcome multiple barriers and are often used for in vivo delivery of mRNA vaccine due to many benefits, such as enhanced mRNA stability, long circulating time, high fusion capacity with cellular membranes and endosome escape. The rapid approval of

LNP-formulated mRNA vaccines against COVID19 has witnessed the rapid development, robust protection efficacy and good safety profile of such vaccines. Nevertheless, there are minor side effects reported, which may be attributed to the inflammatory nature of the lipid components in the vaccine. With further investigation, lipid-based barriers may facilitate the development of more effective mRNA vaccines and therapeutic drugs while avoiding unwanted side effect.

2.2 Polymer-based delivery

Polymers are also widely used for mRNA delivery and offer similar advantages of lipids. Polyethylenimine (PEI), polyamidoamine (PAMAM) dendrimer and polysaccharide are the most commonly used polymers for mRNA delivery (Chahal et al., 2016; Moghimi et al., 2005). The PEI-packaged HIV gp120 mRNA vaccine can trigger high levels of specific antibodies against HIV infection in mice, but the in vivo toxicity impedes its clinically application (Li et al., 2016; Moyo et al., 2019). The PAMAM dendrimer with the self-amplifying mRNAs vaccine can protect mice from Ebola, H1N1 and Zika virus infection (Chahal et al., 2016). In addition, the poly(aspartamide)s conjugated with ionizable aminoethylene, which can be charged at acidic pH, also lead to a high mRNA delivery and rapid release of the mRNA from endosome to cytosol (Kim et al., 2019). Overall, polymer-based mRNA delivery showed good delivery efficiency, but their use is limited by obvious toxicity. Further studies may need to focus on the structure modification or combination with other helper materials to mitigate the polymer toxicity without sacrificing the delivery efficacy.

2.3 Peptide-based delivery

Peptides has also been applied for mRNA delivery owing to the cationic or amphipathic amine groups, such as arginine, which can electrostatically bind to the negatively charged mRNA and form nanocomplex (McCarthy et al., 2014). The RALA peptide, which contains repetitive arginine-alanine-leucine-alanine (RALA) motifs, can condense mRNA into nanoparticle, and transfect DCs and elicit T cell-mediated immune response (Li et al., 2004). Anionic peptides conjugated to positively charged polymers was also used for delivering mRNA into antigen presenting cells efficiently and specifically, and triggered antigen presenting cells activation (Lou et al., 2019). In addition, protamine was also used as cationic peptide for mRNA delivery, as it can protect mRNA from RNase degradation and active TLR7 as an adjuvant (Kallen et al., 2013; Skold et al., 2015). The mRNA vaccines which based on protamine-formulated delivery have been in clinical trials for treating multiple cancer types including prostate cancer (CV9103 and CV9104) and non-small cell lung cancer (CV9201), but the in vivo immune response was too weak for tumor control (Kubler et al., 2015; Papachristofilou et al., 2019; Weide et al., 2009). Therefore, further development of peptide-based mRNA vaccine is needed to improve the magnitude of in vivo immune response.

2.4 Pseudovirus-based particles

Pseudovirus can be utilized for mRNA delivery which refers the RNA virus structures (Lundstrom, 2020). The pseudovirus with mRNA shows high delivery efficiency to cell cytosol thanks to multiple cell entry pathways by the virus (Usme-Ciro et al., 2013). Moreover, some virus structure in the particles can trigger innate immune response, thereby serving as self-adjuvant purpose. The pseudovirus mRNA vaccines have been applied for preventing dengue virus infection and inhibiting tumor growth in mice (White et al., 2013), and the pseudovirus-based mRNA HIV vaccines have been under clinical evaluation (Fuchs et al., 2015). However, the viral-based mRNA delivery depends on packaging cell lines and needs to follow a special manufacturing process, which limits its large-scale production (Rauch et al., 2018). More importantly, the viral-based mRNA particles may induce unwanted antibody production as the virus structure can be recognized as antigen, which was reported in several clinical trials, such as the vaccine for cytomegalovirus and therapeutic cancer vaccine for patients with advanced cancer (Bernstein et al., 2009; Morse et al., 2010). Therefore, future development must focus on the streamline preparation and minimizing the anti-viral structure immunity.

Taken together, there are multiple types of carriers for mRNA delivery, which shows different benefits and shortages (Table 1). So far, lipid-based carriers have been most widely used with excellent delivery efficiency and great safety profile, but long-term monitoring of side effect is necessary to avoid any potential risk.

3. mRNA vaccines for clinical applications

Battling the global pandemic of SARS-COV-2, mRNA vaccines have gone from a small field to a widespread application, making a remarkable contribution to saving many lives. Compared to traditional vaccines,

Table 1 Summary of commonly used mRNA delivery carriers.

Delivery format	Advantages	Challenges	Clinical application
Lipid-based carriers	mRNA stability Blood circulation Delivery efficacy Reproduction Scale up	Potential side effects	Approved for COVID-19 vaccine/ Clinical trials for others
Polymer-based carriers	mRNA stability Delivery efficacy	Toxicity	Preclinical mouse model
Peptide-based carriers	mRNA stability Adjuvant activity	Low immune response	Clinical trial/Preclinical mouse model
Pseudovirus-based carriers	mRNA stability Deliver efficacy	Scale up/ Antibody to viral vectors	Clinical trials

mRNA vaccines enable precise antigen design and production of proteins with "native-like" presentation, stabilized in more immunogenic conformations, expose key antigenic sites or protein antigens from different pathogens that results in a single vaccine targeting multiple targets (Gebre et al., 2021). Due to their outstanding advantages in delivering antigens and eliciting immune responses, mRNA vaccines have great potential in the treatment of many diseases. This section will focus on the applications and summarize the progress in preclinical and clinical trials of mRNA vaccines in recent years.

The application of mRNA vaccines can be roughly divided into three directions. First, it is applied to resist SARS-COV-2. The outbreak of SARS-COV-2 in the end of 2019 has caused a worldwide pandemic. At the forefront of the fight against Covid-19, mRNA vaccine products from different manufacturing companies have been used clinically or are still in clinical trials. The FDA-approved mRNA-1273 vaccine comes from Moderna Inc. and encodes transmembrane prefusion SPIKE protein of Covid-19 by nucleoside-modified mRNA (Corbett et al., 2020). The SPIKE protein on SARS-CoV-2 surface is an essential component for virus binding and uptake into mammalian cells (Letko et al., 2020). Vaccination with mRNA-1273 induced remarkable type 1 helper T cell (Th1)-biased CD4 T cell responses but weak Th2 or CD8 T cell responses and efficient neutralizing antibodies, which provides effective protection against SARS-CoV-2 (Baden et al., 2021; Corbett et al., 2020). Another mRNA vaccine that achieved colossal success clinically is BNT162b2 that manufactured by BioNTech and Pfizer (Tauzin et al., 2021). BNT162b2, a nucleoside-modified mRNA encapsulated in LNP that encodes SARS-CoV-2 SPIKE glycoprotein, can elicit a strong responses of IFNγ+ or IL-2+ CD8+ T cells and CD4+ Th1 cells and induce high neutralizing antibody titres in serum from immunized individuals (Sahin et al., 2021). Other companies also have designed different mRNA vaccines. CVnCoV from CureVac uses unmodified mRNA to express full-length SPIKE protein and simultaneously activate innate immunity (Kremsner et al., 2021). Nevertheless, the clinical trial outcomes were not as good as the previous two mRNA vaccines, and the effective rate against Covid-19 was only 47%, according to their company statement. ARCT-021 from Arcturus Inc. uses self-amplifying mRNA to encode the full length of SPIKE protein, which also induced a strong Th1-dominant immune response and a robust viral antigen-specific CD8+ T lymphocyte response (de Alwis et al., 2021). LNP-nCoVsaRNA developed by Imperial College London and Acuitas Therapeutics uses self-amplifying mRNA to encode the SPIKE protein to provoke Th1-biased response against SARS-CoV-2 (McKay et al., 2020). ARCoV from Walvax biotechnology and Abogen Inc. uses unmodified mRNA to encode secreted spike receptor-binding domain (RBD) to elicit neutralizing antibodies against SARS-CoV-2 as well as a Th1-biased cellular response in preclinical experiments (Zhang et al., 2020). So far, the two mRNA vaccines using LNP-delivered mRNA with nucleoside modification by Moderna Inc. and BioNtech Inc. exhibited robust efficacy and received FDA approval in an unprecedented speed.

In addition to SARS-COV-2, the prevention effect of mRNA vaccines was also tested against other infectious diseases, including cytomegalovirus, Zika, human metapneumovirus, respiratory syncytial virus, Influenza A, Chikungunya virus (CHIKV), and Rabies in preclinical animal experiments or clinical trials (Alberer et al., 2017; Chahal et al., 2016; Chaudhary et al., 2021; Kose et al., 2019; Magini et al., 2016; Richner et al., 2017b). Generally, these mRNA vaccines are injected into the skin, muscle, or subcutaneous space to transfer mRNA into antigen presenting cells that finally induce specific CD8+ T cell responses, robust polyfunctional Th1 cells, and antibody responses to resist virus (Chaudhary et al., 2021). The mRNA vaccine against CHIKV encodes a ultrapotent neutralizing human monoclonal antibody to treat CHIKV infection (Kose et al., 2019). Apart from fighting virus infection, mRNA vaccines have been reported to elicit immune responses against *Plasmodium* species or *Streptococci* infection

(Baeza Garcia et al., 2018; Maruggi et al., 2017). *Plasmodium* express an ortholog of the mammalian cytokine macrophage migration inhibitory factor (PMIF), which modulates the host inflammatory response to malaria (Cordery et al., 2007; Dobson et al., 2009). The mRNA vaccine based on an RNA replicon encoding PMIF could reduce the expression of the Th1-associated inflammatory markers TNFα, IL-12, IFNγ, and augment T follicular helper cells and germinal center responses to increase anti-*Plasmodium* antibody titers (Baeza Garcia et al., 2018). In conclusion, the above results indicate that the mRNA vaccine has substantial application value in preventing infectious diseases caused by viruses, bacteria, and eukaryotic microorganisms.

Another vast potential application scenario for mRNA vaccines is cancer immunotherapy. Since 1995, there has been an attempt to use mRNA vaccine encoding carcinoembryonic antigen to elicit tumor antigen-specific immune responses in mice (Conry et al., 1995). Nowadays, multiple mRNA vaccines targeting different solid tumors like non-small-cell lung cancer, colorectal cancer, pancreatic adenocarcinoma, melanoma, triple-negative breast cancer, ovarian cancer, and gastrointestinal cancer are in clinical trials (Burris et al., 2019; Cafri et al., 2020; Hou et al., 2021; Sahin et al., 2020b; Sebastian et al., 2014). With the fast development of nucleotide sequencing technology, individualized sequencing results from tumor patients have revealed many potential neoantigens. Neoantigens are tumor-specific and highly immunogenic antigens generated by genetic mutations on tumor cells and have great potential value for the design of tumor vaccines (Schumacher and Schreiber, 2015). For example, an mRNA vaccine called mRNA-4157 encoding up to 34 neoantigens was designed to induce neoantigen-specific T cells and associated anti-tumor responses to treat unresectable (locally advanced or metastatic) solid tumors and resected cutaneous melanoma (Burris et al., 2019). The novel mRNA vaccine against gastrointestinal cancer encoding defined neoantigens, predicted neoepitopes, and mutations of driver genes in a single mRNA construct can elicit mainly CD4+, and not CD8+, T cell-specific responses to inhibit tumor growth (Cafri et al., 2020). BNT111, an intravenously administered liposomal RNA vaccine for melanoma, can induce strong antigen-specific CD4+ and CD8+ T cell immunity by precisely and effectively targeting dendritic cells in various lymphoid compartments accompanied by TLR7-triggered interferon (IFN) α production (Kranz et al., 2016b; Sahin et al., 2020b).

To modify the very immunosuppressive nature of tumor tissue microenvironment, mRNA can also be used to express proinflammatory cytokines to tailor tumor microenvironment besides expressing tumor antigens. In a clinical trial (NCT03739931), mRNA 2752 encoding human OX40L, IL-23, and IL-36γ was administrated to induce a proinflammatory tumor microenvironment and to promote T cell expansion, infiltration, and memory responses to prevent tumor progression (Bauer et al., 2019; Patel et al., 2020). Recently, another fascinating study reported an mRNA-based chimeric antigen receptor (CAR) T cell therapy treating heart failure. By using mRNA complexed with CD5-targeted LNP to express the fibroblast activation protein alpha-CAR protein in mice, the researchers observed specific and strong killing of fibroblasts, reduce fibrosis and restored cardiac function after injury (Rurik et al., 2022).

The ability of encoding functional proteins has given mRNA therapies enormous potential in treating genetic diseases, autoimmune diseases and cancer immunotherapies. Altogether, evidenced by the success of mRNA vaccines for controlling SARS-COV-2 global pandemic, mRNA vaccines hold promise to revolutionize future vaccines for treating infectious diseases, cancer, and beyond (Chaudhary et al., 2021).

References

Alberer, M., Gnad-Vogt, U., Hong, H.S., Mehr, K.T., Backert, L., Finak, G., Gottardo, R., Bica, M.A., Garofano, A., Koch, S.D., Fotin-Mleczek, M., Hoerr, I., Clemens, R., von Sonnenburg, F., 2017. Safety and immunogenicity of a mRNA rabies vaccine in healthy adults: an open-label, non-randomised, prospective, first-in-human phase 1 clinical trial. Lancet 390, 1511–1520.

Baden, L.R., El Sahly, H.M., Essink, B., Kotloff, K., Frey, S., Novak, R., Diemert, D., Spector, S.A., Rouphael, N., Creech, C.B., McGettigan, J., Khetan, S., Segall, N., Solis, J., Brosz, A., Fierro, C., Schwartz, H., Neuzil, K., Corey, L., Gilbert, P., Janes, H., Follmann, D., Marovich, M., Mascola, J., Polakowski, L., Ledgerwood, J., Graham, B.S., Bennett, H., Pajon, R., Knightly, C., Leav, B., Deng, W., Zhou, H., Han, S., Ivarsson, M., Miller, J., Zaks, T., COVE Study Group, 2021. Efficacy and safety of the mRNA-1273 SARS-CoV-2 vaccine. N. Engl. J. Med. 384, 403–416.

Baeza Garcia, A., Siu, E., Sun, T., Exler, V., Brito, L., Hekele, A., Otten, G., Augustijn, K., Janse, C.J., Ulmer, J.B., Bernhagen, J., Fikrig, E., Geall, A., Bucala, R., 2018. Neutralization of the Plasmodium-encoded MIF ortholog confers protective immunity against malaria infection. Nat. Commun. 9, 2714.

Baiersdorfer, M., Boros, G., Muramatsu, H., Mahiny, A., Vlatkovic, I., Sahin, U., Kariko, K., 2019. A facile method for the removal of dsRNA contaminant from in vitro-transcribed mRNA. Mol. Ther. Nucleic Acids 15, 26–35.

Bauer, T., Patel, M., Jimeno, A., Wang, D., McDermott, J., Zacharek, S., Randolph, W., Johansen, L., Hopson, K., Frederick, J., Zaks, T., Meehan, R.S., 2019. Abstract CT210: a phase I, open-label, multicenter, dose escalation study of mRNA-2752, a lipid nanoparticle encapsulating mRNAs encoding human OX40L, IL-23, and IL-36γ, for intratumoral injection alone and in combination with immune checkpoint blockade. Cancer Res. 79, CT210.

Beissert, T., Perkovic, M., Vogel, A., Erbar, S., Walzer, K.C., Hempel, T., Brill, S., Haefner, E., Becker, R., Tureci, O., Sahin, U., 2020. A trans-amplifying RNA vaccine strategy for induction of potent protective immunity. Mol. Ther. 28, 119–128.

Bernstein, D.I., Reap, E.A., Katen, K., Watson, A., Smith, K., Norberg, P., Olmsted, R.A., Hoeper, A., Morris, J., Negri, S., Maughan, M.F., Chulay, J.D., 2009. Randomized, double-blind, phase 1 trial of an alphavirus replicon vaccine for cytomegalovirus in CMV seronegative adult volunteers. Vaccine 28, 484–493.

Brenner, S., Jacob, F., Meselson, M., 1961. An unstable intermediate carrying information from genes to ribosomes for protein synthesis. Nature 190, 576–581.

Bresson, S.M., Conrad, N.K., 2013. The human nuclear poly(a)-binding protein promotes RNA hyperadenylation and decay. PLoS Genet. 9, e1003893.

Brito, L.A., Chan, M., Shaw, C.A., Hekele, A., Carsillo, T., Schaefer, M., Archer, J., Seubert, A., Otten, G.R., Beard, C.W., Dey, A.K., Lilja, A., Valiante, N.M., Mason, P.W., Mandl, C.W., Barnett, S.W., Dormitzer, P.R., Ulmer, J.B., Singh, M., O'Hagan, D.T., Geall, A.J., 2014. A cationic nanoemulsion for the delivery of next-generation RNA vaccines. Mol. Ther. 22, 2118–2129.

Burris, H.A., Patel, M.R., Cho, D.C., Clarke, J.M., Gutierrez, M., Zaks, T.Z., Frederick, J., Hopson, K., Mody, K., Binanti-Berube, A., Robert-Tissot, C., Goldstein, B., Breton, B., Sun, J., Zhong, S., Pruitt, S.K., Keating, K., Meehan, R.S., Gainor, J.F., 2019. A phase I multicenter study to assess the safety, tolerability, and immunogenicity of mRNA-4157 alone in patients with resected solid tumors and in combination with pembrolizumab in patients with unresectable solid tumors. J. Clin. Oncol. 37, 2523.

Cafri, G., Gartner, J.J., Zaks, T., Hopson, K., Levin, N., Paria, B.C., Parkhurst, M.R., Yossef, R., Lowery, F.J., Jafferji, M.S., Prickett, T.D., Goff, S.L., McGowan, C.T., Seitter, S., Shindorf, M.L., Parikh, A., Chatani, P.D., Robbins, P.F., Rosenberg, S.A., 2020. mRNA vaccine-induced neoantigen-specific T cell immunity in patients with gastrointestinal cancer. J. Clin. Invest. 130, 5976–5988.

Chahal, J.S., Khan, O.F., Cooper, C.L., McPartlan, J.S., Tsosie, J.K., Tilley, L.D., Sidik, S.M., Lourido, S., Langer, R., Bavari, S., Ploegh, H.L., Anderson, D.G., 2016. Dendrimer-RNA nanoparticles generate protective immunity against lethal Ebola, H1N1 influenza, and toxoplasma gondii challenges with a single dose. Proc. Natl. Acad. Sci. U. S. A. 113, E4133–E4142.

Chaudhary, N., Weissman, D., Whitehead, K.A., 2021. mRNA vaccines for infectious diseases: principles, delivery and clinical translation. Nat. Rev. Drug Discov. 20, 817–838.

Conry, R.M., LoBuglio, A.F., Wright, M., Sumerel, L., Pike, M.J., Johanning, F., Benjamin, R., Lu, D., Curiel, D.T., 1995. Characterization of a messenger RNA polynucleotide vaccine vector. Cancer Res. 55, 1397–1400.

Corbett, K.S., Flynn, B., Foulds, K.E., Francica, J.R., Boyoglu-Barnum, S., Werner, A.P., Flach, B., O'Connell, S., Bock, K.W., Minai, M., Nagata, B.M., Andersen, H., Martinez, D.R., Noe, A.T., Douek, N., Donaldson, M.M., Nji, N.N., Alvarado, G.S., Edwards, D.K., Flebbe, D.R., Lamb, E., Doria-Rose, N.A., Lin, B.C., Louder, M.K., O'Dell, S., Schmidt, S.D., Phung, E., Chang, L.A., Yap, C., Todd, J.M., Pessaint, L., Van Ry, A., Browne, S., Greenhouse, J., Putman-Taylor, T., Strasbaugh, A., Campbell, T.A., Cook, A., Dodson, A., Steingrebe, K., Shi, W., Zhang, Y., Abiona, O.M., Wang, L., Pegu, A., Yang, E.S., Leung, K., Zhou, T., Teng, I.T., Widge, A., Gordon, I., Novik, L., Gillespie, R.A., Loomis, R.J., Moliva, J.I., Stewart-Jones, G., Himansu, S., Kong, W.P., Nason, M.C., Morabito, K.M., Ruckwardt, T.J., Ledgerwood, J.E., Gaudinski, M.R., Kwong, P.D., Mascola, J.R., Carfi, A., Lewis, M.G., Baric, R.S., McDermott, A., Moore, I.N., Sullivan, N.J., Roederer, M., Seder, R.A., Graham, B.S., 2020. Evaluation of the mRNA-1273 vaccine against SARS-CoV-2 in nonhuman Primates. N. Engl. J. Med. 383, 1544–1555.

Cordery, D.V., Kishore, U., Kyes, S., Shafi, M.J., Watkins, K.R., Williams, T.N., Marsh, K., Urban, B.C., 2007. Characterization of a Plasmodium falciparum macrophage-migration inhibitory factor homologue. J Infect Dis 195, 905–912.

de Alwis, R., Gan, E.S., Chen, S., Leong, Y.S., Tan, H.C., Zhang, S.L., Yau, C., Low, J.G.H., Kalimuddin, S., Matsuda, D., Allen, E.C., Hartman, P., Park, K.J., Alayyoubi, M., Bhaskaran, H., Dukanovic, A., Bao, Y., Clemente, B., Vega, J., Roberts, S., Gonzalez, J.A., Sablad, M., Yelin, R., Taylor, W., Tachikawa, K., Parker, S., Karmali, P., Davis, J., Sullivan, B.M., Sullivan, S.M., Hughes, S.G., Chivukula, P., Ooi, E.E., 2021. A single dose of self-transcribing and replicating RNA-based SARS-CoV-2 vaccine produces protective adaptive immunity in mice. Mol. Ther. 29, 1970–1983.

Dobson, S.E., Augustijn, K.D., Brannigan, J.A., Schnick, C., Janse, C.J., Dodson, E.J., Waters, A.P., Wilkinson, A.J., 2009. The crystal structures of macrophage migration inhibitory factor from Plasmodium falciparum and Plasmodium berghei. Protein Sci. 18, 2578–2591.

Elroy-Stein, O., Fuerst, T.R., Moss, B., 1989. Cap-independent translation of mRNA conferred by encephalomyocarditis virus 5′ sequence improves the performance of the vaccinia virus/bacteriophage T7 hybrid expression system. Proc. Natl. Acad. Sci. U. S. A. 86, 6126–6130.

Eygeris, Y., Patel, S., Jozic, A., Sahay, G., 2020. Deconvoluting lipid nanoparticle structure for messenger RNA delivery. Nano Lett. 20, 4543–4549.

Fuchs, J.D., Frank, I., Elizaga, M.L., Allen, M., Frahm, N., Kochar, N., Li, S., Edupuganti, S., Kalams, S.A., Tomaras, G.D., Sheets, R., Pensiero, M., Tremblay, M.A., Higgins, T.J., Latham, T., Egan, M.A., Clarke, D.K., Eldridge, J.H., HVTN 090 Study Group and the National Institutes of Allergy and Infectious Diseases HIV Vaccine Trials Network, Mulligan, M., Rouphael, N., Estep, S., Rybczyk, K., Dunbar, D., Buchbinder, S., Wagner, T., Isbell, R., Chinnell, V., Bae, J., Escamilla, G., Tseng, J., Fair, R., Ramirez, S., Broder, G., Briesemeister, L., Ferrara, A., 2015. First-in-human evaluation of the safety and immunogenicity of a recombinant vesicular stomatitis virus human immunodeficiency Virus-1 gag vaccine (HVTN 090). Open Forum Infect Dis 2. ofv082.

Gebre, M.S., Brito, L.A., Tostanoski, L.H., Edwards, D.K., Carfi, A., Barouch, D.H., 2021. Novel approaches for vaccine development. Cell 184, 1589–1603.

Gohin, M., Fournier, E., Dufort, I., Sirard, M.A., 2014. Discovery, identification and sequence analysis of RNAs selected for very short or long poly a tail in immature bovine oocytes. Mol. Hum. Reprod. 20, 127–138.

Gouy, M., Gautier, C., 1982. Codon usage in bacteria: correlation with gene expressivity. Nucleic Acids Res. 10, 7055–7074.

Grudzien-Nogalska, E., Jemielity, J., Kowalska, J., Darzynkiewicz, E., Rhoads, R.E., 2007. Phosphorothioate cap analogs stabilize mRNA and increase translational efficiency in mammalian cells. RNA 13, 1745–1755.

Gustafsson, C., Govindarajan, S., Minshull, J., 2004. Codon bias and heterologous protein expression. Trends Biotechnol. 22, 346–353.

Holtkamp, S., Kreiter, S., Selmi, A., Simon, P., Koslowski, M., Huber, C., Tureci, O., Sahin, U., 2006. Modification of antigen-encoding RNA increases stability, translational efficacy, and T-cell stimulatory capacity of dendritic cells. Blood 108, 4009–4017.

Hou, X., Zaks, T., Langer, R., Dong, Y., 2021. Lipid nanoparticles for mRNA delivery. Nat Rev Mater, 1–17.

Kallen, K.J., Heidenreich, R., Schnee, M., Petsch, B., Schlake, T., Thess, A., Baumhof, P., Scheel, B., Koch, S.D., Fotin-Mleczek, M., 2013. A novel, disruptive vaccination technology: self-adjuvanted RNActive((R)) vaccines. Hum. Vaccin. Immunother. 9, 2263–2276.

Kariko, K., Buckstein, M., Ni, H., Weissman, D., 2005. Suppression of RNA recognition by toll-like receptors: the impact of nucleoside modification and the evolutionary origin of RNA. Immunity 23, 165–175.

Kariko, K., Muramatsu, H., Welsh, F.A., Ludwig, J., Kato, H., Akira, S., Weissman, D., 2008. Incorporation of pseudouridine into mRNA yields superior nonimmunogenic vector with increased translational capacity and biological stability. Mol. Ther. 16, 1833–1840.

Kauffman, K.J., Dorkin, J.R., Yang, J.H., Heartlein, M.W., DeRosa, F., Mir, F.F., Fenton, O.S., Anderson, D.G., 2015. Optimization of lipid nanoparticle formulations for mRNA delivery in vivo with fractional factorial and definitive screening designs. Nano Lett. 15, 7300–7306.

Kauffman, K.J., Mir, F.F., Jhunjhunwala, S., Kaczmarek, J.C., Hurtado, J.E., Yang, J.H., Webber, M.J., Kowalski, P.S., Heartlein, M.W., DeRosa, F., Anderson, D.G., 2016. Efficacy and immunogenicity of unmodified and pseudouridine-modified mRNA delivered systemically with lipid nanoparticles in vivo. Biomaterials 109, 78–87.

Kim, H.J., Ogura, S., Otabe, T., Kamegawa, R., Sato, M., Kataoka, K., Miyata, K., 2019. Fine-tuning of hydrophobicity in amphiphilic Polyaspartamide derivatives for rapid and transient expression of messenger RNA directed toward genome engineering in brain. ACS Cent Sci 5, 1866–1875.

Knop, K., Hoogenboom, R., Fischer, D., Schubert, U.S., 2010. Poly(ethylene glycol) in drug delivery: pros and cons as well as potential alternatives. Angew. Chem. Int. Ed. Engl. 49, 6288–6308.

Koltover, I., Salditt, T., Radler, J.O., Safinya, C.R., 1998. An inverted hexagonal phase of cationic liposome-DNA complexes related to DNA release and delivery. Science 281, 78–81.

Kormann, M.S., Hasenpusch, G., Aneja, M.K., Nica, G., Flemmer, A.W., Herber-Jonat, S., Huppmann, M., Mays, L.E., Illenyi, M., Schams, A., Griese, M., Bittmann, I., Handgretinger, R., Hartl, D., Rosenecker, J., Rudolph, C., 2011. Expression of therapeutic proteins after delivery of chemically modified mRNA in mice. Nat. Biotechnol. 29, 154–157.

Kose, N., Fox, J.M., Sapparapu, G., Bombardi, R., Tennekoon, R.N., de Silva, A.D., Elbashir, S.M., Theisen, M.A., Humphris-Narayanan, E., Ciaramella, G., Himansu, S., Diamond, M.S., Crowe Jr., J.E., 2019. A lipid-encapsulated mRNA encoding a potently neutralizing human monoclonal antibody protects against chikungunya infection. Sci Immunol 4.

Kowalska, J., Lewdorowicz, M., Zuberek, J., Grudzien-Nogalska, E., Bojarska, E., Stepinski, J., Rhoads, R.E., Darzynkiewicz, E., Davis, R.E., Jemielity, J., 2008. Synthesis and characterization of mRNA cap analogs containing phosphorothioate substitutions that bind tightly to eIF4E and are resistant to the decapping pyrophosphatase DcpS. RNA 14, 1119–1131.

Kranz, L.M., Diken, M., Haas, H., Kreiter, S., Loquai, C., Reuter, K.C., Meng, M., Fritz, D., Vascotto, F., Hefesha, H., Grunwitz, C., Vormehr, M., Husemann, Y., Selmi, A., Kuhn, A.N., Buck, J., Derhovanessian, E., Rae, R., Attig, S., Diekmann, J., Jabulowsky, R.A., Heesch, S., Hassel, J., Langguth, P., Grabbe, S., Huber, C., Tureci, O., Sahin, U., 2016a. Systemic RNA delivery to dendritic cells exploits antiviral defence for cancer immunotherapy. Nature 534, 396–401.

Kranz, L.M., Diken, M., Haas, H., Kreiter, S., Loquai, C., Reuter, K.C., Meng, M., Fritz, D., Vascotto, F., Hefesha, H., Grunwitz, C., Vormehr, M., Hüsemann, Y., Selmi, A., Kuhn, A.N., Buck, J., Derhovanessian, E., Rae, R., Attig, S., Diekmann, J., Jabulowsky, R.A., Heesch, S., Hassel, J., Langguth, P., Grabbe, S., Huber, C., Türeci, Ö., Sahin, U., 2016b. Systemic RNA delivery to dendritic cells exploits antiviral defence for cancer immunotherapy. Nature 534, 396–401.

Kremsner, P.G., Mann, P., Kroidl, A., Leroux-Roels, I., Schindler, C., Gabor, J.J., Schunk, M., Leroux-Roels, G., Bosch, J.J., Fendel, R., Kreidenweiss, A., Velavan, T.P., Fotin-Mleczek, M., Mueller, S.O., Quintini, G., Schönborn-Kellenberger, O., Vahrenhorst, D., Verstraeten, T., Alves de Mesquita, M., Walz, L., Wolz, O.O., Oostvogels, L., 2021. Safety and immunogenicity of an mRNA-lipid nanoparticle vaccine candidate against SARS-CoV-2: A phase 1 randomized clinical trial. Wien. Klin. Wochenschr. 133, 931–941.

Krieg, P.A., Melton, D.A., 1984. Functional messenger RNAs are produced by SP6 in vitro transcription of cloned cDNAs. Nucleic Acids Res. 12, 7057–7070.

Kubler, H., Scheel, B., Gnad-Vogt, U., Miller, K., Schultze-Seemann, W., Vom Dorp, F., Parmiani, G., Hampel, C., Wedel, S., Trojan, L., Jocham, D., Maurer, T., Rippin, G., Fotin-Mleczek, M., von der Mulbe, F., Probst, J., Hoerr, I., Kallen, K.J., Lander, T., Stenzl, A., 2015. Self-adjuvanted mRNA vaccination in advanced prostate cancer patients: a first-in-man phase I/IIa study. J. Immunother. Cancer 3, 26.

Kuhn, A.N., Diken, M., Kreiter, S., Selmi, A., Kowalska, J., Jemielity, J., Darzynkiewicz, E., Huber, C., Tureci, O., Sahin, U., 2010. Phosphorothioate cap analogs increase stability and translational efficiency of RNA vaccines in immature dendritic cells and induce superior immune responses in vivo. Gene Ther. 17, 961–971.

Letko, M., Marzi, A., Munster, V., 2020. Functional assessment of cell entry and receptor usage for SARS-CoV-2 and other lineage B betacoronaviruses. Nat. Microbiol. 5, 562–569.

Li, W., Nicol, F., Szoka Jr., F.C., 2004. GALA: a designed synthetic pH-responsive amphipathic peptide with applications in drug and gene delivery. Adv. Drug Deliv. Rev. 56, 967–985.

Li, M., Zhao, M., Fu, Y., Li, Y., Gong, T., Zhang, Z., Sun, X., 2016. Enhanced intranasal delivery of mRNA vaccine by overcoming the nasal epithelial barrier via intra- and paracellular pathways. J. Control. Release 228, 9–19.

Lima, S.A., Chipman, L.B., Nicholson, A.L., Chen, Y.H., Yee, B.A., Yeo, G.W., Coller, J., Pasquinelli, A.E., 2017. Short poly(A) tails are a conserved feature of highly expressed genes. Nat. Struct. Mol. Biol. 24, 1057–1063.

Linares-Fernandez, S., Moreno, J., Lambert, E., Mercier-Gouy, P., Vachez, L., Verrier, B., Exposito, J.Y., 2021. Combining an optimized mRNA template with a double purification process allows strong expression of in vitro transcribed mRNA. Mol Ther Nucleic Acids 26, 945–956.

Lou, B., De Koker, S., Lau, C.Y.J., Hennink, W.E., Mastrobattista, E., 2019. mRNA Polyplexes with post-conjugated GALA peptides efficiently target, transfect, and activate antigen presenting cells. Bioconjug. Chem. 30, 461–475.

Lundstrom, K., 2018. Self-replicating RNA viruses for RNA therapeutics. Molecules 23.

Lundstrom, K., 2020. Self-amplifying RNA viruses as RNA vaccines. Int. J. Mol. Sci. 21.

Magini, D., Giovani, C., Mangiavacchi, S., Maccari, S., Cecchi, R., Ulmer, J.B., De Gregorio, E., Geall, A.J., Brazzoli, M., Bertholet, S., 2016. Self-amplifying mRNA vaccines expressing multiple conserved influenza antigens confer protection against homologous and Heterosubtypic viral challenge. PLoS One 11, e0161193.

Maier, M.A., Jayaraman, M., Matsuda, S., Liu, J., Barros, S., Querbes, W., Tam, Y.K., Ansell, S.M., Kumar, V., Qin, J., Zhang, X., Wang, Q., Panesar, S., Hutabarat, R., Carioto, M., Hettinger, J., Kandasamy, P., Butler, D., Rajeev, K.G., Pang, B., Charisse, K., Fitzgerald, K., Mui, B.L., Du, X., Cullis, P., Madden, T.D., Hope, M.J., Manoharan, M., Akinc, A., 2013. Biodegradable lipids enabling rapidly eliminated lipid nanoparticles for systemic delivery of RNAi therapeutics. Mol. Ther. 21, 1570–1578.

Malone, R.W., Felgner, P.L., Verma, I.M., 1989. Cationic liposome-mediated RNA transfection. Proc. Natl. Acad. Sci. U. S. A. 86, 6077–6081.

Maruggi, G., Chiarot, E., Giovani, C., Buccato, S., Bonacci, S., Frigimelica, E., Margarit, I., Geall, A., Bensi, G., Maione, D., 2017. Immunogenicity and protective efficacy induced by self-amplifying mRNA vaccines encoding bacterial antigens. Vaccine 35, 361–368.

McCarthy, H.O., McCaffrey, J., McCrudden, C.M., Zholobenko, A., Ali, A.A., McBride, J.W., Massey, A.S., Pentlavalli, S., Chen, K.H., Cole, G., Loughran, S.P., Dunne, N.J., Donnelly, R.F., Kett, V.L., Robson, T., 2014. Development and characterization of self-assembling nanoparticles using a bio-inspired amphipathic peptide for gene delivery. J. Control. Release 189, 141–149.

McKay, P.F., Hu, K., Blakney, A.K., Samnuan, K., Brown, J.C., Penn, R., Zhou, J., Bouton, C.R., Rogers, P., Polra, K., Lin, P.J.C., Barbosa, C., Tam, Y.K., Barclay, W.S., Shattock, R.J., 2020. Self-amplifying RNA SARS-CoV-2 lipid nanoparticle vaccine candidate induces high neutralizing antibody titers in mice. Nat. Commun. 11, 3523.

Miao, L., Li, L., Huang, Y., Delcassian, D., Chahal, J., Han, J., Shi, Y., Sadtler, K., Gao, W., Lin, J., Doloff, J.C., Langer, R., Anderson, D.G., 2019. Delivery of mRNA vaccines with heterocyclic lipids increases anti-tumor efficacy by STING-mediated immune cell activation. Nat. Biotechnol. 37, 1174–1185.

Midoux, P., Pichon, C., 2015. Lipid-based mRNA vaccine delivery systems. Expert Rev. Vaccines 14, 221–234.

Mockey, M., Goncalves, C., Dupuy, F.P., Lemoine, F.M., Pichon, C., Midoux, P., 2006. mRNA transfection of dendritic cells: synergistic effect of ARCA mRNA capping with poly(a) chains in cis and in trans for a high protein expression level. Biochem. Biophys. Res. Commun. 340, 1062–1068.

Moghimi, S.M., Symonds, P., Murray, J.C., Hunter, A.C., Debska, G., Szewczyk, A., 2005. A two-stage poly(ethylenimine)-mediated cytotoxicity: implications for gene transfer/therapy. Mol. Ther. 11, 990–995.

Morse, M.A., Hobeika, A.C., Osada, T., Berglund, P., Hubby, B., Negri, S., Niedzwiecki, D., Devi, G.R., Burnett, B.K., Clay, T.M., Smith, J., Lyerly, H.K., 2010. An alphavirus vector overcomes the presence of neutralizing antibodies and elevated numbers of Tregs to induce immune responses in humans with advanced cancer. J. Clin. Invest. 120, 3234–3241.

Moyo, N., Vogel, A.B., Buus, S., Erbar, S., Wee, E.G., Sahin, U., Hanke, T., 2019. Efficient induction of T cells against conserved HIV-1 regions by mosaic vaccines delivered as self-amplifying mRNA. Mol Ther Methods Clin Dev 12, 32–46.

Natalizio, B.J., Wente, S.R., 2013. Postage for the messenger: designating routes for nuclear mRNA export. Trends Cell Biol. 23, 365–373.

Nelson, J., Sorensen, E.W., Mintri, S., Rabideau, A.E., Zheng, W., Besin, G., Khatwani, N., Su, S.V., Miracco, E.J., Issa, W.J., Hoge, S., Stanton, M.G., Joyal, J.L., 2020. Impact of mRNA chemistry and manufacturing process on innate immune activation. Sci. Adv. 6, eaaz6893.

Papachristofilou, A., Hipp, M.M., Klinkhardt, U., Fruh, M., Sebastian, M., Weiss, C., Pless, M., Cathomas, R., Hilbe, W., Pall, G., Wehler, T., Alt, J., Bischoff, H., Geissler, M., Griesinger, F., Kallen, K.J., Fotin-Mleczek, M., Schroder, A., Scheel, B., Muth, A., Seibel, T., Stosnach, C., Doener, F., Hong, H.S., Koch, S.D., Gnad-Vogt, U., Zippelius, A., 2019. Phase Ib evaluation of a self-adjuvanted protamine formulated mRNA-based active cancer immunotherapy, BI1361849 (CV9202), combined with local radiation treatment in patients with stage IV non-small cell lung cancer. J. Immunother. Cancer 7, 38.

Passmore, L.A., Coller, J., 2022. Roles of mRNA poly(A) tails in regulation of eukaryotic gene expression. Nat. Rev. Mol. Cell Biol. 23, 93–106.

Patel, M.R., Bauer, T.M., Jimeno, A., Wang, D., LoRusso, P., Do, K.T., Stemmer, S.M., Maurice-Dror, C., Geva, R., Zacharek, S., Laino, A.S., Sun, J., Frederick, J., Zhou, H.,

Randolph, W., Cohen, P.S., Meehan, R.S., Sullivan, R.J., 2020. A phase I study of mRNA-2752, a lipid nanoparticle encapsulating mRNAs encoding human OX40L, IL-23, and IL-36γ, for intratumoral (iTu) injection alone and in combination with durvalumab. J. Clin. Oncol. 38, 3092.
Persano, S., Guevara, M.L., Li, Z., Mai, J., Ferrari, M., Pompa, P.P., Shen, H., 2017. Lipopolyplex potentiates anti-tumor immunity of mRNA-based vaccination. Biomaterials 125, 81–89.
Pichlmair, A., Reis e Sousa, C., 2007. Innate recognition of viruses. Immunity 27, 370–383.
Polack, F.P., Thomas, S.J., Kitchin, N., Absalon, J., Gurtman, A., Lockhart, S., Perez, J.L., Perez Marc, G., Moreira, E.D., Zerbini, C., Bailey, R., Swanson, K.A., Roychoudhury, S., Koury, K., Li, P., Kalina, W.V., Cooper, D., Frenck Jr., R.W., Hammitt, L.L., Tureci, O., Nell, H., Schaefer, A., Unal, S., Tresnan, D.B., Mather, S., Dormitzer, P.R., Sahin, U., Jansen, K.U., Gruber, W.C., C4591001 Clinical Trial Group, 2020. Safety and efficacy of the BNT162b2 mRNA Covid-19 vaccine. N. Engl. J. Med. 383, 2603–2615.
Pollard, C., Rejman, J., De Haes, W., Verrier, B., Van Gulck, E., Naessens, T., De Smedt, S., Bogaert, P., Grooten, J., Vanham, G., De Koker, S., 2013. Type I IFN counteracts the induction of antigen-specific immune responses by lipid-based delivery of mRNA vaccines. Mol. Ther. 21, 251–259.
Rauch, S., Jasny, E., Schmidt, K.E., Petsch, B., 2018. New vaccine technologies to combat outbreak situations. Front. Immunol. 9, 1963.
Richner, J.M., Himansu, S., Dowd, K.A., Butler, S.L., Salazar, V., Fox, J.M., Julander, J.G., Tang, W.W., Shresta, S., Pierson, T.C., Ciaramella, G., Diamond, M.S., 2017a. Modified mRNA vaccines protect against Zika virus infection. Cell 169, 176.
Richner, J.M., Himansu, S., Dowd, K.A., Butler, S.L., Salazar, V., Fox, J.M., Julander, J.G., Tang, W.W., Shresta, S., Pierson, T.C., Ciaramella, G., Diamond, M.S., 2017b. Modified mRNA vaccines protect against Zika virus infection. Cell 168 (1114–1125), e1110.
Rurik, J.G., Tombácz, I., Yadegari, A., Méndez Fernández, P.O., Shewale, S.V., Li, L., Kimura, T., Soliman, O.Y., Papp, T.E., Tam, Y.K., Mui, B.L., Albelda, S.M., Puré, E., June, C.H., Aghajanian, H., Weissman, D., Parhiz, H., Epstein, J.A., 2022. CAR T cells produced in vivo to treat cardiac injury. Science 375, 91–96.
Sabnis, S., Kumarasinghe, E.S., Salerno, T., Mihai, C., Ketova, T., Senn, J.J., Lynn, A., Bulychev, A., McFadyen, I., Chan, J., Almarsson, O., Stanton, M.G., Benenato, K.E., 2018. A novel amino lipid series for mRNA delivery: improved endosomal escape and sustained pharmacology and safety in non-human Primates. Mol. Ther. 26, 1509–1519.
Sahin, U., Kariko, K., Tureci, O., 2014. mRNA-based therapeutics—developing a new class of drugs. Nat. Rev. Drug Discov. 13, 759–780.
Sahin, U., Oehm, P., Derhovanessian, E., Jabulowsky, R.A., Vormehr, M., Gold, M., Maurus, D., Schwarck-Kokarakis, D., Kuhn, A.N., Omokoko, T., Kranz, L.M., Diken, M., Kreiter, S., Haas, H., Attig, S., Rae, R., Cuk, K., Kemmer-Bruck, A., Breitkreuz, A., Tolliver, C., Caspar, J., Quinkhardt, J., Hebich, L., Stein, M., Hohberger, A., Vogler, I., Liebig, I., Renken, S., Sikorski, J., Leierer, M., Muller, V., Mitzel-Rink, H., Miederer, M., Huber, C., Grabbe, S., Utikal, J., Pinter, A., Kaufmann, R., Hassel, J.C., Loquai, C., Tureci, O., 2020a. An RNA vaccine drives immunity in checkpoint-inhibitor-treated melanoma. Nature 585, 107–112.
Sahin, U., Oehm, P., Derhovanessian, E., Jabulowsky, R.A., Vormehr, M., Gold, M., Maurus, D., Schwarck-Kokarakis, D., Kuhn, A.N., Omokoko, T., Kranz, L.M., Diken, M., Kreiter, S., Haas, H., Attig, S., Rae, R., Cuk, K., Kemmer-Brück, A., Breitkreuz, A., Tolliver, C., Caspar, J., Quinkhardt, J., Hebich, L., Stein, M., Hohberger, A., Vogler, I., Liebig, I., Renken, S., Sikorski, J., Leierer, M., Müller,

V., Mitzel-Rink, H., Miederer, M., Huber, C., Grabbe, S., Utikal, J., Pinter, A., Kaufmann, R., Hassel, J.C., Loquai, C., Türeci, Ö., 2020b. An RNA vaccine drives immunity in checkpoint-inhibitor-treated melanoma. Nature 585, 107–112.

Sahin, U., Muik, A., Vogler, I., Derhovanessian, E., Kranz, L.M., Vormehr, M., Quandt, J., Bidmon, N., Ulges, A., Baum, A., Pascal, K.E., Maurus, D., Brachtendorf, S., Lörks, V., Sikorski, J., Koch, P., Hilker, R., Becker, D., Eller, A.K., Grützner, J., Tonigold, M., Boesler, C., Rosenbaum, C., Heesen, L., Kühnle, M.C., Poran, A., Dong, J.Z., Luxemburger, U., Kemmer-Brück, A., Langer, D., Bexon, M., Bolte, S., Palanche, T., Schultz, A., Baumann, S., Mahiny, A.J., Boros, G., Reinholz, J., Szabó, G.T., Karikó, K., Shi, P.Y., Fontes-Garfias, C., Perez, J.L., Cutler, M., Cooper, D., Kyratsous, C.A., Dormitzer, P.R., Jansen, K.U., Türeci, Ö., 2021. BNT162b2 vaccine induces neutralizing antibodies and poly-specific T cells in humans. Nature 595, 572–577.

Schumacher, T.N., Schreiber, R.D., 2015. Neoantigens in cancer immunotherapy. Science 348, 69–74.

Sebastian, M., Papachristofilou, A., Weiss, C., Früh, M., Cathomas, R., Hilbe, W., Wehler, T., Rippin, G., Koch, S.D., Scheel, B., Fotin-Mleczek, M., Heidenreich, R., Kallen, K.J., Gnad-Vogt, U., Zippelius, A., 2014. Phase Ib study evaluating a self-adjuvanted mRNA cancer vaccine (RNActive®) combined with local radiation as consolidation and maintenance treatment for patients with stage IV non-small cell lung cancer. BMC Cancer 14, 748.

Skold, A.E., van Beek, J.J., Sittig, S.P., Bakdash, G., Tel, J., Schreibelt, G., de Vries, I.J., 2015. Protamine-stabilized RNA as an ex vivo stimulant of primary human dendritic cell subsets. Cancer Immunol. Immunother. 64, 1461–1473.

Stepinski, J., Waddell, C., Stolarski, R., Darzynkiewicz, E., Rhoads, R.E., 2001. Synthesis and properties of mRNAs containing the novel "anti-reverse" cap analogs 7-methyl(3'-O-methyl)GpppG and 7-methyl (3'-deoxy)GpppG. RNA 7, 1486–1495.

Tan, X., Wan, Y., 2008. Enhanced protein expression by internal ribosomal entry site-driven mRNA translation as a novel approach for in vitro loading of dendritic cells with antigens. Hum. Immunol. 69, 32–40.

Tauzin, A., Nayrac, M., Benlarbi, M., Gong, S.Y., Gasser, R., Beaudoin-Bussieres, G., Brassard, N., Laumaea, A., Vezina, D., Prevost, J., Anand, S.P., Bourassa, C., Gendron-Lepage, G., Medjahed, H., Goyette, G., Niessl, J., Tastet, O., Gokool, L., Morrisseau, C., Arlotto, P., Stamatatos, L., McGuire, A.T., Larochelle, C., Uchil, P., Lu, M., Mothes, W., De Serres, G., Moreira, S., Roger, M., Richard, J., Martel-Laferriere, V., Duerr, R., Tremblay, C., Kaufmann, D.E., Finzi, A., 2021. A single dose of the SARS-CoV-2 vaccine BNT162b2 elicits Fc-mediated antibody effector functions and T cell responses. Cell Host Microbe 29 (1137–1150), e1136.

Thess, A., Grund, S., Mui, B.L., Hope, M.J., Baumhof, P., Fotin-Mleczek, M., Schlake, T., 2015. Sequence-engineered mRNA without chemical nucleoside modifications enables an effective protein therapy in large animals. Mol. Ther. 23, 1456–1464.

Usme-Ciro, J.A., Campillo-Pedroza, N., Almazan, F., Gallego-Gomez, J.C., 2013. Cytoplasmic RNA viruses as potential vehicles for the delivery of therapeutic small RNAs. Virol. J. 10, 185.

Vogel, A.B., Kanevsky, I., Che, Y., Swanson, K.A., Muik, A., Vormehr, M., Kranz, L.M., Walzer, K.C., Hein, S., Guler, A., Loschko, J., Maddur, M.S., Ota-Setlik, A., Tompkins, K., Cole, J., Lui, B.G., Ziegenhals, T., Plaschke, A., Eisel, D., Dany, S.C., Fesser, S., Erbar, S., Bates, F., Schneider, D., Jesionek, B., Sanger, B., Wallisch, A.K., Feuchter, Y., Junginger, H., Krumm, S.A., Heinen, A.P., Adams-Quack, P., Schlereth, J., Schille, S., Kroner, C., de la Caridad Guimil Garcia, R., Hiller, T., Fischer, L., Sellers, R.S., Choudhary, S., Gonzalez, O., Vascotto, F., Gutman, M.R., Fontenot, J.A., Hall-Ursone, S., Brasky, K.,

Griffor, M.C., Han, S., Su, A.A.H., Lees, J.A., Nedoma, N.L., Mashalidis, E.H., Sahasrabudhe, P.V., Tan, C.Y., Pavliakova, D., Singh, G., Fontes-Garfias, C., Pride, M., Scully, I.L., Ciolino, T., Obregon, J., Gazi, M., Carrion Jr., R., Alfson, K.J., Kalina, W.V., Kaushal, D., Shi, P.Y., Klamp, T., Rosenbaum, C., Kuhn, A.N., Tureci, O., Dormitzer, P.R., Jansen, K.U., Sahin, U., 2021. BNT162b vaccines protect rhesus macaques from SARS-CoV-2. Nature 592, 283–289.

Weide, B., Pascolo, S., Scheel, B., Derhovanessian, E., Pflugfelder, A., Eigentler, T.K., Pawelec, G., Hoerr, I., Rammensee, H.G., Garbe, C., 2009. Direct injection of protamine-protected mRNA: results of a phase 1/2 vaccination trial in metastatic melanoma patients. J. Immunother. 32, 498–507.

Weissman, D., Pardi, N., Muramatsu, H., Kariko, K., 2013. HPLC purification of in vitro transcribed long RNA. Methods Mol. Biol. 969, 43–54.

White, L.J., Sariol, C.A., Mattocks, M.D., Wahala, M.P.B.W., Yingsiwaphat, V., Collier, M.L., Whitley, J., Mikkelsen, R., Rodriguez, I.V., Martinez, M.I., de Silva, A., Johnston, R.E., 2013. An alphavirus vector-based tetravalent dengue vaccine induces a rapid and protective immune response in macaques that differs qualitatively from immunity induced by live virus infection. J. Virol. 87, 3409–3424.

Wolff, J.A., Malone, R.W., Williams, P., Chong, W., Acsadi, G., Jani, A., Felgner, P.L., 1990. Direct gene transfer into mouse muscle in vivo. Science 247, 1465–1468.

Wu, H.Y., Ke, T.Y., Liao, W.Y., Chang, N.Y., 2013. Regulation of coronaviral poly(A) tail length during infection. PLoS One 8, e70548.

Zeng, C., Zhang, C., Walker, P.G., Dong, Y., 2020. Formulation and delivery technologies for mRNA vaccines. Curr. Top. Microbiol. Immunol.

Zhang, N.N., Li, X.F., Deng, Y.Q., Zhao, H., Huang, Y.J., Yang, G., Huang, W.J., Gao, P., Zhou, C., Zhang, R.R., Guo, Y., Sun, S.H., Fan, H., Zu, S.L., Chen, Q., He, Q., Cao, T.S., Huang, X.Y., Qiu, H.Y., Nie, J.H., Jiang, Y., Yan, H.Y., Ye, Q., Zhong, X., Xue, X.L., Zha, Z.Y., Zhou, D., Yang, X., Wang, Y.C., Ying, B., Qin, C.F., 2020. A thermostable mRNA vaccine against COVID-19. Cell 182 (1271–1283), e1216.

Zhang, H., You, X., Wang, X., Cui, L., Wang, Z., Xu, F., Li, M., Yang, Z., Liu, J., Huang, P., Kang, Y., Wu, J., Xia, X., 2021. Delivery of mRNA vaccine with a lipid-like material potentiates antitumor efficacy through toll-like receptor 4 signaling. Proc. Natl. Acad. Sci. U. S. A. 118.

Zhu, X., Tao, W., Liu, D., Wu, J., Guo, Z., Ji, X., Bharwani, Z., Zhao, L., Zhao, X., Farokhzad, O.C., Shi, J., 2017. Surface De-PEGylation controls nanoparticle-mediated siRNA delivery in vitro and in vivo. Theranostics 7, 1990–2002.

FSC
www.fsc.org
MIX
Papier | Fördert gute Waldnutzung
FSC® C083411

Kolibri 360 GmbH
Ferdinand-Jühlke-Straße 7
99095 Erfurt, Deutschland
produktsicherheit@kolibri360.de